Animal Cell Bioreactors

BIOTECHNOLOGY SERIES

Animal Cell Bioreactors

Edited by

Chester S. Ho

Daniel I.C. Wang

Butterworth–Heinemann

Boston   London   Singapore   Sydney   Toronto   Wellington

Copyright © 1991 by Butterworth–Heinemann, a division of Reed Publishing (USA) Inc.
All rights reserved.

Recognizing the importance of preserving what has been written, it is the policy of Butterworth–Heinemann to have the books it publishes printed on acid-free paper, and we exert our best efforts to that end.

Editorial and production supervision by Science Tech Publishers, Madison, WI 53705.

Library of Congress Cataloging-in-Publication Data
Animal cell bioreactors / edited by Chester S. Ho, Daniel I. C. Wang.
    p. cm. — (Biotechnology ; 17)
   Includes bibliographical references and index.
   ISBN 0-409-90123-7
   1. Animal cell biotechnology. 2. Bioreactors. I. Ho, Chester
S., 1950. II. Wang, Daniel I-chyau, 1936– III. Series.
TP248.27.A53A54 1991
660'.6—dc20   90-2654
                                            CIP

British Library Cataloguing in Publication Data
Animal cell bioreactors.
   1. Biotechnology. Applications of animal cultured cells
   I. Ho, Chester S. II. Wang, Daniel I. C. III. Series
   660.6

   ISBN 0-409-90123-7

Butterworth–Heinemann
80 Montvale Avenue
Stoneham, MA 02180

10 9 8 7 6 5 4 3 2 1

Printed in the United States of America

CONTRIBUTORS

Georges Belfort
Bioseparations Research Center
Department of Chemical
  Engineering
Rensselaer Polytechnic Institute
Troy, New York

Harvey W. Blanch
Chemical Engineering
  Department
University of California-Berkeley
Berkeley, California

David Broad
Celltech Limited
Slough, Berkshire
United Kingdom

Matthew S. Croughan
Genentech Inc.
South San Francisco, California

Bruce E. Dale
Department of Chemical
  Engineering
Texas A & M University
College Station, Texas

Simon Gardiner
Celltech Limited
Slough, Berkshire
United Kingdom

Robert J. Gillies
Department of Biochemistry
College of Medicine
Arizona Health Sciences Center
University of Arizona
Tucson, Arizona

Michael W. Glacken
SmithKline Beecham
  Pharmaceuticals
King of Prussia, Pennsylvania

Bryan Griffiths
Center for Applied Microbiology
  and Research
Porton, Salisbury
Wiltshire, United Kingdom

Brian S. Hampson
Research and Development
  Department
Charles River Laboratories
Wilmington, Massachusetts

Carole A. Heath
Thayer School of Engineering
Dartmouth College
Hanover, New Hampshire

Chester S. Ho
Charles River Laboratories
Wilmington, Massachusetts, and
BioChet Corporation
Boston, Massachusetts

Kazuaki Kitano
Microbiology Research
  Laboratories
Takeda Chemical Industries, Ltd.
Osaka, Japan

Denis Looby
Division of Biologics
PHLS CAMR
Porton Down, Salisbury,
Wiltshire, United Kingdom

N. Maroudas
University of Surrey
Guildford, Surrey
United Kingdom

Otto-W. Merten
Institut Pasteur
Paris, France

William M. Miller
Chemical Engineering
    Department
Northwestern University
Evanston, Illinois

Mary L. Nicholson
Massachusetts Department of
    Public Health
Massachusetts Center for Disease
    Control
Boston, Massachusetts

Christopher P. Prior
Rhône-Poulenc Rorer
King of Prussia, Pennsylvania

Aleš Prokop
Department of Chemical
    Engineering
Washington University
St. Louis, Missouri

Gordon G. Pugh
Charles River Laboratories
Wilmington, Massachusetts

Malcolm Rhodes
Celltech Limited
Slough, Berkshire
United Kingdom

Peter W. Runstadler, Jr.
Verax Corporation
Lebanon, New Hampshire

Winfried Scheirer
Sandoz Research Institute
Vienna, Austria

Raymond E. Spier
University of Surrey
Guildford, Surrey
United Kingdom

Michiyuki Tokashiki
Teijin, Tokyo Research Center
Tokyo, Japan

John N. Vournakis
Genmap Inc.
New Haven, Connecticut

Daniel I.C. Wang
Massachusetts Institute of
    Technology
Cambridge, Massachusetts

Joseph K. Welply
Monsanto Company
St. Louis, Missouri

CONTENTS

ix

PART IV.   ANIMAL CELL BIOREACTOR DESIGN, OPERATION, AND CONTROL

PREFACE

Biotechnology is entering a new era. To date, many new biopharmaceutical products developed with this new technology are being sold commercially and approximately 100 are in clinical trials. According to industrial sources, over 250 more are in the research or preclinical development stage. Animal cell culture, in particular, is responsible for a wide range of medically important products, such as vaccines, antibodies, hormones, cytokines, growth factors, and enzymes. As these products advance toward commercialization, the biotechnology industry must rapidly increase the capacity requirement for large-scale in vitro production of animal cell products.

This book addresses the underlying principles and strategies in the in vitro cell culture biotechnology. It addresses the engineering aspects such as mass transfer, instrumentation, and control ensuring successful design and operation of animal cell bioreactors. The purpose of the book is to provide a comprehensive analysis and review in the advancement of the bioreactor systems for large-scale animal cell cultures.

Animal cell culture is here to stay. A broad spectrum of subjects regarding animal cell bioreactors will be useful to researchers and practitioners with different scientific backgrounds. Progress in this area will depend, to a great extent, on a concerted and continued effort by all. It is fair to state that much work remains to be done and that it certainly will tax the ingenuity of researchers in the years to come. Indeed, the future of animal cell culture is a bright one, and we certainly hope *Animal Cell Bioreactors* will play a role in this exciting march into the future.

We thank the contributing authors for their enthusiasm, encouragement, and efforts in making this project possible. To the staff at Butterworth–Heinemann, we are grateful for their patience and assistance during the preparation and publication of this book.

Chester S. Ho
Daniel I.C. Wang

PART
I

Historical Development of
Animal Cell Bioreactors

An Overview of Animal Cell Biotechnology: The Conjoint Application of Science, Art, and Engineering

R.E. Spier

Much has been made of the influence of "science" on society, whereupon the discussion often reverts to considerations of humans in space or of the value of nuclear reactors. In such situations it is clear that it is the practical applications of the knowledge base within a social context that is under examination. This latter activity may be more properly designated as engineering. In view of the confusion it is worthwhile to restate some primary definitions.

Science: A mental activity associated with the data base of knowledge contained in the mind. There are three kinds of such knowledge, the first is knowledge of the composition of the items in the world outside the self, the second is understanding how those external elements interact to produce the observed phenomena, and the third is the series of recipes or protocols that enable us to manipulate, influence, or modify the world external to ourselves. To "do" science is to increase the scope or depth and certainty (by testing) of the mental data base by increasing the knowledge available for assimilation. Such knowledge is then made available to other minds by the use of communication processes.

3

Art: The physical activity differentiated from basic bodily functions which normally involves the manipulation of nonbody materials, or when body manipulations are involved they encompass more than the physical activities necessary for survival. Other words that can be used synonymously with art are craft, skill, capability, technique, or manufacture. It is not possible to effect art without at the same time using some of the mental data base, or science. In a similar sense, it is not possible to "do" science without the application of art. We can now put the word *technology* into its literal context as a *science*; for when it is translated literally it means the *study of techniques*.

Engineering: This word relates to the word ingenious and in its usage may be seen to involve the conjoint application of science and art "in the interest of the society." Thus engineering embraces the three dimensions of mind, body, and society (Spier 1989a). The pianist at practice uses science and art. When the pianist performs in public he operates as an engineer; a communication engineer, transmitting the message of the author of the work, after interpretation, to the minds of the observers. Other performers, painters, sculptors, actors, and dancers, are likewise engaged in information transfer.

How then does the animal cell biotechnologist fit into this picture? There is little doubt that such individuals make use of an extensive "on-board," (or in the mind), data base coupled with the physical ability applied to make their processes work. There is also the additional dimension that all such science and art is directed to improving the lot of society by increasing its standard of health, decreasing the weight of infection and suffering, and making available therapeutic materials that can save and promote life. In addition, we have to consider the social implications of some of the molecules that can now be made in abundance by the animal cell biotechnologist, which can be used to enhance, extend, and improve the already healthy body by further developing particular bodily functions beyond the range commonly experienced in extant populations. Such functions could include brain size, muscle development, longevity, and the number and disposition of organs and limbs. Faced with such a potential, the animal cell biotechnologist has a responsibility to society to act in a manner conducive to the well-being of society as a whole. This means that the manner in which such putative improvements are introduced has to be done with care and with a desire to educate the society in such a way that it can most aptly delineate situations that are to its advantage and disadvantage. In this sense, the animal cell biotechnologist must be a fully fledged engineer and accept the duties and responsibilities that engineers have traditionally accepted as part of their professionalism. There is also a second theme implicit in this introduction. This is the determination of how science and art interact in progressing the field of animal cell biotechnology. If we consider a particular aspect of the subject, such as the development of a cell line that can thrive in a serum-free medium, then we can see how mental, (scientific), and bodily (art), activities interact to achieve the desired ends.

The following is an example of how an engineer might achieve these ends.

- Mentally define the objective.
- Analyze the objective using collateral information regarding the nature of cells and their workings as to its feasibility.
- Use the data base of reported work to attempt to answer the feasibility question.
- Decide on the protocols for the practical work.
- Physically implement these protocols.
- Physically observe and record the effects of the implemented protocols.
- Mentally relate the observations (now incorporated in the scientific data base in the mind), to the original objectives.
- If the original objectives were not achieved, then change the practical protocols according to some mental concepts as to how to achieve the objectives (this could be to use more of the extant data base that relates to this specific project or to a slightly different view of the nature and workings of the cell as influenced by the observations that resulted from the previous experimentation).
- A second basis for changing the protocols could be changing the practical techniques themselves, such as the way solutions were made, the way solutions were dispensed, the origin of the cells, how the cells were handled, the way in which the cells and medium were incubated, and the nature of the container in which the incubation took place, or the way in which the cells were observed at the end of the experiment, etc.
- (Indeed, it would be important to repeat the initial observations according to the initial protocol several times to determine the within- and between-experiment variation, so that an estimate can be made of the effect of the experimental variables on the system as opposed to the system variables).
- Redesign the practical protocols and attempt the experiment again.
- This loop is repeated until the objectives are achieved.

There is clearly an interaction between the intellectual and the physical aspects of this process such that the successful achievement of the objective cannot be obtained without the appropriate interaction between these two facets of endeavor. (It could be argued that the mental activity is but a manifestation of the physical since thought is dependent on the interaction of nerve cells, chemical messengers, and electrical effects. Therefore, the mind/body dualism as carried through to the science/art dualism is not an accurate description of what actually happens. There does seem to be a difference, however, between the physical activity that goes on in the mind viz a viz the physical activity perpetrated by the muscles, which makes useful the distinction between mental and physical activities as extended to the science/art dichotomy).

The mind and its science sets up the practical work and the results of the practical work revert back to the mind for interpretation and for decisions as to the way ahead. (The quality of the practical work, or art, determines whether the subsequent mental activity can be used effectively to progress the achievement of the objectives). The data base of specific examples is also used as is the data base of condensed, summarized, or generalized experience, which is often given the name *theory*. A theory can be generalized numerically or it can be rationalized as a series of directives with an "if ... then do ..." form. Thus science in the form of the mental data base and the generalizations that enable the effective use of that data base and art in the form of physical skills must both be applied to achieve both simple and complex objectives. In the subsequent sections of this introductory chapter I shall seek to explore this relationship further with some specific examples of work in progress that seeks to make animal cell biotechnology processes as productive on a cost per unit of product basis as that achieved by other microbial systems.

1.1 SHEAR AND BUBBLES

There are lessons to be learned from the history of the development of animal cell biotechnology as it has been practiced in industry since about 1955 (Spier and Griffiths 1985). The first industrial processes were designed in the mid-1950s to produce polio vaccine from African Green Monkey kidney cells grown as an adherent culture in either stationary flasks or in rolling bottles. The scale up of such processes simply involved increasing the number of bottles and did not require developing new protocols other than the handling of more bottles and manipulating fluids in larger collection vessels for inactivating the killed (Salk) vaccine or for adding stabilizing agents to the live (Sabine) vaccine. As an industrial process it did not require new techniques and since the ability of the multiple process to supply adequate quantities of vaccine was not challenged, the process was maintained in this form until the development of the microcarrier culture. This new technique, initiated by Van Wezel in 1967, had been developed to the point where both the Dutch Instituut voor de Volksgezondheid (in the late 1970s) and the Institute Merieux in France (in the early 1980s) continued to use this unit process for producing polio vaccine. It was during this period of changing to a scaled-up unit process that difficulties were experienced. Such difficulties were often attributed to the damage of the cells by the hydrodynamic shear forces in the bioreactor which was caused by stirring the microcarriers, and also by bringing them into contact with the air/liquid interface that exists at the surface of a bubble of air or oxygen, which was used to aerate the deep tank cultures. However, there were parallel developments in the scale up of another animal-cell based vaccine that cast doubt on these facile assertions explaining the causes of the problems of microcarrier scale up.

During the early 1960s it was shown that the foot-and-mouth (FMD) virus could be grown in primary bovine cells grown in culture dishes. This was followed by the demonstration that FMD would grow in the continuous attachment-dependent cell line derived from the tissues of baby hamster kidneys (BHK). Within one year of this demonstration, Capstick and Chapman succeeded in growing the anchorage-dependent cell in suspension culture. This led directly into the large-scale, deep-stirred tank culture of animal cells because, unlike the polio vaccine situation, the demand for the double or triple yearly vaccinations of cattle with FMD vaccine (a single dose of vaccine containing three types of FMD virus would require from 3–10 ml of cell culture fluid for its manufacture), coupled with the need to vaccinate several hundred millions of cattle, placed a considerable strain on the production techniques. Vaccination campaigns also were limited by the supply of vaccine. During the scaleup of this deep-stirred tank culture system many of the problems of dealing with allegedly sensitive, cell-wall-free, delicate cells were met and overcome (Telling and Radlett 1970). It was found that with the appropriate materials in the culture medium, it was possible to both stir the cells with the turbine-type impellers, which were used for bacteria and fungi, and that it was possible to control both pH and dissolved oxygen levels by bubbling air through the culture medium to maintain a required pH between 7.0–7.4. The key ingredients of the culture medium that enabled this regimen to operate were the 5–10% of adult bovine serum coupled with the addition of the polyol Pluronic F68. Having arrived at a successful modus operandi, the scaleup of these cultures to the 1,000 liter and later the 4,000 liter scales presented few new problems. It is instructive to ask how theory and practice had interacted to achieve this effective industrial scale process. The key discoveries were made as a result of a "go–look–see" approach to the growth of the FMD virus in the cell cultures, which were becoming available in bottle form. The thrust behind this endeavor was that the supply of tongue tissue (used in the Frankle process) was both limited and not very easy to control. Would cells in culture substitute? The screening of a number of cell preparations resulted in the choice of the BHK21 clone 13 cell line. The concept of screening cell lines had been well established in the antibiotic industry; it was now necessary to adapt that idea to the animal-cell culture practices of the time in order to arrive at the appropriate cell line. Suspension, or anchorage-independent cell growth was also well used for growing mouse L and HeLa cells prior to attempts to make the BHK monolayer cell line capable of growing in suspension culture. Again, one can envisage a direct transfer of both the concepts and the techniques. Once growth in suspension had been demonstrated, it was a short step to scale up when the standard microbial bio-reactors were used and the conditions most propitious for the animal cells were found. Because the animal cells were regarded as being shear sensitive, a number of viscosity enhancers were assayed to decrease the effects of local hydrodynamic stress. Whether the relationship between viscosity and hy-drodynamic stress was known to the investigators at that time is uncertain, particularly when the material that seemed to provide the greatest benefits, the Pluronic F68, did not affect the viscosity of the medium but seemed rather to express its hydrodynamic properties as a foam stabilizer, (Handa et al. 1987). Nevertheless, screening a number of materials did yield a com-ponent that has been found to be of considerable significance.

The discovery of Pluronic can be argued to have been arrived at by empirical means. An assumption of this type requires that an experimenter does not have any preconceived notions about how the material will react. Rather, there is a commitment to unbiased observation regardless of the effects achieved. Indeed, we still do not have a clearly defined mechanism whereby Pluronic exerts its effects. In this case our data base has cause-and-effect knowledge, but knowledge about the mechanism of action is clearly unnecessary for success at this stage. Therefore, a trial-and-error approach worked in the discovery of Pluronic. The trials may have been polarized by a theoretical bias toward viscosity increasers, but the discursive activity resulted in material that has little if any effects on viscosity, but does enable cell culture in deep-stirred tank bioreactors. Thus the mechanism of action of the Pluronic remains to be determined. When such knowledge does be-come available, it may be possible to increase its effectiveness by enhancing those aspects of the molecule that are active in achieving its beneficial effects. In summary, this example has demonstrated that screening systems can yield practical materials. Further developments of the so-discovered ma-terials can be achieved either by further screening techniques that concen-trate on molecules with similar structures or by discovering the mechanism of action of the active part of the molecule. One can then either effect random changes or make changes that have a theoretical basis.

The work with BHK suspension cells clearly demonstrated that neither shear itself nor the actual bubbling of gas through the culture medium nec-essarily caused the destruction of cells. Such cultures were also achieved for the once-thought-delicate hybridoma cell lines as well as the alpha-inter-feron-producing Namalwa cell line. In all these cases, the cell lines selected for scaleup were "bioreactor hardened". That is, care was taken to select cultures that had performed well in the suspension mode at the bench scale. From such cultures amplified innocula were prepared that could be used in the large-scale operation. The characteristics of the bioreactor-hardened cul-tures have only been defined in an operational sense.

Having the ability to operate the suspension cultures on a large scale for some time permitted a series of studies to allay the suspicions that shear and bubbles were the cause of the problem. The work on shear showed that with suboptimal medium and a weak cell line while using stirrer speeds beyond those necessary for mixing, animal cells in culture could be damaged by the hydrodynamic environment. This work was followed by examinations of cell survivability in the highly defined shear environment that exists in the annulus of the Couette viscometer. In this environment, cells were shown to survive shear rates of 200/sec for over 15 hours (Smith et al. 1987; Wudtke and Schugere 1987; Meilhoc et al. 1987). To make this information useful, it is necessary to relate the conditions of the viscometer to what occurs in a stirred bioreactor. For this, the concept of the average shear effect must be applied for we would not expect the cells to stay in the zone of maximum shear for extended periods of time as they would naturally be swept away from such regions with maximum speed. However, techniques for assessing the average shear rely on creating the same effect as a known shear rate, which has been obtained in a defined system. Such work with animal cells in stirred bioreactors is yet lacking.

In addition, it was becoming clear from the application of the concept of the Kolmogorov radius or length (Croughan et al. 1987) that while it was possible to obtain conditions in stirred tanks whereby the Kolmogorov radius could be made of comparable size to the diameter of a suspension cell (a size that would be expected to induce damage to the cell), such conditions need not occur in cell cultures that relied on the stirrer for the mere mixing of the cells and for the fluids that were admitted into the vessel for pH control.

Both the experimental data and the theoretical considerations consolidated those elements of the existing data base, which included the concept that although animal cells in stirred tanks could be damaged by hydrodynamic effects, it was not necessary to generate such extreme conditions.

A similar scenario applies to the putative damage of animal cells by the air/liquid interfaces expressed at the surfaces of bubbles. In Handa's initial experiments, bubble-induced damage in a fully formulated medium was found to be difficult. Indeed, the inclusion of Pluronic F68 in the medium reduced the damage observed in unfortified media to levels indistinguishable from unbubbled control cultures. Further investigation of this effect focused attention on the area of bubble disengagement as the position where the cells were subjected to violent oscillations in the absence of the protective agent. Again, the video microscopy used in these studies showed that Pluronic exerted a "calming" influence by stabilizing a layer of foam on the top of the culture. The retained liquid film between the bubbles was sufficiently stable to allow sufficient time for the cells to leave the zone of bubble breakage before the bubbles actually disintegrated. A corollary to this study was the demonstration that the amount of bubble damage sustained was directly proportional to the area available for bubble disengagement. This means that the size limitation of the reactor in this zone will ameliorate the amount of damage done in a situation that was unprotected by Pluronic or equivalent materials.

Again, this study of the influence of bubbles on the cells reaffirms the view that bubbles per se do not damage cells and that situations can be engineered so that the zone where bubble disengagement occurs can either be stabilized with agents or decreased in size to an innocuous level. Conventional practice arrived at this position and additional studies provided new knowledge that explained the success of those practices. Clearly, additional development work should be done on the tuning of the protective agents, but major improvements are not to be expected.

The caged aerator was developed as a spin-off from the mistaken belief that bubbles did damage cells (Spier and Whiteside 1983). This was used successfully for microcarrier cultures. However, the reason for the success of this system was not the separation of the air/liquid interface from the cells but rather from the decrease in the amount of foam that formed. In microcarrier cultures, the carriers concentrate in the foam layer and are thus held in an environment that neither supplies adequate nutrients nor removes the toxic products of metabolism. Such carriers and their cells die, resulting in decreased yields. The caged aerator eliminates such effects and enables the bubble aeration of cultures at a commercial scale of operation. Many of the commercial companies now provide such caged aerators, or variants of them, as one of the options available in the configuration of the purchased bioreactor.

In this example, a mistaken concept led to a beneficial effect in an unexpected direction. However, once the practical consequences of the implementation of the caged aerator had been observed, it was possible to make a conceptual shift and recognize the new aerator as operating more in the form of a foam breaker than a device that separates bubbles from cells. In this sense, even incorrect science can lead to advancement providing that the basic thrust of the work is progressive and seeks to improve on present-day performance.

1.2 HIGH CELL CONCENTRATION SYSTEMS

During the 1980s, there has been a radical change in the types of animal cell bioreactor systems available. Prior to the late 1970s, bioreactors in common use were the stirred tank or the air-lift variant for suspension cells. There were also bioreactors that were modified stirred-tank reactors or static-bed systems for growing anchorage-dependent cells. Such reactors were operated with local cell concentrations of $0.5-5 \times 10^6$ cells/ml of bioreactor volume. More recently developed bioreactors are designed to operate at local cell concentrations of $0.5-5 \times 10^8$ cells/ml. (See Griffiths 1988 for a more detailed listing of almost 60 different bioreactor types reported.) There are three points to note when considering the latest bioreactor types. Although the local cell concentration has been given as one of the characteristic parameters, the actual concentration of product molecules may not reflect the high local cell concentration. This results from the need to irrigate the cell mass with a volume of medium that is closer to the total cell number than the volume in which the cells are held. On the other hand, when the expressed product is produced in a manner not dependent on the growth of the cells, it is also possible to restrict the amount of medium used in the irrigation and thus generate product at a high concentration. The second point to note is that some products require the cells to be growing at the maximum rate in order to obtain maximum yields of product. Such a product could be a lytic virus such as polio, FMD, or vaccinia. Under these circumstances it may be more cost-effective to operate at low, local-cell concentration and arrange for optimal conditions to achieve rapid cell growth. The third point is that there are many possible advantages of operating at high, local-cell concentrations. Among these advantages are the improved prospects of being able to keep the cells alive and productive on a medium that is both devoid of serum and other high molecular weight proteins (Tyo and Spier 1987). This and the requirement of generating product at higher concentrations leads to the prospect of more efficient downstream process purification operations. However, the cost of obtaining such high performance has to be increased because the high, local-cell concentrations require more complex systems for monitoring and controlling the process. There are also added complications when attempting to scale up such bioprocesses to the industrial level. This problem of the scaleup of such high cell concentration systems is my next focus because in this area, science, art, and engineering interact in a multifaceted and diverse manner.

There is little doubt that most of the systems in use today have not relied on any great depth of theory for their manifestation. If we consider the hollow fiber case then perhaps the first theoretical principle which was brought to bear was that of Knazek (Anonymous 1986), who in his early work likened his hollow fiber capillary system to the capillaries of the body, which by the in vitro system could be set up to serve the same function. This work was further developed by Endotronics and others (Anonymous 1986; Amicon 1985; CD Medical Inc. 1988; Schonherr et al. 1987) into practical units where it was recognized that one of the main limitations for using such systems was the polarization of the hollow fiber cartridge such that the end where the fresh medium was applied would benefit from more favorable conditions than the end where the spent medium exited. To some extent this problem, having been recognized conceptually, was overcome by the method of medium feeding, which required that for part of a cycle all the medium fed to the cartridge would pass across the entire surface of the hollow fiber membranes, through the cell mass in the extracapillary space, and collect in a balance chamber. The second half of the cycle would then involve expressing the medium back through the cell mass and through the membrane of the hollow fiber and from there back into the lumen of the hollow fiber and on to the exit of the cartridge.

While it is clearly possible to juggle with the molecular weight cut-off of the hollow fiber membrane, the manufacturers of the system were more prone to take the hollow fiber cartridges that were available and to "see if they would work." If that was to be the case, then the system would be based on that apparatus. Alternatively, if a low molecular weight cut off was used, then it was thought that some of the growth-promoting molecules in the serum would not be available to the cells so that it became necessary to provide serum directly to the cell-side of the hollow fiber cartridge. Having reached this stage of the development, there were three further considerations that soon became operational. The first concerned the ability to supply the required amounts of nutrients, the second involved the removal of toxic waste materials, and the third involved the provision of oxygen.

There are two approaches to the supply and removal problems. The first is to rely on a system that provides nutrients on demand. This is assessed by sensors or off-line measurements and is related to standard concentrations, which are held to be necessary for the system to thrive. The second approach is to calculate from the growth rate of the cells and the number of cells, what the expected nutrient demand is likely to be and to arrange the flow of medium to meet that demand. (It has to be recognized that the latter approach is fraught with problems since the utilization rate of metabolites by the cells depends on the state of the cells. Since it is generally not possible to measure the cell concentrations directly in these systems, the calculated addition of nutrients could be inappropriate for the demand). Therefore most of the calculations relate to the milliliters of space available for the cells, which if multiplied by 10^9 should give an indication of the maximum number of cells in the system. This would then indicate the rate at which medium would have to be fed at the maximum rate because, as a rough rule of thumb, it requires about 1 liter of fully formulated medium to produce 10^9 cells from an inoculum of one-tenth the size. It is not yet clear how much medium is required or the minimum composition of such a medium that is needed to maintain already grown cells in a state where product is generated without further increase in biomass.

Although the supply-side of the operation is more or less well understood in sufficient depth to enable practical systems, the possible toxic effects of the material discharged from the cells has not been as thoroughly determined. The two primary candidate toxic materials are lactic acid and ammonia. Many studies have been performed to demonstrate the effects of these metabolic end products and, indeed, particular cell populations have been inhibited in their growth when the level of either or both of these parameters rises above benign levels. If such levels are known, it is possible to adjust the media flow rates to effectively wash out these inhibitors. However, this could be a costly option when compared to the alternative of either selecting or adapting the cell population to tolerate the levels of waste metabolite likely to be encountered in the culture. As yet, bioreactors that work in conjunction with the equivalent of artificial kidneys are not commercially available, although it is likely that this situation will be rectified soon.

Recent research in the authors' laboratory and elsewhere, (Chou et al. 1974), has shown that the cell-produced inhibitor situation does not rest with lactic acid and ammonia molecules. Rather, it appears that there is a range of molecules with molecular weights between 10 and 150 kDa that cells make in culture, several of which are cell growth inhibitors. This has implications that are far from understood when considering the scale-up of concentrated cell cultures. Where the cells are separated from the nutrient supply and waste-product removal stream by a membrane, the molecular weight cut-off of the membrane could well be set to concentrate such large molecular weight growth inhibitors on the cell side of the membrane with consequences that could be deleterious for biomass production but which may or may not impede product generation. It should again be noted that many such systems are in operation without any clear idea as to how the cell-derived molecules influence the generation of product. This is a case, perhaps, where the art of cell culture is driving the system forward and the science is dragging along behind. It is of course hoped that when science does catch up, it will lead to improvements in the performance of the on-going art.

The situation with regard to the oxygenation of concentrated cell cultures is less retarded than that involving the large molecular weight growth inhibitors. The problems of oxygen demand and supply have been with the cell culturist since the early days of cell culture scaleup (Spier and Griffiths 1984). The primary limitation has been the amount of oxygen that will dissolve in a given volume of medium. Clearly, the principle question to be asked is how far away can the cell be from a source of oxygen before the cells become oxygen limited? The answer to this depends on the cell's rate of respiration, the local concentration of cells, and the geometric parameter under consideration. There is also the prospect of taking some basic physiological data and analytically determining the thickness of the cell sheet. This scenario would ensue if all the cells of the body were evenly layered on an area equivalent to the surface area of all the body's capillaries (McCullough and Spier 1990). Some 20 or so cell diameters has been calculated for this parameter. Further theoretical calculations can be effected where it can be shown for a model situation of diffusion with reaction (the cells are using up the oxygen as it is diffusing through the cell layers) that enough oxygen can be made available from that dissolved in the medium (equilibrated with air at normal pressure) to provide for the oxygenation of a cell sheet approximately 20 to 30 cells thick. This gives the value of the geometric dimension of 0.2–0.3 mm. Further work in the authors' laboratory has shown that in cells immobilized in agar gels, that the theoretical considerations described above do hold in some measure in experimental situation (Wilson and Spier 1988; Murdin et al. 1987). Such a dimension is critical for the design of cell bioreactors where the cells are held at high concentrations. It means that the distance between capillaries in the hollow fiber cartridge should not be greater than about 0.5–1 mm apart. This same dimension applies when the cells are to be provided with nutrients across flat sheets of planar membranes or when cells are held entrapped or immured in spherical particles or microcarriers. It could be argued that if the cells are held in particles greater than 1 mm in diameter, then the cells in the middle of the particle would not survive for lack of oxygen. Do such dead cells have a negative influence on the productivity of the culture? After all, making cells merely to die is wasting the resources of medium and space in the bioreactor. On the other hand, the dead cells may promote productivity in ways that are not yet known. Again, we do not know whether the cells that die are replaced by cells that use the liberated materials from the dead cells for further growth.

In spite of the many questions and loose ends in our understanding of the way in which concentrated cell cultures may be oxygenated, the practice of such cultures has proceeded with scant regard for the niceties of theory. Whether the theoretical consideration as discussed above will be of value in the design of second- or third-generation high cell concentration bioreactors remains to be determined. Nevertheless, such calculations do validate current practice and could help convince those who doubt the feasibility of operating such system, that high cell concentrations are not only feasible but could well be desirable in the appropriate circumstances.

1.3 MODERN BIOPROCESS ENGINEERING FOR ANIMAL-CELL CONTAINING SYSTEMS

Animal cells in culture are predicted to be the source of products whose value could mount to more than half the value of all the products from the new biotechnology combined (Spier 1988). The processes by which such products are made are basically simple and consist of the following:

- The formulation of a growth medium.
- The addition of the cell inoculum to the growth medium in a container.
- The growth stage incubation.
- The removal of the spent growth medium (which could itself be the crude product material).
- The addition of the medium into which the product is to be secreted or discharged.
- The product stage incubation.
- The discharge of the product-containing medium.
- The concentration, purification, and, if necessary, modification of the product molecules.

- The bottling, labeling, packing, storage, and dispatch of the final product.

While the outline delineated above represents the "process biotechnology" schema for an animal-cell-based bioprocess, there is a much larger body of work that encompasses the basic process to ascertain the state of the process and the materials both in it and those made by it. Thus quality control (tests that determine if a material has achieved a particular control value) and quality assurance (measurements made to ascertain the probability that a certain parameter will be met) operations have become exercises in their own right in recent times. Although quality control has largely grown up to meet the requirements of the regulatory agencies, it has led to the tightening up of all practices, which has resulted in increased reliability with which processes have been brought into productive operation.

In so far as the regulatory agencies, and through them the society at large, have insisted on the implementation of practices that lead to the safest, most efficacious products made by processes that are as controllable as is possible, the animal cell biotechnologist has enjoyed the role of the engineer who applies both science and art to the achievement of his productive goals.

It is clear that given the constraints alluded to above, the goal of the animal cell biotechnologist is to arrive at the most cost-effective way of manufacturing a product. To achieve this end, the researcher would consider the following:

- The medium has to be inexpensive.
- The medium has to be as free from protein as possible (to improve the yield and performance of the downstream concentration and purification operations).
- The cells should be as productive as possible.
- The process for product generation should be in a container operated under conditions to arrive at the most product for the lowest cost (bearing in mind that the same or a modified container should also be usable for the next, different product that might be scheduled).
- The downstream operation should be contain as few stages as possible and each stage should be as effective as is practicable.

Given these five guiding principles, the biotechnologist then approaches the selected process. There is some knowledge that relates some of the principles with each other. For example, to produce a particular product in a given cell line it may not be possible to formulate a medium low in protein. Or, the most efficient bioprocess could be based on a suspension cell where the products have to be more extensively processed in the downstream area to ensure that the possibility of making a cancer-inducing product is reduced to the lowest level. Or, the lowest protein-containing medium has to be fortified with additives that make it one of the most expensive media. Or, the genetically engineered cell line that costs a great deal to produce is difficult to scale up in the available bioreactors. Such conundra are the bane of the animal cell biotechnologist.

Presented with such a welter of imponderables the usual approach has been to begin at the beginning with the bioprocess when it is in its bottle phase. At this scale of operations much discursive work can be done to determine the sensitivity of the system to the removal of the proteinaceous components of the medium. This work can also determine whether the cell population one is working with is the most productive possible. There are levels of productivity that can be recognized as maximal for the cell, which are based on the rate at which cells can produce the proteinaceous biomass needed for their own replication (Spier 1989b). Thus equipped with the most efficacious small-scale culture system, the remaining questions concern the scale-up. This latter operation is conditioned by the amount of product for which there is a market, which is itself determined by the price at which that material is offered for sale. (Clearly some materials are more price sensitive than others depending upon the existing state of the market, particularly with regard to competing products and the perceived value of the material to the purchaser.) A further consideration is the estimate of the time it takes to scale up with a system that is already in use compared to a system that seems to offer advantages over in-house capability but is as yet untried. All these considerations have to be compared to the on-going background activity of interactions with the regulatory agencies, which are often a source of directives (actual or conjectured), such as which particular culture system should or should not be used or more usually which system is less likely to cause problems with these authorities. Similar considerations obtrude into the determination of the way in which the product is concentrated and purified and subsequently processed.

We thus develop the process by a mixture of concepts and practical or discursive experimentation, the results of which have to be interpreted before decisions are made regarding how to proceed. Additional thought is necessary to "second guess" how the regulatory authorities are likely to respond to a given presentation of a process. All these activities, in an industrial context, are used in a race against the clock, necessitated by the threat of competition or the demands of the shareholders. It is thus that science in the form of the data base that enables us to approach practical problems in conjunction with those arts that enable the manifestation of the process and the presentation of that process to the society at large via the regulatory authorities, is engineered into an operation that brings the people of our societies materials to improve their health and well-being. Science, art, or engineering alone could not achieve this end, but when applied conjointly, valuable progress is assured.

1.4 CONCLUSIONS

The relationship between the mental, physical, and social aspects of the discipline of animal cell biotechnology have been examined in general principles and some specific instances. Before work can commence, the instigator has some mental concept of what is to be achieved and some ideas about the way to proceed. Such thoughts are based more on the data base, which is made of past practical experiences and on concepts of the nature of a cell and the way cells work in their more usual environments, than upon any numerical model or deep-seated theory of the nature of the cellular state. The latter approach has been adopted after it has been shown practically that it was possible to achieve a particular mode of cell culture whereupon the numerical analysis of the existing performance enabled the boundaries of the system to be discerned. This is often of considerable importance for it determines the necessity for either new techniques or gives an indication when it is no longer cost-effective to continue improving a particular operational performance. Also, numerical analysis of the physicochemical parameters that affect the performance of cells has delineated the boundaries of hydrodynamic conditions within which it is prudent to maintain the cultures. As yet, mathematical models have neither been used to extend existing systems beyond current performance capabilities nor have they predicted new phenomena. Also, they have not led us to discover new materials that can be derived from cells in culture. We still have a long way to go to account for the 100,000 molecules that may be part of an animal cell and to be able to recognize the 1,000 or so molecules they excrete. When this has been achieved it may prove useful to build models of how animal cells work in order to project the existence of either new molecules or new ways in which the known molecules must interact to comply with the features predicated in the theoretical construct. Even when such a realization is achieved, the practical manifestation of the consequences of the concepts generated from such models has to be effected and applied to benefit society. In the future, like the present, we will have conjointly applied science, art, and engineering in the area of animal cell biotechnology for progress.

REFERENCES

Amicon Applications (1985) Amicon Applications Bulletin I-254.

Anonymous (1986) in 1986 Endotronics applications data bulletin no. 4513, Endotronics, Inc., Coon Rapids, MN.

CD Medical Inc, Dow Chemical Company (1988) Jencons Scientific, Ltd., Leaflet Ref. 7/88, Cherrycourt Way Ind. Estate, Leighton Buzzard, England.

Chou, H.N., Black, P.H., and Roblin, R.O. (1974) *Proc. Natl. Acad. Sci. U.S.A.* 71, 1748.

Croughan, M.S., Hamel, J.-F., and Wang, D.I. (1897) *Biotechnol. Bioeng.* 29, 130–141.

Griffiths, J.B. (1988) in *Animal Cell Biotechnology* Vol 3, (Spier, R.E., and Griffiths, J.B., eds.), pp. 179–221 Academic Press, London.

Handa, A., Emery, A.N., and Spier, R.E. (1987) *Dev. Biol. Stand.* 66, 241–254.

McCullough, K., and Spier, R.E. (1990) in pp. 265–315, Cambridge University Press, Cambridge, England.

Meilhoc, E., McQueen, A., and Bailey, J.E. (1987) *Flow effects on viability and metabolism of suspended mammalian cells,* in *194th Meeting of the American Chemical Society,* New Orleans (Abstract MBDT 56).

Murdin, A., Wilson, R., Kirkby, N.F., and Spier, R.E. (1987) in *Modern Approaches to Animal Cell Technology* (Spier, R.E. and Griffiths, J.B., eds.), pp. 353–364, Butterworths, Guildford, England.

Schonherr, O.T., van Gelder, P.T.J.A., van Hees, P.J., van Os, A.M.J.M., and Roelofs, H.W.M. (1987) *Develop. Biol. Stand.* 66, 211–220.

Smith, C.G., Greenfield, P.F., and Randerson, D.H. (1987) in *Modern Approaches to Animal Cell Technology* (Spier, R.E., and Griffiths, J.B., eds.) pp. 316–327, Butterworths, Guildford, England.

Spier, R.E. (1988) *Chimicaoggi* Sept., pp. 51–55.

Spier, R.E. (1989a) *Enzyme Microbial Technol.*, 11, 319.

Spier, R.E. (1989b) in *Advances in Animal Cell Biology and Technology for Bioprocesses* (Spier, R.E., Griffiths, J.B., Stephenne, J., and Crooy, P., eds.) pp. 32–43, Butterworths, Guildford, England.

Spier, R.E., and Griffiths, J.B. (1984) *Devel. Biol. Stand.* 55, 81–92.

Spier, R.E., and Griffiths, J.B. (1985) in *Animal Cell Biotechnology* Vol. I (Spier, R.E., and Griffiths, J.B., eds.) pp. 3–13, Academic Press, London.

Spier, R.E., and Whiteside, J.P. (1983) *Dev. Biol. Stand.* 55, 151–152.

Telling, R.C., and Radlett, P.J. (1970) *Large-Scale Cultivation of Mammalian Cells,* *Advances in Applied Microbiology* Vol. 13, pp. 91–119, Academic Press, London.

Tyo, M., and Spier, R.E. (1987) *Enzyme Microbial Technol.* 9, 514–520.

Wilson, R., and Spier, R.E. (1988) *Enzyme Microbial Technol.* 10, 161–164.

Wudtke, M., and Schugerl, K. (1987) in *Modern Approaches to Animal Cell Technology* (Spier, R.E., and Griffiths, J.B., eds.) p. 297–315, Butterworths, Guildford, England.

Implications of Cell Biology
on Bioreactor Operation

Implications of Cell Biology
in Animal Cell Biotechnology

Aleš Prokop

The purpose of this article is to review the current status of mammalian cell biology with a special relevance to molecular and physical signaling. Cell signaling is viewed as an interaction of three components. First, the cell, its biochemical and mechanical status, including different signal-response cascades; second, the status of the extracellular environment, including cell extracellular matrix and fluid-flow properties; and third, the time scale or rates of biological and physical phenomena involved, including the dynamics of response. Schematically, the interaction of the above three components is depicted in Figure 2–1. The article's structure very closely follows the individual components subheading from this figure. The implications for bioreactor design are interdispersed in the text when relevant.

2.1 CELL STATUS

2.1.1 Lipid Bilayer and Surface Receptors

Mammalian cells are surrounded by a phospholipid bilayer (plasma membrane) embedded with enzymes and structural proteins, which mediate communication between the cell and the environment. These cells lack an outer

This paper was supported in part by a BRSG S07 RR070S4-22 grant awarded by NIH.

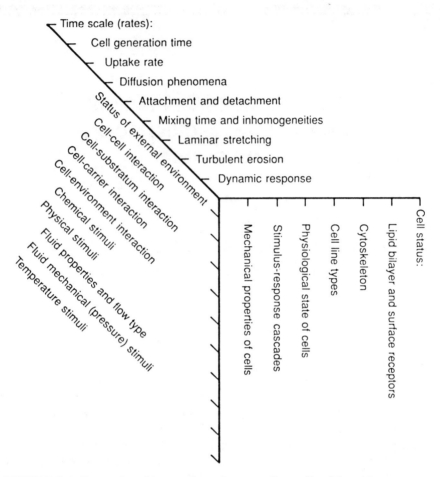

FIGURE 2-1 Dynamics of interaction of mammalian cells with environment.

cell wall and as a result are highly sensitive to environmental stimuli such as osmotic changes, hydrodynamic forces, and pH and nutrient changes. The lipid bilayer membranes are essentially elastic (deformable) structures. Some specialized proteins (receptors) in membrane are involved in accepting different external stimuli such as chemical and physical (osmotic, mechanical), etc., for example, ion channel proteins (Guharay 1984; Lansman et al. 1987). Some basic plasma membrane functions are listed in Table 2-1. Most membrane proteins (receptors) are attached to microfilament proteins of the cytoskeleton (see below) such as actin and spectrin. This attachment limits receptor lateral mobility in the membrane (Salisbury et al. 1983) and their internalization upon presentation of an external stimuli (chemical ligand, mechanical) for certain classes of receptors. This inter-

TABLE 2–1   Functions of Plasma Membrane (Lipid Bilayer) in Mammalian Cells (Adapted with permission from Kotyk et al. 1988)

Type of Function	Example of Cells
Transport of small molecules and ions	All mammalian cells
Transport of large molecules	Macrophages
Fusion	All mammalian cells in an appropriate environment
Ligand binding	
hormones	Numerous mammalian cells
antigens	Lymphocytes

nalization is accomplished through endocytosis (receptor mediated endocytosis, RME) and leads either to receptor recycling to the cell surface or to their degradation. De novo receptor synthesis can also occur. Besides, receptors can be internalized and recycled in the absence of a ligand (constitutive recycling). The present understanding of a receptor pathway is schematically depicted in Figure 2–2. The primary function of endocytosis is to provide a cell with nutrients, growth factors and hormones, and to carry protein processing and clearance (Stahl and Schwartz 1986).

In Table 2–2, the effect of chemical (and mechanical) stimuli on individual steps of a receptor cycle is listed for three types of receptors (generally a signal detection system, SDS; for more details see Section 2.1.5 in this chapter). The mechanical stimuli are discussed in Section 2.3.2. Goldstein et al. (1985) classified receptor mediated endocytosis on the basis of whether receptor and ligand can be recycled or/and degraded.

The components of a lipid bilayer membrane exhibit a variety of movements: (1) rotation and bending of C–C bonds; (2) rotation of lipids around an axis perpendicular to the membrane; (3) lateral diffusion in the membrane plane; and (4) flip-flop movement between the monolayers of the bilayer membrane. Movements 2 through 4 are depicted in Figure 2–3. Of those, a lateral diffusion is important and is discussed later. Membrane proteins and receptors, in analogy to membrane lipids, also exhibit a variety of movements. Some may be relevant to the signal transduction and to the receptor capping (endocytosis). The lateral diffusion of proteins is in the order of 10^{-12}–10^{-16} m²/s, close (or mostly smaller) to the corresponding diffusion of lipids (10^{-11}–10^{-15} m²/s). It is, perhaps, the nonhomogeneity of lipid membranes and voids in the lipid membrane created by lateral lipid movement that enable proteins to move around. The tortuosity of protein movement in membranes and cytoskeleton involvement may explain their somewhat small diffusion coefficients. Beck (1987) and Kotyk et al. (1988) listed some lateral diffusion coefficients of lipids and membrane proteins. Wallach (1987) presented a relationship between lipid and protein lateral mobility.

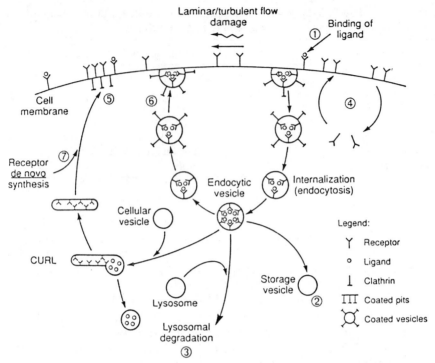

FIGURE 2–2 The cell surface receptor pathway following ligand occupancy (for the most part). (1) Binding of ligand; (2) receptor/ligand internalization, endocytosis and receptor down-regulation; (3) receptor/ligand degradation; (4) constitutive receptor recycling to membrane surface; (5) receptor recycling; (6) receptor/ligand recycling; and (7) receptor de novo synthesis. CURL is compartment of uncoupling of receptor and ligand (adapted with permission from Limbird 1987).

2.1.2 Cytoskeleton

There are three types of cytoplasmic fibers in mammalian cells that make up the cell cytoskeleton and are embedded in a dense, highly viscous cytoplasmic gel (Fulton 1984; Darnell et al. 1986): the microtubules, which are 25 nm in diameter and possibly hollow; the intermediate filaments (10 nm in diameter); and the microfilaments (6 nm in diameter). The total number of cytoskeleton-associated proteins is on the order of 1,000.

A rigid framework of microtubules, composed of tubulin, is responsible for maintaining the reciprocal position of cytoplasmic components, particularly of lysosomes and mitochondria. Microtubular network is usually close to the nucleus and anchored to material associated with the centriole. A contractile system is made up of microfilaments, composed of spectrin, actin, and other proteins. Filaments, usually in the form of thick bundles (stress fibers) are localized on the inner side of the plasma membrane across the cell and connect the membrane to the nucleus. The attachment to the

TABLE 2–2 Possible Effects of Chemical Ligand and Mechanical Stimuli on Surface Receptor Cycle (for Types of SDS, *A* through *C*, See Figure 2–6 and Table 2–4)

Effect	SDS Type:	Chemical Stimuli A	B	C	Mechanical Stimuli and Ligand[1,2] A/B	C
Surface receptor binding (1)[3]						
Ligand		Yes	Yes	Yes	ENH	NA
Damage		Yes	Yes	Yes	Yes	Yes
Receptor and ligand internalization/endocytosis and receptor down-regulation (storage vesicles) (2)		Some	Yes	No	ENH	No
Receptor and ligand internalization and degradation (3)		Some	Yes	No	ENH	No
Constitutive receptor recycling to membrane surface (in the absence of ligand) (4)		Some	Yes	No	?	No
Receptor recycling (5)		Some	Yes	No	?	No
Receptor and ligand recycling (6)		Some	Yes	No	?	No
De novo receptor synthesis (7)		Some	Yes	No	?	?

ENH, enhancement; NA, not available.
[1]Sprague et al. 1987.
[2]Davies et al. 1984.
[3]Numbers in parentheses refer to pathway numbers in Figure 2–2.

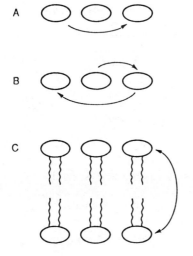

FIGURE 2–3 Three types of lipid (protein) mobility in lipid bilayer membrane. (A) Axial rotation. (B) Lateral diffusion. (C) Flip-flop movement.

plasma membrane phospholipid bilayer is via specialized proteins. A pseudopodial extensions of plasma membrane (microvilli), mediated through stress fibers (Langarger et al. 1986), provide better surface area to volume ratio for nutrient uptake, reception, attachment, and movement. Both spectrin (actin) and tubulin are formed by polymerization of protein subunits and their dissociation and reassembly are closely regulated in the cell (Cleveland 1982). Intermediate filaments are linked to cell differentiation and may hold the cell nucleus in position. They are the most stable components of the cytoskeleton and the least soluble. However, the intermediate filament network of *Drosophila* was shown to be very sensitive to number of stresses (e.g., heat), and collapses around the nucleus shortly after a heat shock. The normal morphology is resumed when cells are allowed to recover from the stress (Welch and Suhan 1986).

2.1.3 Cell Line Types

A primary culture results from subculturing and cloning of an originally very heterogeneous mixture of different cell types (from tissue) and has a finite life span. In vitro cell line transformation (see below) gives rise to a continuous cell line. A comparison of properties of finite and continuous cell lines is shown in Table 2–3 .

Cells in a primary (secondary) cell culture are typically adherent (anchorage-dependent). The initial contact between cells or between cells and a substrate (a surface for adhesion) is via, among others, a morphological pattern known as a tight junction, which may develop into gap junction (Griffiths and Riley 1985). Junctions provide a cell-to-cell communication and the exchange of small molecules between cells. Similar cell junctions,

TABLE 2–3 **Properties of Finite and Continuous Cell Lines (Adapted with permission from Freshney 1987)**

Type of Cells (life span)	Finite	Continuous
Transformation	No	Yes (to a different degree)
Tumorigenicity	No	Yes
Anchorage dependence	Yes	No (slight)
Contact inhibition	Yes	No
Density limitation of growth	Yes	No (or less so)
Mode of growth	Monolayer	Suspension or mono-and multi-layer
Maintenance	Cyclic	Continuous possible
Serum requirement	High	Low
Surface markers	Tissue specific	Chromosomal
Growth rate	Slow	Rapid

called adhesion plaques (patches) enable cell adhesion to a substrate. The attachment is mediated via anchor glycoprotein (e.g., fibronectin in fibroblasts) capping, formed as a result of cell surface cross-linking by multivalent substrate (Griffiths and Riley 1985). With the help of cytoskeletal elements attached to glycoproteins, the membrane fluidity is decreased, the lateral movement of anchor glycoproteins is lowered, and cell rounding is prevented. A final outcome of cell adhesion is thus cell spreading (flattening). For negatively charged cells and positively charged substrates, divalent ions usually participate in protein-mediated adhesion. Adhesion is discussed further in Section 2.2.2.

Many nonmalignant cells, particularly fibroblasts, exhibit a typical saturation pattern when grown attached to a substrate. The density-dependent inhibition of growth (topoinhibition) results in monolayer formation with limited overlapping and multilayering in dense cultures. A detached cell, usually due to mitosis, will lose its ability to grow and proliferate unless reattached (Ben-Ze'ev et al. 1980). Detached cells are round and arrested in a certain stage of the cell cycle (G_1).

Transformed and genetically engineered cells are different from those of normal cells in many ways (Freshney 1987). They exhibit rounded shape, have an increased life-span, an increased lateral mobility of membrane proteins, a decreased number of membrane proteins and receptors, and alterations in cytoskeletal elements. Because of the last two changes they usually lose the ability to adhere (anchorage-independent cells). A lower number of calcium ion channels in transformed cells was noted by Chen et al. (1988) and the disappearance from the cell surface of transformed cells of cell adhesion molecules (CAMs) (See Section 2.3.1) was correlated with a decreased intercellular adhesion (Brackenburry et al. 1984). Besides non-adherent transformed cells, other lines are in an intermediate state: i.e., they are able to grow either attached or in suspension. When attached they grow in multiple layers beyond the confluent monolayer stage. This is particularly apparent for those lines capable of moderate attachment to supports like microcarriers, as observed for some genetically engineered fibroblast cultures. Since the transformation is a multistep process (Quintanilla et al. 1986) it is conceivable that the type of growth (ranging from a normal to a mostly attaching type to a completely nonattaching type with several intermediate states) would result from a degree of the cell transformation (i.e., of the incorporation of a foreign viral genome). On the other hand, cell types that would normally not attach because of their genetic make-up (lymphocytes and derived hybridomas) can in some cases be adapted for attachment. The cell's adaptation is explained on the basis of availability of CAMs (Barnes 1987). Both transformed and engineered cell lines typically exhibit enhanced growth and proliferation as compared to normal cell lines, a property of great advantage for biotechnological purposes.

Hybridomas result from cell hybridization of B or T lymphocytes with suitable indefinite life-span, cell-like lymphomas, myelomas, etc. (transformed cells). The product of fusion is a cell having nuclei from both parent cells (heterocaryon), immediately undergoing mitosis, and yielding a mononucleated hybrid cell. In hybrid cells many genes of either parent continue to be expressed (Eshhar 1985).

Lymphocytes (and transformed cells) typically have a depolymerized microfilamental cytoskeleton. They are also suspected to have the other two major components of cytoskeleton in a less polymerized state (Varani et al. 1983; Mely-Goubert and Bellgran 1981). The same applies to hybrid cells or hybridomas (Gowingt et al. 1984). The hybrids are expected to have morphology of an intermediate type between those of the parents. A limited amount of information is available regarding the cytoskeleton organization in hybridoma cells (Chen et al. 1985). Some information on cytoskeleton of lymphocytes resulted from work on response to mechanical stimuli (Pasternak and Elson 1985; Mazur and Williamson 1987). Little is known about the interaction between the three major filament systems of the cytoskeleton caused by any imposed stimuli (chemical or mechanical).

In summary, mammalian cells' bilayer membrane lacks rigidity and thus does not provide satisfactory protection against outside disturbances. Surface proteins and receptors, embedded in the membrane structure, provide a means of communication with the external environment by accepting all kinds of stimuli. Because of an intimate association of surface receptors with the cell cytoskeleton, a response to external stimuli may result in cytoskeleton modification. The cells' recovery from any stress is accompanied by a reforming of normal morphology. In hybridomas, the cytoskeleton network is less distinguished compared to a typical mammalian cell, provides a limited support to keep the cells' shape, and contributes to anchorage-independence of such cells.

2.1.4 Physiological State of Cells

Two physiological states are important in relation to practical bioprocessing goals: the growing state and a maintenance state. The growing state is typically encountered under the cultivation conditions where a suspension culture technique is used. Exceptions may include a high-density culture in microcapsules or one mechanically entrapped in hollow fibers in a ceramic or gel matrix or in a specially constructed membrane-sandwiched culture. Under such a situation cells may initially grow vigorously and cease to do so when they reach a high density. Similarly, anchorage-dependent cells may grow vigorously before reaching a confluency on a support (be it microcarrier or highly macroporous matrix). At this point, proliferation is considerably reduced and a culture will reach a maintenance status, provided a convenient media switch is provided (from serum to serum-free media). The nutrient uptake is usually reduced to a minimum (as evidenced by radiolabeled amino acid or nucleotide(s) uptake, oxygen, and glucose uptake). However, no systematic studies are available to distinguish between an intensively proliferating culture and that of the maintenance status (non-growing yet product-generating state), such as a simple comparison (through nutrient uptake) between a cell population on microcarriers before and after the confluency. Also, the comparison between a low- and high-density hybridoma culture is warranted. The "maintenance bioreactor" of Monsanto/Invitron Static Maintenance Reactor (SMR) represents the extreme end of this spectrum (Tolbert et al. 1985). Thus, there can be different degrees of maintenance between the two extreme ends: very low density (usually with high-growth rates) and high-density (usually with a slowly growing population) approaching a tissue density. The arrest at a maintenance status is possible at G_1 or G_0 state (Figure 2–4). The stimuli leading to such a state are not known but can be possibly a combination of both chemical and mechanical ones. The involvement of chemical stimuli (e.g. depletion of growth factors or serum) has been studied by Schneider et al. (1988).

Some new in situ techniques are available now, such as NMR spectroscopy, to monitor the physiological state of cells. The concentrations of intermediates of cellular metabolism, particularly those of energy metabolism (ATP, phosphorylated compounds, etc.) can be estimated via this tech-

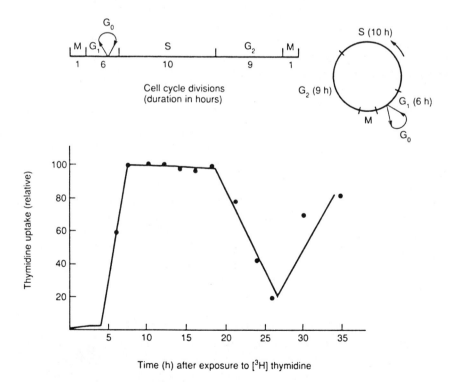

FIGURE 2–4 G_1 or G_0 represents cell cycle states with minimal nutrient uptake (adapted with permission from Baserga and Weibel 1969).

nique. However, it requires a special culture mode to fit into the instrument (Fernandez and Clark 1987; Fernandez et al. 1988). Nutrient uptake rates can be estimated in situ similarly.

Another way of characterizing the cells status is by fluorescence-activated cell sorting (FACS) (Altshuler et al. 1986). A double label of cells through DNA and RNA (Darzynkiewicz et al. 1980) enables an estimation of the cell cycle position and, consequently, to answer the question of whether a cell is proliferating or not (G_0).

2.1.5 Stimulus-Response Cascades (SRC)

An information transfer in mammalian cells in general can be described as in Figure 2–5. A stimulus enters the system, is processed by a sensory processor, and results in a specific cell response. The sensory processor is composed of a signal (stimulus) detector, followed by modulation, transduction, translation, and amplification. There are several generic information systems used by cells, also called SDSs (Figure 2–6, Table 2–4). The G-protein (GP) pathway exists in two types, details of which are in Figures 2–7 and 2–8. The family of GPs (guanine nucleotide-binding proteins) has a heterotrimeric structure consisting of α, β, and γ subunits, all involved in a cyclic manner in an actual pathway. Note that the adenylate cyclase GP pathway intercepts both negative and positive signals, and the phospholipase C/ inositol phosphate pathway only a positive one. The amplification is understood as an event giving rise to multiple molecules (second messengers)

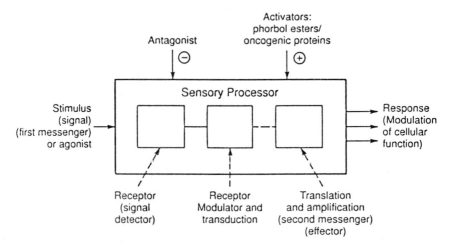

FIGURE 2–5 Schematics of an information transfer system (an alternative terminology is in parentheses). The agonist is an activator that mimics hormone action, binds to the receptor, and causes the normal function. The antagonist (blocker) binds to the receptor but does not activate normal function.

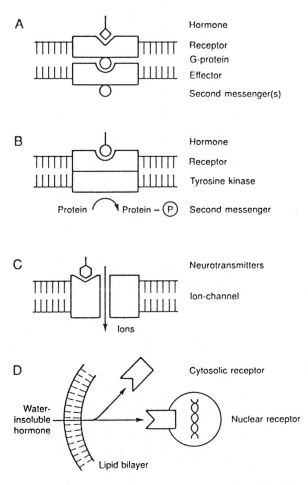

FIGURE 2–6 Four generic SDS (pathways): (A) G-protein (GP); (B) tyrosine-kinase (TK); (C) ion-channel (IC); and (D) nuclear and/or cytosolic membrane (NCM) (adapted with permission from Rawls 1987). Note that some ion-channels (e.g., Ca^{2+} channels) can be directly or indirectly also modulated by GP's (Rosenthal et al. 1988).

from one molecule that starts a cascade [e.g., many PIP_2s with one GP activated phospholipase C, resulting from one molecule of the first messenger (a hormone) binding to the R_p receptor (R_p is receptor protein)]. Such molecular specificity will assure the fidelity of signal reception as well as its amplification to achieve a desired cellular response. A limited number of cell-signalling mechanisms with a limited number of second messengers can, however, achieve a myriad of cellular responses. This is perhaps possible through many further types of covalent modification of hormones, receptors,

TABLE 2–4 Signal Detection Systems (SDS)

Signal Detection	Signaling Compounds	Major Effect
GP Plasma membrane receptor-G-protein-effector-second messenger (SDS A)	Water-soluble hormones, e.g. transferrin, catecholamines, some thyroid hormones, and lipid soluble prostaglandins and gonadotropins	Short-term specific cellular responses (metabolism, secretion) via second messenger in cytosol
TK Plasma membrane receptor-tyrosine kinase (SDS B)	Some hormones and growth factors (insulin, IGF, EGF, PDGF, CSF, TGF)	Short-term specific responses (cell proliferation) or cellular transformation
IC Ion-channel (some could be G-protein modulated)[1] (SDS C)	Some neurotransmitters (catecholamines) and others	Ion uptake, excitatory response, and memory
NCM Nuclear and/or cytosolic receptor following signaling compound diffusion via plasma membrane (SDS D)	Lipid-soluble steroid hormones and some thyroid hormones	Transcriptional control, mRNA stability, growth, and differentiation

EGF, epidermal growth factor; PDGF, platelet-derived growth factor; TGF, transforming growth factor; IGF, insulin-like growth factor; CSF colony stimulating factor.
[1]Scott and Dolphin 1987.

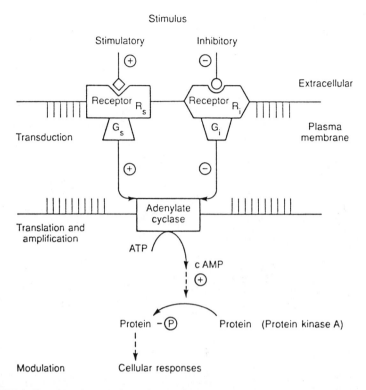

FIGURE 2–7 Adenylate cyclase GP pathway. Note that the scheme may not represent the actual physical configuration of molecules in the cell membrane.

and effector molecules as well as through protease processing (Rodbell 1985; Lichtstein and Rodbard 1987). Each form may have a different function (programmable messengers). For example, the biological activity is lost upon the deglycosylation of choriogonadotropin hormone (Sairam and Bhargavi 1985). Other signaling pathways (see Figures 2–6B through 2–6D) are less complicated and explanations in the legends to Figure 2–6 should suffice. Examples of signaling compound effects of the GP pathways, such as cAMP (adenylate cyclase), calmodulin/PIP_2/DAG/Ca^{2+}/protein kinase C, and of the growth factor/tyrosine kinase pathway are listed in Wallach (1987) (see also Berridge 1988). There are probably few signaling pathways still to be discovered. Among those known, some interactions have been noted. One pathway may antagonise (inhibit) or positively affect the others (Kikkawa and Nishizuka 1986).

The membrane proteins (and receptors) have been categorized according to extracellular, transmembrane, and cytosolic domain typology (Figure 2–9). The corresponding effector molecules are either not known or form an integral part of the receptor molecule. For example, class B involves the

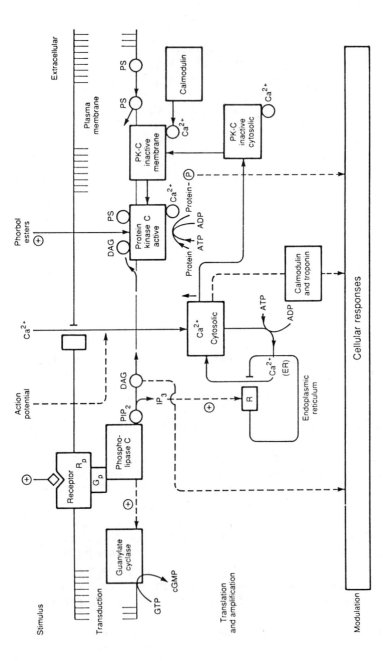

FIGURE 2–8 Phospholipase C/inositol phosphate(s) GP pathway. In addition to IP_3, some other inositol-based messengers seem to be involved. PIP_2, phosphatidylinositol-4,5-biphosphate; DAG, 1,2-diacylglycerol; IP_3, inositol-1,4,5-triphosphate; PS, phosphatidyl-serine. Note that the associated arachidonic acid cascade pathway, encompassing several potent agonists that promote cell growth (prostaglandins) is not depicted in this GP pathway. Also, the cGMP (cyclic GMP) effects are not included here. The scheme may not represent the actual physical configuration of molecules in the cell membrane.

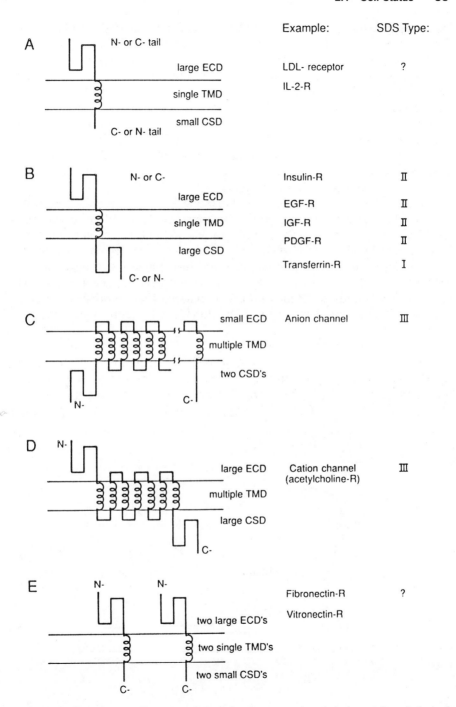

FIGURE 2–9 Classes of membrane proteins (receptors) and their topology (adapted from Wallach (1987). ECD, extracellular domain; TMD, transmembrane domain; CSD, cytosolic domain.

inserts of tyrosine kinase (SDS II) (Yarden and Ullrich 1988a and 1988b). Such categorization does not fit into the above scheme of types of the SDS.

In many SDS, the number of receptors is under some regulation at the level of gene expression. In many other cases, receptor activity is modulated by covalent modification: phosphorylation, glycosylation, etc. In the adenylate cyclase system, receptor is modified by the coupling protein (GP), others are under an endocytosis cycle regulation. Some receptors may be modulated at sites different from the ligand binding site. Such inhibitions can interfere with the free receptor and the receptor/ligand complex. The direct attachment of a modulator to the ligand binding site can completely inhibit the receptor function (antagonist) (Wallach 1987).

Another control is exercised via hormone (e.g., insulin) binding to its receptor. This triggers a process known as down-regulation (desensitization), resulting in the decrease of the number of receptors at the cell surface (Kikkawa and Nishizuka 1986; Czech et al. 1988). The down-regulation consists of two components: RME and degradation. In the absence of degradation, a dynamic equilibrium will usually result between the surface and internal receptor pool (Goldstein et al. 1985). The dynamics in the presence of degradation is more complex (for the LDL—low-density lipoprotein—receptor see Figure 2–10).

One function of receptors is to bring their ligands to intracellular sites for recycling. In addition to cytosolic or nuclear membrane receptors (type III SDS pathway), this function is carried out for thyroid or steroid hormones through RME (Figure 2–11). Examples of RME uptake include epidermal growth factor (EGF) (in fibroblasts), insulin (in many types of cells),

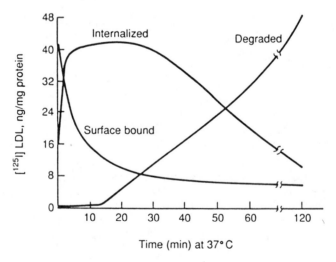

FIGURE 2–10 Dynamics of receptor degradation (adapted with permission from Brown and Goldstein 1979).

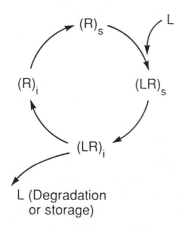

FIGURE 2–11 Schematics of RME cycle (adapted with permission from Stahl and Schwartz 1986). Note that this is a part of a complex fate of receptors, as shown in Figure 2–2. L, Ligand; $(R)_s$, $(R)_i$, $(LR)_s$, and $(LR)_i$ represent surface or internal receptors (R) and surface or internal ligand-receptor complex (LR), respectively.

LDL protein (in fibroblasts), transcobalamin (in hepatocytes), and transferrin (in erythrocytes) (Stahl and Schwartz 1986).

Another control over receptor function results from interaction of a class of proteins (encoded by cellular and retrovirus oncogene) with respective receptors. Thus, EGF and other growth factors appear to employ the TK detection system (Table 2–4). A number of oncogenes can interfere by disturbing this growth control pathway and lead to continuous cell proliferation. The erb B oncogene encodes a protein identical to a part of the EGF receptor (partial homology) and can thus antagonize the EGF function. Several other protein products of oncogene can affect the phospholipase C/inositol phosphate GP pathway or NCM pathway (Berridge 1985; McCormick 1989). A negative control of cell proliferation has been also indicated via anti-oncogene products (Whyte et al. 1988; Green 1989). A clear understanding of the relationship between these genes, the proteins they express, and the way in which cell growth is affected remains to be achieved (Spier 1988).

The surface protein function control and signaling is thus exercised through several mechanisms. Among these, some function to terminate signal transduction processes. A listing of different termination reactions can be found in Mooibroek and Wang (1988).

The identification of signals involved in eucaryotic gene control is far from complete. Many signalling pathways have not been identified, those that have been identified represent the most dramatic effects. There is much to learn about the actual mechanisms involved. At this point, the use of new knowledge on stimulus-response cascades in practical cell culture technology has been limited to a few growth factors and oncogenic applications (Spier 1988). In a rather exceptional example, cGMP was found to stimulate antibody production in hybridomas (Dalili and Ollis 1988). With further progress, we will see more deliberate manipulation of cell cultures via these factors.

2.1.6 Mechanical Properties of Cells

A full understanding of the mechanical properties of cells should be based on understanding the structure or function of individual components, their interaction as well as the behavior of whole cells. Thus, mechanicochemical properties of cell membranes, of the cortical submembrane layer, and of isolated cytoskeletal proteins in different topological association with the above membranes should be sought. The microtrabecular cytosol background (not yet identified) should also be included.

One major group of mechanical properties include viscoelastic properties. Animal cells are viscoelastic when they recover their original shape following a brief deformation. Cellular viscosity is a major controlling parameter in this process. Due to the lack of deeper understanding, a phenomenological model, linear standard viscoelastic solid model, is used. It can yield moduli of cellular viscosity and elasticity (Elson 1988). Simply put, viscoelastic properties can be understood as a combination (in series and in parallel) of springs and a dashpot (Figure 2–12). The spring represents the elastic force (Hooke's law) and the dashpot is the viscosity (Newtonian viscosity). The series spring k_2 allows an immediate elastic response, the parallel k_1 permits eventual recovery. The above approach also allows for a time-varying exposure to stresses or strains.

Extensive data on actin rheology and erythrocytes deformability have been summarized by Elson (1988). One conclusion of actin data is of importance to us: the actin filament network, an actin cross-linked by another protein, can behave simultaneously as rigid (solid) and fluid elements (Sato et al. 1987), in contrast to pure actin solutions. The most appropriate explanation for this is a combination of depolymerization and repolymerization: a sol to gel transition. In a rapidly deforming environment (turbulent flow), actin stiffens and behaves like a viscoelastic solid. When deformed over longer time it behaves as a liquid (deforms and recovers). Fibroblasts and other cells behave similarly.

The viscoelastic properties of erythrocytes are better understood and characterized partly because of their simple structure (Waugh and Hochmuth 1987; Skalak et al. 1989). Different contributions caused by membrane shear, expansion, and bending can be easily resolved. Data on nonerythroid animal cells (leukocytes and adherent cells) are rare (Elson 1988). In spite

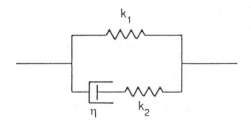

FIGURE 2–12 Linear standard model for a viscoelastic solid.

of this, these cells also attest to their importance to cytoskeleton and plasma membrane. Table 2–5 presents some important mechanicochemical properties of mammalian cells, with emphasis on nonerythroid cells. No attempt is made to discuss differences between erythroid and nonerythroid cells. The importance and function of each property is also briefly stated in Table 2–5. An extensive tabulation of cellular viscosities is found in Bereiter-Hahn (1987).

The membrane shear elastic modulus is considerably higher compared to that of actin itself. Thus, the membrane is the dominant source of viscous dissipation during the recovery process. Moduli k_1 and k_2 (see Figure 2–12) are also listed in Table 2–5. Note that $G = (k_1 + k_2)/2$ for a rapidly deforming field. The upper value of G holds for T lymphocytes, which indicates that these cells are more resistant to deformation. It is perhaps due to their high actin content as compared to B-lymphocytes (Mely-Goubert and Bellgran 1981). Lymphoid (transformed) cells should be thus even more resistant.

The shear membrane viscosity may reflect a real situation in a bioreactor with a fluctuating flow field. The membrane viscosity (and also other mechanicochemical properties) can increase several times as a result of appropriate cross—linking of surface proteins embedded in the plasma membrane (Chasis et al. 1985). These chemical stimuli are typically present in media and thus modulate the cell stiffness. Cell-cell interaction and cell-extracellular matrix (ECM) interaction can also affect it (for ECM see section 2.2.2) (Elson 1988).

The wide range of cytosolic viscosities only reflects the nonnewtonian nature of cytoplasm: As the shear rate goes down, the measured viscosity goes up (Valberg and Albertini 1985).

When small forces are applied for a certain period of time, the result is a permanent cell deformation, a process called a creep and force relaxation. The magnitude of the permanent deformation is proportional to the total time of extension and the level of the applied force. For forces beyond a certain level (yield stress), the cell fragments and is disrupted. The estimate of yield stress of lymphocytes is not available (see Table 2–5). Hiramoto (1987) recently reviewed methods for measuring cytomechanical properties.

Very little is known about viscoelastic behavior of adherent cells. The deformability of such cells has been recently probed by a "cell poker" (Peterson et al. 1982; Duszyk et al. 1989); imposing a small force on a cell and following its response.

The forces involved at the cell adhesion/detachment (anchorage-dependent cells) can be also conveniently measured with the help of a cell poker (Francis et al. 1987). The knowledge of the adhesion and/or detachment forces is of critical importance in microcarrier technology. Guarnaccia and Schnaar (1982) measured the adhesion forces by allowing the cells to attach to plastic wells, then inverted and centrifuged the wells to determine the centrifugal force necessary to dislocate the cells. Crouch et al. (1985) sug-

TABLE 2-5 Mechanico- and Physicochemical Properties of Mammalian Cells and Their Components

Property	Value	Function	Cell Type	Reference
Shear elastic modulus (G)	500–800 dyne/cm^2	Resistance to deform (extend) cell membrane without changing surface area	B and T lymphocytes	1,2
Elastic modulus k_1	100–500 dyne/cm^2	Resistance to deform plasma membrane and cortical cytoplasm	B and T lymphocytes	1,2
(Maxwell) fluid element k_2	300–1000 dyne/cm^2	Resistance to deform ungelled actin in cytoplasm	B and T lymphocytes	1,2
Equilibrium shear elastic modulus of actin	14 dyne/cm^2	Resistance to deform actin in solution	Actin (amoeba)	3
Membrane area compressibility modulus	600 dyne/cm	Tension to change membrane area (expand/contract)	B lymphocytes	1
Bending elastic modulus	10^{-12} dyne × cm	Membrane resistance to bending	Erythrocytes	4
Shear (dynamic) membrane viscosity	100–200 dyne × s/cm^2	Frictional resistance to oscillating flow (rate of deformation)	B and T lymphocytes	1,2
Low shear apparent cytosolic viscosity	10^4 dyne × s/cm^2	Frictional resistance to very low flow	Macrophage	5

Property	Value	Description	Context	Ref.
High shear apparent cytosolic viscosity	0.01 dyne $\times$ s/cm^2	Frictional resistance to high flow	—	—
Yield stress in membrane	NA	Stress in membrane beyond the point of reversible recovery	Lymphocytes	—
Yield stress	7.5 dyne/cm^2	Stress beyond the point of reversible recovery	Rabbit muscle actin	6
Yield stress	800 dyne/cm^2	"	Cross-linked cytoskeleton/macrophage	7
Protein diffusion coefficient	10^{-8} cm^2/s	Mass transfer	12–440 kDa proteins in fibroblasts	8
"	10^{-7} cm^2/s	"	In water	8
Membrane thickness	10–20 nm	Signaling	—	9
Outer surface (zeta) potential	0–(−15mV)	—	—	9

References are as follows:
1, Schmidt-Schonbeim 1987; 2, Sung et al. 1986; 3, Sato et al. 1987; 4, Hochmuth and Waugh 1987; 5, Valberg and Albertini 1985; 6, Sato et al. 1985; 7, Zaner 1986; 8, Jain and Cohen 1981; 9, Stryer 1988.
NA, not available.

gested another way to quickly assess a degree of attachment on glass and plastic surfaces in a shear force gradient apparatus. For different combination of cell-growth surface, the attachment forces varied in the range of 0.1–0.9 dyne/cm², the detachment ones in the range of 30–55 dyne/cm² for standard growth conditions. The effect of pH, serum, and attachment factors can be conveniently measured in this way. From the above numbers it appears that once cells have adhered and spread onto a surface, they are three orders of magnitude less sensitive to shear forces. Also, wherever possible, they should be permitted to attach and spread before being exposed to a shear.

To summarize, the understanding of cellular mechanics can be obtained from measurements of cellular deformability. The extensive data available for simple erythrocyte cells are of little use for complex nucleated and adherent cells because of their anisotropicity. More explanatory models of mechanicochemical behavior of cells coupled with convenient measurements may answer questions about mechanism, structure, and function, some of which have importance for reactor design and operation.

2.2 STATUS OF EXTRACELLULAR ENVIRONMENT

2.2.1 Cell-Cell Interactions

Cell differentiation in vivo and in vitro depends on various signals from adjoining and/or distant cells. These signals maintain and regulate gene expression. The specificity of the response is cell-type dependent, however. Signaling between cells includes largely soluble factors and is determined by the secretion and target site. For endocrine signaling, a signaling compound (e.g., a hormone) acts on distant target cells. Paracrine signaling involves factors (e.g., catecholamines and other neurotransmitters) that act upon adjacent cells, and as autocrine signaling cells respond to substances (e.g., growth factor) they release themselves. All three categories of signal molecules are usually present in typical cell in vitro culture media (Barnes 1987).

In addition to soluble factors, the cell-cell interaction may include insoluble signals mediated by cell surface molecules, denoted as cell adhesion molecules (CAMs) (Obrink 1986; Edelman 1988). At least six different CAMs have now been identified and all are large intrinsic cell surface glycoproteins that are mobile in the plane of the membrane. Since many CAMs are expected to be linked to the cell surface by a phosphatidyl inositol (PI) (and perhaps to phospholipase C) (Low 1987), they may resemble a receptor type II. A cytoskeleton interaction via a given CAM chain is also involved. Several models of homophillic binding between cells, as mediated by CAMs have been proposed. The simplest one is depicted in Figure 2–13A. They involve both calcium-dependent as well as independent types. Thus, PI, cytoskeleton, calcium, as well as chemical modulation (carbohydrate), prev-

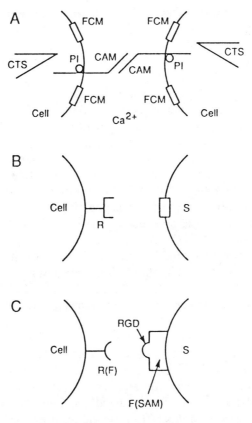

FIGURE 2–13 Examples of cell-cell and cell-ECM interaction. (A) Homophillic (cell-cell), (B) direct heterophillic (cell-ECM), and (C) indirect heterophillic (cell-SAM-ECM). CAM, Cell adhesion molecule; PI, phosphoinositol; CTS, cytoskeleton; R, receptor; F, fibronectin; SAM, substrate adhesion molecule; RGD, Arg-Gly-Asp sequence; FCM, focal contact molecule; S, substratum.

alence modulation (charge), etc. may alter the signal processing and consequently the cell-binding strength and cell shape.

2.2.2 Cell-Substratum (Extracellular Matrix) Interaction

Other morphoregulatory insoluble molecules involved in signaling between cell and substratum are substrate adhesion molecules (SAMs), their receptors on respective cells, focal contact molecules (FCMs), and cell junction molecules (CJMs). The adhesion with a substratum via SAMs involves interaction between cell-surface receptors and one or more extracellular molecules. Some SAM molecules, such as fibronectin, possess a multiplicity of cellular binding sites as well as binding sites for other extracellular matrix molecules (heparin). Fibronectin is of special importance to fibroblast cells. The initial binding (heterophillic) of a cell to a substratum may be via one or two SAM molecules (see Figure 2–13C) or directly to a substratum through substratum "cell receptor" (see Figure 2–13C). The SAM family includes as many as 30 molecules, among them collagen, laminin, fibronectin, vitronectin, etc. (Kornblihtt and Gutman 1988). The attachment of fibronectin to cells is mediated via Arg-Gly-Asp (RGD) sequence located within the cell-binding region of the fibronectin molecule receptor (Ruoslahti 1988). The second cell binding region has been also implicated (McDonald 1988). Also, fibronectin has a heparan sulfate binding domain, the second SAM involved. The definitive proof of linking heparan sulfate with matrix is still lacking. Various SAM molecules are frequently added to plastic substrata in the form of surface coating to facilitate specific molecular interactions since the specificity of the substratum cell receptor is not high (Kleinman et al. 1987). In the presence of serum, the adhesion to substrate appears to be mediated by vitronectin and its cell-surface receptor rather than by fibronectin (McDonald 1988), implying an inhibition of fibronectin receptor by serum proteins (Pitt et al. 1986). Sulfated glycolipids can also participate negatively in the cell-adhesion process (Roberts and Ginsburg 1988).

The initial binding of a cell to a substratum is quickly strengthened via a build-up of adhesion molecules as well as by the cytoskeleton coupling on the other side. The discrete, specialized regions of the first contact have been denoted as areas of focal contact, with characteristic FCMs. A successive or parallel formation of specialized cell junctions then completes the adhesion process: adhesion junctions (spot desmosomes), gap junctions, and tight junctions, each featuring specific CJMs (McClay and Ettensohn 1987; Burridge et al. 1988). Such structures and molecules add mechanical strength to the cell-matrix aggregate. The spreading represents the last stage in which the initial spherical cell flattens out, perhaps as a result of separate stimuli. The whole stimulus-response process involved in the adhesion and spreading somewhat resembles that of the signal detection mechanism because of effects of calcium ionophores and phorbol esters (phospholipase C/inositol phosphate pathway) (Curtis 1987).

Many cells possess multiple types of surface receptors and can bind to several extracellular matrix components simultaneously (SAMs). The receptors recognize the specific sites on the matrix molecules. A family of complex surface glycoproteins, called integrins, have been shown to act as a receptor for fibronectin with some affinity for laminin (Buck and Horwitz 1987). The integrin receptor is also bound via an intracellular (peripheral) protein to the cytoskeleton complex, the degree of association depending on the level of integrin phosphorylation. The opposite holds for transformed cells with altered morphology and adhesive properties: phosphorylation is enhanced at the cell transformation (Buck and Horwitz 1987).

A thermodynamic approach has been suggested to measure the above non-polar interactions between cells and substratum, based on interfacial free energy balance (Schakenraad et al. 1988). The interfacial free energy of adhesion (in erg/cm^2) is defined as:

$$\Delta F_{adh} = \gamma_{sc} - \gamma_{sl} - \gamma_{cl} < 0$$

where the terms on the right side of the above equation are, respectively, the substratum-cell, the substratum-liquid, and the cell-liquid interfacial free energies. The individual contributions can be assessed experimentally, based on a contact angle measurement (Absolom et al. 1983), leading to the surface free energy of a substratum. A high value for this parameter has been shown to be positively correlated with cell adhesion and spreading (Schakenraad et al. 1986). More general thermodynamic theory can also account for additional polar (electrostatic) interactions (van Oss 1989).

In addition to specific molecular forces, nonspecific forces contribute to cell adhesion. The major nonspecific force is that of electrostatic attraction due to opposite charges between cells and substrata. The cells' overall charge is negative (Sherbet et al. 1972) as it is for plastic or glass substrates, unless they are surface coated or derivatized (e.g., by polylysine, DEAE-dextran). The second contributing force could be the van der Waals forces between ions and dipoles. Several complicating facts should also be mentioned. First, the cell surface charge is not uniform; and second, substratum takes on many of the properties of the adsorbed proteins resulting from the presence of proteins in media (usually sera). The above discussion has been more extensively presented by Barngrover (1986). Rutter and Vincent (1988) elaborated on physicochemical background theory to explain microbial adhesion to surfaces in general. The contribution of the nonspecific physical forces to the overall mechanism of cell adhesion remains unclear. The distinction between nonspecific physical and specific molecular interactions is purely semantic, since on the molecular level, chemical interactions do involve physical forces (in addition to van der Waals and dipolar interactions there are also hydrogen-binding and hydrophobic interactions). The physical forces may be of initial importance in specific molecular forces strengthening the relatively weak attraction. The resolution of these two forces or determining their ratio remains one of the challenges of this field.

In addition to the morphology control, ECM status also has an effect on gene expression. It has been postulated that the transmembrane proteins (perhaps receptors for SAM), associated cytoskeletal components, and, in turn, polyribosomes, mRNA stability, and transcription (gene expression) may be influenced (Bissell and Aggeler 1987; Reid et al. 1988). A model proposed for this complex interaction would involve physical forces generated by tensional continuity through the structural (cytoskeletal) components (Ingber and Jamieson 1984). Thus, through ECM effects on cell shape, generated physical forces might control cellular gene expression. To rephrase this statement: signals from cell-cell and/or cell-matrix interaction can regulate gene expression with a high specificity. As a consequence, in vitro culture conditions (media) must incorporate appropriate signals (hormones, growth factors, extracellular matrix components) derived from cell-cell or cell-matrix interactions normally present in vivo in order to retain most of the cells' usual repertoire of gene expression (Reid et al. 1988). For example, hormonally defined media may differ considerably depending on the type of support matrix used (collagen, plastic). The type of ECM is determined to a large degree by a cell ECM. Transformed cell lines do not assemble a normal ECM and thus their media requirements differ considerably (Marsillo et al. 1984).

2.2.3 Cell-Carrier Interaction

This type of interaction represents a special type of cell-ECM interaction, involving electrostatic charges in addition to SAMs, FMCs, and CJMs as above. The most frequently used microcarriers are small beads of a convenient diameter (usually in the range of 150–250 μm) made of natural or synthetic polymers. Recent reviews by Butler (1987) and Fleischaker (1987) provide several examples. The most common are dextran-based microcarriers of Cytodex type (Pharmacia) with ion exchange groups. Cytodex 1 has DEAE-groups throughout the whole matrix, with a high retention of mammalian cell products—proteins—as well as serum proteins in the bead matrix. Cytodex 2 avoids this undesirable protein adsorption to the bead interior by having only a surface derivatization of dextran in the form of an outer surface layer. Cytodex 3 is collagen (SAM) coated, thus providing cells with a natural microenvironment. The ratio of nonspecific (physical charge) to specific (molecular) forces involved in the cell-microcarrier interaction is not known. A coating with a synthetic peptide (having RGD sequence) would enforce specific molecular forces (Pierschbacher and Ruoslahti 1984). A recently introduced dextran-based microcarrier (Schulz et al. 1986, Pfeifer & Langen 1987) allows a fine adjustment of bead charge surface relative to the cell type.

Several surface topologies have been observed on microcarriers: monolayer, multilayer, and bridging. They either reflect differences in cell surface characteristics (transformed versus nontransformed, with many intermediate stages between) or a culture development in time. In general, nontransformed cell lines will form a monolayer whereas transformed and genetically engineered cells (if attaching) will form multilayers. An analogy can perhaps be drawn between the cell-cell interaction in tissues (and involvement of gap junction proteins) with bead bridging mediated by cells. A search for specific CJMs involved at bead bridging might be warranted. The types of surface topologies also have a profound effect on gene expression. The specific productivity (in units per cell and time) of human beta-interferon by mouse L-cells attached to microcarriers was highest for a monolayer culture and decreased significantly by factor of 10 when cells formed a trilayer on microcarriers (Schulz et al. 1987). Any generalization would be premature since more data on different cell lines and products are needed.

Highly porous macrocarriers represent another ECM support material. Verax Corporation beads (WS-IMMO) made of modified and cross-linked, sponge-like collagen with inserted weighting of heavy metal particles is the best available. The WS-IMMO bead is suitable for both anchorage-dependent as well as independent cells, either attached or entrapped in the interconnected pore structure of beads (Dean et al. 1987). Excellent nutrient distribution within the matrix, an appropriate size (500 μm), and specific weight make it very suitable for fluidized reactor culturing. Details of the bead manufacturing technology are not clear. One distinct difference from microcarriers is that the cells in porous macrocarriers are protected from shear (fluid-bead) and collision (bead-bead) phenomena, limiting the operational range of microcarriers application (usually the volume fraction of beads has an upper and relatively low optimum). Recently, another highly porous matrix was introduced, based on collagen-glycosamino-glycan (GAG) species, possibly providing better chemical signals than collagen itself (Cahn 1989).

Ceramic matrix is occasionally used to culture both anchorage-dependent and independent cells. Charles River Biotechnical Services, Inc., markets a nonporous matrix for adherent cells and a porous one for nonadherent cells and adherent cells when harvesting is not required. In none of this systems is a surface treatment (e.g., by SAM) provided (Lydersen 1987).

2.3 CELL-ENVIRONMENT INTERACTION

2.3.1 Chemical Stimuli

Having defined some needs for ECM components and soluble signals previously it is now quite easy to discuss the chemical environment requirements. Table 2–6 enumerates typical classes of media nutrients and supplements as they appear in the serum media as well as in hormonally defined serum-free media. The first four classes have been already discussed in relation to signal transduction and cell adhesion. Others are not relevant to the present discussion. The elimination of the undesired serum components allows for the precise control of the extracellular environment and for shunting of cellular metabolism from growth-related pathway to product synthesis. It may also be possible to control anchorage versus suspension culture by the introduction or removal of adhesion proteins (Barnes 1987). The mechanism of gene regulation by steroid hormones was recently discussed at length (Beato 1989).

2.3.2 Physical Stimuli

2.3.2.1 Fluid Properties and Flow Type   In most cases, media can be characterized as low viscosity newtonian fluids. Higher viscosities are encountered in the case of polymer additions. Their effect is to reduce the

TABLE 2–6   Some Basal Nutrients and Supplements of a Cell Culture Media

Chemical Class	Function
Metal binding proteins	
Transferrin	Iron transport
Ceruloplasmin	Copper transport
Lipid carriers	
Lipoproteins	Lipid transport
Albumin	Lipid binding, vitamin transport
Lipoprotein particles	
Low- and high-density lipoprotein	Cholesterol uptake
Growth factors	Proliferation
Insulin, somatomedins, glucocorticoids, Sex hormones, triiodothyronine, glucagon, Vasopresin, PDGF, EGF, FGF, hydrocortisone, Transformation factors, etc.	
Attachment factors and ECM components	Attachment/spreading
Fibronectin	
Some spreading factors	
Extracellular enzymes and protease inhibitors	
Thrombin	Growth-stimulatory
Catalase	
Protease inhibitors (e.g., antitrypsin proteins)	
Lipids and components	
Lipid-soluble vitamins (E)	Antioxidant
Fatty acids (linoleic, oleic)	Bind to albumin
Ethanolamine	Precursor of prostaglandins
Polyamines (putrescine)	
Amino acids and poly-L-lysine (PLL)	Cell surface charge modification (PLL)
Purines and pyrimidines	
Water-soluble vitamins	
Ascorbic acid	Collagen synthesis antioxidant
Selenium	Co-factor for glutathione peroxidase
Trace elements	

Kolmogoroff length and protect cells from turbulent shear (Shintani et al. 1988). Similar effects (besides biological ones) are attributed to a protective action of high serum concentrations. The other physicochemical properties are not relevant.

The fluid flow type is related to the reactor design and mixing conditions. Generally, there are two extreme cases of fluid flow: laminar and turbulent.

A more extensive treatment is discussed in Prokop and Rosenberg (1989), so only a brief discussion follows. Laminar flow with prevailing viscous forces dissipation is typical of low energy input mixing. As examples of bioreactor, with predominant laminar flow and shear stress an axial type of flow mixer (marine impeller), a static mixer with anchorage-dependent cells attached to it (Grabner and Paul 1981) or horizontal loop type bioreactor (Fiechter 1978) can serve. In a laboratory, laminar flow conditions can be controlled to a high degree in a rotational viscometer (Taylor 1935).

In most other cell culturing devices, the turbulent flow (and Reynolds stress) is the more prevailing situation. The viscous (laminar) component is usually negligible. Higher oxygen transfer and/or a need for removal of solid (microcarriers, macroporous carriers) inhomogeneities results in higher energy inputs and consequently in turbulent conditions. They are characterized by spatial and temporal variations of fluid velocities (and pressure fluctuations). The complexity of the flow pattern in a bioreactor and the difficulty of quantifying the magnitude of shear stress makes it difficult to ascertain the effects on cells and their metabolism.

Another type of stress is encountered when a cell meets with another object (be it another cell, bioreactor wall, or mixer blade). Although there is a general belief that under such circumstances two objects never touch (because the process of draining liquid film between two objects takes a finite time; Das et al. 1987) there could be a substantial level of interaction: a deformation of one or both objects could occur, depending on their elasticity.

2.3.2.2 Fluid Mechanical (Pressure) Stimuli An interaction similar to that above may occur at turbulent mixing when a cell "hits" a turbulent eddy (laminar shear leads to a cell extension by opposite streamlines or streamlines with different velocities). Thus, two processes may have a similar mechanism resulting from pressure or mechanical stimuli. Cells have evolved to accept such stimuli and seem to possess a special category of pressure-sensitive ion channels. The presence of such channels (also called stretch- or mechano-activated) have been demonstrated with only a few cell types and will undoubtedly be found to be true for all cell types. The progress in this field is rapidly developing.

One possible mechanism for transduction of a fluid mechanic signal is that shear stress opens up a mechanicosensitive ion channel, thus allowing for increased potassium leakage, which causes the membrane potential to become more negative. This enhanced cell membrane polarization allows certain voltage-sensitive calcium (Ca^{2+}) channels to open, resulting in an elevation of the intracellular calcium concentration (see Figure 2–8). The influx of Ca^{2+} is accompanied by a depolarization of the membrane potential followed by alternations in ECM, cytoskeleton, and cell function (Nerem and Levesque 1987; Chittur et al. 1988). Stretch-activated ion channels have been detected in endothelial and fibroblast cells (Lansman et al. 1987; Christensen 1987; Olesen et al. 1988; Stockbridge and French 1988) and voltage-sensitive calcium channels in fibroblasts by Chen et al. (1988). The recent data of Ando et al. (1988), however support an alternative view: Calcium release from intracellular reservoirs (such as the endoplasmic reticulum) is a result of low shear stress. The authors observed a two-order magnitude increase in cytoplasmic Ca^{2+} under low external Ca^{2+} conditions and 1–20 dynes/cm^2 shearing of bovine endothelial cells. The monitoring of the intracellular Ca^{2+} was accomplished through cells loading with fluorescent Ca^{2+}-sensitive dye Fura 2. Studies combining Ca^{2+} monitoring with blockage of the calcium ion channels with calcium antagonists may provide insights into the intrinsic mechanism (external versus intracellular Ca^{2+} mobilization) (Levesque et al. 1989).

In addition to a high specificity of pressure reception by cells as discussed above, another possibility of less specific transduction mechanism exists. The cytoplasm contains several structures with discrete and continuous morphological and molecular conformations. One of these is the cytoskeletal complex, described earlier. The second is the microtrabecular network of the cytoplasm, representing a tensile continuity through which physical forces can be transmitted from one point to another. Both systems represent a physical connection from the plasma membrane into the nucleus (and complement the above chemical transducing system of the second messengers) (Packard 1986). The role of cytoskeleton in signal transduction has been postulated earlier (Bissell and Aggeler 1987). The details of both transduction mechanisms are not available. However, it is conceivable that the pressure-sensitive ion channel pathway as well as that of a tensile continuity may have much in common with chemical transducing pathways in terms of sharing common messengers and resulting altered cell functions. In addition, shear stress induced membrane perturbations could mimic a ligand-receptor interaction and induce the SRC, or direct activation of an effector may be possible (Frangos et al. 1988).

One speculation about the end-product of a shear-triggered pathway is the possibility of synthesis of shear specific proteins. Goochee et al. (1987) and Rosenberg (1987) demonstrated that such a specific protein class may exist in mammalian as well as in plant cells grown in suspension as a result of appropriate shear stress (level and exposure). Recent work in the author's laboratory is devoted to plant stress proteins, their characterization, elucidation of their control (transcription, translation), and to promoter characterization.

2.3.2.3 Temperature Stimuli and Other Environmental Factors
Temperature fluctuations may have a consequence similar to pressure fluctuations discussed above. An improper bioreactor design in terms of heat transfer (cooling/heating coils, jacket, low mixing) may cause temperature gradients

and inhomogeneities within the bioreactor. Under such conditions cells may be exposed to fluctuating temperature (e.g., that of jacket and bulk liquid). Similarly, the response to nonisothermal temperature fluctuations due to poor control design is also important. A temperature shock (usually several degrees Celsius above the optimum and for a time period of 1–2 hours) results in the production of heat shock proteins (HSPs) at the expense of normal protein synthesis (e.g., a rDNA product) (Piper 1987). It is not known, however, whether temperature fluctuations can cause a similar response.

Similar to temperature shocks (shifts), drastic change in the dissolved oxygen concentration may provoke a similar situation (anaerobic stress proteins).

2.4 DYNAMICS OF INTERACTION OF CELL AND ENVIRONMENT

One possible way to analyze the interaction between a cell and the environment is to employ systems analysis/synthesis and/or regime analysis (Prokop 1982; Roels 1982; Sweere et al. 1987). Prokop (1982) advanced a notion of systemic variables for the description of complex, ill-defined biological/biotechnological systems. The systemic variable is defined as a decisive variable for each level of hierarchy of biological systems (subcellular, cellular, population, and reactor), analogous to the notion of a limiting step in reaction engineering. The dynamic hierarchy of time characteristics, i.e., the relaxation times of individual subsystems involved, is useful in this respect. The relaxation time (occasionally a term such as characteristic time, time constant, half-time, etc. are used) is a measure of the rate of mechanism; time needed by that mechanism to smooth out a change to a certain fraction. A low value for a characteristic time means a fast mechanism; a high value means a slow mechanism (Sweere et al. 1987). Other than the characteristic time of biological subsystems, one can use the cell generation time, rate of substrate consumption, and rate of oxygen uptake (the last two terms can be appropriately converted to time values) on a population level. Also, engineering characteristic times such as mixing time, can be included (Prokop 1982). Figure 2–14 presents two sets of characteristic times for both biological and reactor levels. Some methodology on their estimation can be found in Sweere et al. (1987).

In the following discussion, a comparison is made of the characteristic times of internal biological processes with those characterizing the relevant changes in environmental conditions. The following comments will be useful. (1) An absolute comparison of characteristic times is possible at each level of hierarchy (e.g., at the biological mechanisms level). Such comparison can determine strong interactions, e.g., mRNA stability (long times) will limit the overall protein synthesis. (2) A comparison between adjoining

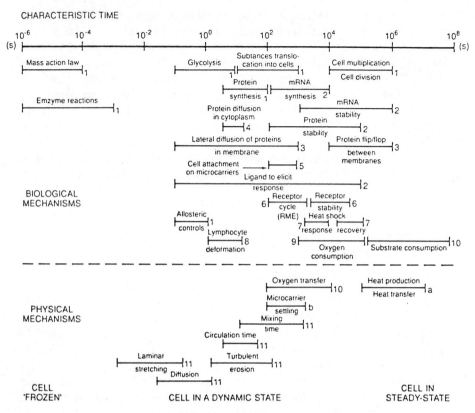

FIGURE 2-14 Characteristic times of biological and physical mechanisms. Numbers attached to bars indicate references. 1, Prokop 1982; 2, Stryer 1988; 3, Kotyk et al. 1988; 4, Jain and Cohen 1981; 5, Hu et al. 1985; 6, Breitfeld et al. 1985, Schlessinger 1988, and Salas et al. 1987; 7, Bergen en Henegouwen and Linnemans 1987; 8, Chien 1987; 9, Fleischaker and Sinskey 1981; 10, Birch et al. 1987; 11, Sweere et al. 1987 (data on air-lift reactor). a, Data of Sweere et al. (1987) shifted by 3 orders of magnitude due to similar shift of oxygen consumption; b, estimation.

levels is possible only for certain associated phenomena between two levels and will yield a weak interaction. Three possibilities are treated further. (2A) The environmental effect is not important when its characteristic times are small (change is too fast) compared to that of the organism. This means that the organism cannot adapt to such a change and is "frozen" in the initial state (Esener et al. 1983). For example, comparing time domains of diffusion, mixing, and oxygen transfer with those of oxygen and substrate consumption indicate that the biological processes should not be affected. Similarly, cells will not attach to microcarriers (they will be frozen in the initial suspension state) when turbulent erosion times are too short compared to the cell attachment. The turbulent erosion represents the Reynolds stresses in the impeller region. (2B) When the environmental processes are too slow (long characteristic times), the organism will be at steady state in relation to the particular environment (Esener et al. 1983). In case the time needed for oxygen transfer to the liquid phase is much larger than that of oxygen consumption, oxygen depletion will occur in the bioreactor. Related to this could be long characteristic times of liquid circulation as well as short ones of heat transfer and oxygen transfer, resulting in temperature and dissolved oxygen gradients. In this case a synthesis of HSP and/or anaerobic stress proteins may also occur. In another situation, the micro-carriers will settle in case of relatively longer characteristic times for mixing and circulation. (2C) For the overlapping time domains between biological and environmental mechanisms, the organism will dynamically respond to a change and will not be directly related to the environmental conditions (Sweere et al. 1987). There may be a phase or frequency shift in the response. This is the most important situation relevant to the topic of this paper. For example, as long as the characteristic times of the turbulent erosion and the laminar stretching on one side and that of lymphocyte deformation are comparable (see Figure 2–14), no damage to cells occurs since the cell can recover from a deformation. The laminar stretching represents viscous stress in the bulk liquid. Figure 2–14 presents some relevant data on dynamics of biological subsystems as well as those on the bioreactor side.

From the knowledge originating from the characteristic time analysis, one can make conclusions about which mechanisms should be further in-vestigated and possibly about the "bottleneck" mechanisms. The analysis of transient behavior and cell dynamics is important for bioreactor control. More fundamental understanding of a largely empirical situation will result in better mechanistic models based on both biological and physical under-lying principles. Another goal is to maintain the same regime at different levels for a reactor scale up. The one major advantage associated with the characteristic time-based scale up is saving time and resources, since pro-duction-scale experiments are always expensive and do not always lead to desired results. This saving is achieved through the investigation of char-acteristic times at the scale-down reactor version.

The data on dynamic response of cellular subsystems are, however, very rare and difficult to obtain. It should be noted that some characteristic times can be estimated on theoretical grounds, others only experimentally. The regime analysis based on characteristic times will no doubt gain more im-portance as the progress in both theoretical and experimental areas contin-ues. At this point, the characteristic times concept provides a useful theo-retical framework for further advancement.

REFERENCES

Absolom, D.R., Lamberti, F.V., Policova, Z., et al. (1983) *Appl. Environ. Microbiol.* 46, 90–97.

Altshuler, G.L., Dilwith, R., Sowek, J., and Belfort, G. (1986) in *Biotechnology and Bioengineering Symposium* No. 17 (Scott, C.D., ed.), pp. 725–736, Wiley, New York.

Ando, J., Komatsuda, T., and Kamiya, A. (1988) *In Vitro Cell Develop. Biol.* 24, 871–877.

Barnes, D. (1987) *Biotechniques* 5, 534–542.

Barngrover, D. (1986) in *Mammalian Cell Technology* (Thilly, W.G., ed.), pp. 131–149, Butterworths, Boston.

Baserga, R., and Weibel, F. (1969) in *International Review of Experimental Pathology*, Vol. 7 (Richter, G.W., and Epstein, M.A., eds.), pp. 1–30, Academic Press, New York.

Beato, M. (1989) *Cell* 56, 335–344.

Beck, K. (1987) in *Cytomechanics. The Mechanical Basis of Cell Form and Structure* (Bereiter-Hahn, J., Anderson, O.R., and Reif, W.-E., eds.), pp. 79–99, Springer, Berlin.

Ben-Ze'ev, A., Farmer, S.R., and Penman, S. (1980) *Cell* 21, 365–372.

Bereiter-Hahn, J. (1987) in *Cytomechanics. The Mechanical Basis of Cell Form and Structure* (Bereiter-Hahn, J., Anderson, O.R., and Reif., W.-E., eds.), pp. 3–30, Springer, Berlin.

Bergen en Henegouwen, van P.M.P., and Linnemans, W.A.M. (1987) *Exp. Cell Res.* 171, 367–375.

Berridge, M.J. (1985) in *The Molecules of Life (Readings from Scientific American)*, pp. 96–106, Freeman, New York.

Berridge, M.J. (1988) *Proc. R. Soc. London B.* 234, 359–378.

Birch, J.R., Lambert, K., Thompson, P.W., Kenney, A.C., and Wood, L.A. (1987) in *Large Scale Cell Culture Technology* (Lydersen, B.J., ed.), pp. 1–20, Hanser, Munich, Germany.

Bissell, M.J., and Aggeler, J. (1987) in *Mechanisms of Signal Transduction by Hormones and Growth Factors* (Cabot, M.C., and McKeehan, W.L., eds.), pp. 251–262, Liss, New York.

Brackenburry, R., Greenberg, M.E., and Edelman, G.M. (1984) *J. Cell Biol.* 99, 1944–1954.

Breitfeld, P.P., Simmons, Jr., C.F., Strous, G.J.A.M., Geuze, H.J., and Schwartz, A.L. (1985) in *International Review of Cytology* Vol. 97 (Bourne, G.H., Danielli, J.F., and Jeon, J.W., eds.), pp. 47–95, Academic Press, New York.

Brown, M., and Goldstein, J. (1979) *Proc. Natl. Acad. Sci. USA* 76, 3330–3337.

Buck, C.A., and Horwitz, A.F. (1987) in *Annual Review of Cell Biology*, Vol. 3 (Palade, G.E., Alberts, B.M., and Spudich, J.A., eds.), pp. 179–205, Annual Reviews, Palo Alto, CA.

Burridge, K., Fath, K., Kelly, T., Nuckolls, G., and Turner, C. (1988) in *Annual Reviews of Cell Biology*, Vol. 4, (Palade, G.E., Alberts, B.M., and Spudich, J.A., eds.), pp. 487–525, Annual Reviews, Palo Alto, CA.

Butler, M. (1987) in *Advances in Biochemical Engineering/Biotechnology*, Vol. 34, (Fiechter, A., ed.), pp. 57–84, Springer, Berlin.

Cahn, F. (1989) *Informatrix™* Porous Microcarriers, Biomat Corp., Belmont, MA.

Chasis, J.A., Mohandas, N., and Shohet, S. (1985) *J. Clin. Invest.* 75, 1919–1926.

Chen, C., Corbley, M.J., Roberts, T.M., and Hess, P. (1988) *Science* 239, 1024–1026.

Chen, L.B., Rosenberg, S., Nodakavukaren, K.K., et al. (1985) in *Hybridoma Technology in the Bioscience and Medicine* (Springer, T.A., ed.), pp. 251–268, Plenum Press, New York.

Chien, S. (1987) in *Annual Review of Physiology*, Vol. 49 (Berne, R.M., and Hoffman, J.F., eds.), pp. 177–192, Annual Reviews, Palo Alto, CA.

Chittur, K.K., McIntire, L.V., and Rich, R.R. (1988) *Biotechnol. Prog.* 4, 89–96.

Christensen, O. (1987) *Nature* 330, 66–68.

Cleveland, D.W. (1982) *Cell* 28, 689–691.

Crouch, C.F., Fowler, H.W., and Spier, R.E. (1985) *J. Chem. Technol. Biotechnol.* 35B, 273–281.

Curtis, A. (1987) *J. Cell Sci.* 87, 609–611.

Czech, M.P., Klarlund, J.K., Yagaloff, K.A., Bradford, A.P., and Lewis, R.E. (1988) *J. Biol. Chem.* 263, 11017–11020.

Dalili, M., and Ollis, M.D.F. (1988) *Biotechnol. Lett.* 10, 781–786.

Darnell, J., Lodish, H., and Baltimore, D. (1986) *Molecular Cell Biology*, Scientific American Books, New York.

Darzynkiewicz, Z., Traganos, F., and Melamed, M.R. (1980) *Cytometry* 1, 98–108.

Das, P., Kumar, R., and Ramkrishna, D. (1987) *Chem. Eng. Sci.* 42, 213–220.

Davies, P.F., Dewey, Jr., C.F.D., Bussolari, S.R., Gordon, E.J., and Gimbrone, Jr., M.A. (1984) *J. Clin. Invest.* 73, 1121–1129.

Dean, Jr., R.C. Karkane, S.B., Phillips, P.G., Ray, N.G., and Rundstadler, Jr., P.W. (1987) in *Large Scale Cell Culture Technology* (Lydersen, B.K., ed.), pp. 145–167, Hanser, Munich, West Germany.

Duszyk, M., Swab III, B., Zahalak, G.I., Qian, H., and Elson, E.L. (1989) *Biophys. J.* 55, 683–690.

Edelman, G.M. (1988) *Biochemistry* 27, 3533–3543.

Elson, E.L. (1988) in *Annual Review of Biophysics and Biophysical Chemistry*, Vol. 17 (Engelman, D.M., Cantor, C.R., and Pollard, T.D., eds.), pp. 397–430, Annual Reviews, Palo Alto, CA.

Esener, A.A., Roels, J.A., and Kossen, N.W.F. (1983) *Biotechnol. Bioeng.* 25, 2803–2841.

Eshhar, Z. (1985) in *Hybridoma Technology in the Biosciences and Medicine* (Springer, T.A., ed.), pp. 3–41, Plenum Press, New York.

Fernandez, E.J., and Clark, D.S. (1987) *Enzyme Microbiol. Technol.* 9, 259–271.

Fernandez, E.J., Mancuso, A., Clark, D.S. (1988) *Biotechnol. Prog.* 4, 173–183.

Fiechter, A. (1978) in *Biotechnology (Proc. 1st Eur. Congr. Biotechnol. 1978), Dechema Monogr.*, Vol. 82, pp. 17–36, Verlag Chemie, Weinheim, Germany.

Fleischaker, R. (1987) in *Large Scale Cell Culture Technology* (Lydersen, B.K., ed.), pp. 59–79, Hanser, Munich, Germany.

Fleischaker, R.J., and Sinskey, A.J. (1981) *Eur. J. Appl. Microbiol. Biotechnol.* 12, 193–197.

Francis, G.W., Fisher, L.R., Gamble, R.A., and Gingell, D. (1987) *J. Cell Sci.* 87, 519–523.

Frangos, J.A., McIntire, L.V., and Eskin, S.G. (1988) *Biotechnol. Bioeng.* 32, 1053–1060.

Freshney, R.I. (1987) *Culture of Animal Cells. A Manual of Basic Technique*, 2nd ed., Liss, New York.

Fulton, A.B. (1984) *The Cytoskeleton. Cellular Architecture and Choreography*, Chapman and Hall, New York.

Goldstein, J.L., Brown, M.S., Anderson, R.G.W., Russell, D.W., and Schneider, W.J. (1985) in *Annual Reviews of Cell Biology*, Vol. 1 (Palade, G.E., Alberts, B.M., and Spudich, J.A., eds.), pp. 1–39, Annual Reviews, Palo Alto, CA.

Goochee, C.F., Passini, C., Lall, R., Morrison, D.R., and Kalmuz, E.E. (1987) in (194th Am. Chem. Soc. Natl. Meeting, New Orleans, LA, August 30–Sept. 4, 1987, Div. Microbial Biochem. Technol. (Abstract no. 52).

Gowingt, L.R., Tellam, R.L., and Banyard, M.R.C. (1984) *J. Cell Sci.* 69, 137–146.

Grabner, R., and Paul, E.L. (1981) U.S. Patent 4,296,204.

Green, M.R. (1989) *Cell* 56, 1–3.

Griffiths, J.B., and Riley, P.A. (1985) in *Animal Cell Biotechnology*, Vol. 1 (Spier, R.E., and Griffiths, J.B., eds.), pp. 17–48, Academic Press, London.

Guarnaccia, S.P., and Schnaar, R.L. (1982) *J. Biol. Chem.* 257, 14288–14292.

Guharay, F. (1984) *J. Physiol.* 352, 685–701.

Hiramoto, Y. (1987) in *Cytomechanics. The Mechanical Basis of Cell Form and Structure* (Bereiter-Hahn, J., Anderson, O.R., and Reif, W.-E., eds.), pp. 31–46, Springer, Berlin.

Hochmuth, R.M., and Waugh, R.E. (1987) in *Annual Review of Physiology* Vol. 49 (Berne, R.M., and Hoffman, J.F., eds.), pp. 209–219, Annual Reviews, Palo Alto, CA.

Hu, W.-S., Giard, D.J., and Wang, D.I.C. (1985) *Biotechnol. Bioeng.* 27, 1466–1476.

Ingber, D.E., and Jamieson, J.D. (1984) in *Gene Expression during Normal and Malignant Differentiation* (Anderson, L.C., Gahmberg, C.G., and Ekblon, P., eds.), pp. 13–32, Academic Press, New York.

Jain, S., and Cohen, C. (1981) *Macromolecules* 14, 759–765.

Kikkawa, U., and Nishizuka, Y. (1986) in *Annual Review of Cell Biology* Vol. 2 (Palade, G.E., Alberts, B.M., and Spudich, J.A., eds.), pp. 149–178, Annual Reviews, Palo Alto, CA.

Kleinman, H.K., Luckenbill-Edds, L., Cannon, F.W., and Sephel, G.C. (1987) *Anal. Biochem.* 166, 1–13.

Kornblihtt, A.R., and Gutman, A. (1988) *Biol. Rev.* 63, 465–507.

Kotyk, A., Janáček, K., and Koryta, J. (1988). *Biophysical Chemistry of Membrane Functions*, Wiley, Chichester, England.

Langarger, B., Moeremans, M., Daneels, G., et al. (1986) *J. Cell Biol.* 102, 200–209.

Lansman, J.B., Hallam, T.J., and Rink, T.J. (1987) *Nature* 325, 811–813.

Levesque, M.J., Sprague, E.A., Schwartz, C.J., and Nerem, R.M. (1989) *Biotechnol. Prog.* 5, 1–8.

Lichtstein, D., and Rodbard, D. (1987) *Life Sci.* 40, 2041–2051.

Limbird, L.E. (1987) *Cell Surface Receptors: A Short Course on Theory and Methods*, Martinus Nijhoff Publ., Boston.

Low, M.G. (1987) *Biochem. J.* 244, 1–13.

Lydersen, B.K. (1987) in *Large Scale Cell Culture Technology* (Lydersen, B.K., ed.), pp. 169–192, Hanser, Munich, Germany.

Marsillo, E., Sobel, M.E., and Smith, B.D. (1984) *J. Biol. Chem.* 259, 1401–1404.

Mazur, M.T., and Williamson, J.R. (1977) *J. Cell. Biol.* 75, 185–199.

McClay, D.R., and Ettensohn, C.A. (1987) in *Annual Reviews of Cell Biology*, Vol. 3 (Palade, G.E., Alberts, B.M., Spudich, J.A., eds.), pp. 319–345, Annual Reviews, Palo Alto, CA.

McCormick, F. (1989) *Cell* 56, 5–8.

McDonald, J.A. (1988) in *Annual Review of Cell Biology*, Vol. 4 (Palade, G.E., Alberts, B.M., and Spudich, J.A., eds.), pp. 183–207, Annual Reviews, Palo Alto, CA.

Mely-Goubert, B., and Bellgran, D. (1981) *J. Immunol.* 127, 399–401.

Mooibroek, M.J., and Wang, J.H. (1988) *Biochem. Cell Biol.* 66, 557–566.

Nerem, R.M., and Levesque, M.J. (1987) *Proc. 2nd Japan-US-China Conf. on Biomechanics* (Hayashi, K., and Seguchi, Y., eds.), Osaka, Japan, Sept. 28–Oct. 2, 1987.

Obrink, B. (1986) in *Frontiers of Matrix Biology*, Vol. 11 (Robert, L., ed.), pp. 123–138, Karger, Basel, Switzerland.

Olesen, S.-P., Clapham, D.E., and Davies, P.F. (1988) *Nature* 331, 168–170.

Packard, B. (1986) *Trends Biochem. Sci.* 11, 154.

Pasternak, C., and Elson, E.L. (1985) *J. Cell Biol.* 100, 860–872.

Petersen, N.O., McConnaughey, W.B., and Elson, E.L. (1982) *Proc. Natl. Acad. Sci. USA* 79, 5327–5331.

Pfeifer & Langen (1987) *Dormacell Microcarriers for Cell Cultures*, Dormagen, Germany.

Pierschbacher, M.D., and Ruoslahti, E. (1984) *Proc. Natl. Acad. Sci. USA* 81, 5985–5988.

Piper, P.W. (1987) *Sci. Progr. Oxford* 71, 531–544.

Pitt, W.G., Park, K. and Cooper, S.L. (1986) *J. Coll. Interf. Sci.* 111, 343–362.

Prokop, A. (1982) *Int. J. Gen. Systems* 8, 7–31.

Prokop, A., and Rosenberg, M.Z. (1989) in *Advances in Biochemical Engineering/Biotechnology* Vol. 39 (Fiechter, A., ed.), pp. 29–71, Springer, Berlin.

Quintanilla, M., Brown, K., Ramsden, M., and Balmain, A. (1986) *Nature* 322, 78–80.

Rawls, R.L. (1987) *Chem. Eng. News* 65, 26–39.

Reid, L.M., Abren, S.L., and Montgomery, K. (1988) in *The Liver Biology and Pathology* 2nd ed. (Arias, I.M., Jakoby, W.B., Popper, H., Schachter, D., and Shafritz, D.A., eds.), pp. 717–737, Raven, New York.

Roberts, D.D., and Ginsburg, V. (1988) *Arch. Biochem. Biophys.* 267, 405–415.

Rodbell, M. (1985) *Trends Biochem. Sci.* 10, 461–464.

Roels, J.A. (1982) *J. Chem. Technol. Biotechnol.* 32, 59–72.

Rosenberg, M.Z. (1987) Doctoral Dissertation, Washington University, St. Louis, MO.

Rosenthal, W., Hescheler, J., Trantwein, W., and Schultz, G. (1988) *FASEB J.* 2, 2784–2790.

Ruoslahti, E. (1988) in *Annual Review of Biochemistry* Vol. 57 (Richardson, C.C., Boyer, P.D., Dawid, I.B., and Meistner, A., eds.), pp. 375–413, Annual Reviews, Palo Alto, CA.

Rutter, P.R., and Vincent, B. (1988) in *Physiological Models in Microbiology* Vol. 2 (Bazin, M.J., Prosser, J.I., and Bazin, M.J., eds.), pp. 87–107, CRC Press, Boca Raton, FL.

Sairam, M.R., and Bhargavi, G.N. (1985) *Science* 229, 65–67.

Salas, P.J.I., Vega-Salas, D.E., and Rodriguez-Boulan, E. (1987) *J. Membrane Biol.* 98, 223–236.

Salisbury, J.L., Condeelis, J.S., and Satir, P. (1983) in *International Review of Experimental Pathology*, Vol. 24 (Richter, G.W., and Epstein M.A., eds.), pp. 1–62, Academic Press, New York.

Sato, M., Leimbach, G., Schwarz, W.H., and Pollard, T.D. (1985) *J. Biol. Chem.* 260, 8585–8592.

Sato, M., Schwarz, W.L., and Pollard, T.D. (1987) *Nature* 325, 828–830.

Schakenraad, J.M., Busscher, H.J., Wildevuur, Ch.R.H., and Arends, J. (1986) in *Advances in Biomaterials*, Vol. 6 (Christel, P., Neunier, A. and Lee, A.J.C., eds.), pp. 263–268, Elsevier, Amsterdam.

Schakenraad, J.M., Busscher, H.J., Wildevuur, Ch. R.H., and Arends, J. (1988) *Cell Biophys.* 13, 75–91.

Schlessinger, J. (1988) *Biochemistry* 27, 3119–3123.
Schmidt-Schonbeim, G.W. (1987) in *Handbook of Bioengineering* (Skalak, R., and Chien, S., eds.), pp. 13.1–13.25, McGraw Hill, New York.
Schneider, C., King, R.M., and Philipson, L. (1988) *Cell* 54, 787–793.
Scott, R.H., and Dolphin, A.C. (1987) *Nature* 330, 760–762.
Schulz, R., Krafft, H., and Lehmann, J. (1986) *Biotechnol. Lett.* 8, 557–560.
Schulz, R., Krafft, H., Piehl, G.W., and Lehmann, J. (1987) *Develop. Biol. Stand.* 66, 489–495.
Sherbet, G.V., Lakshmi, M.S., and Rao, K.V. (1972) *Exp. Cell Res.* 70, 113–123.
Shintani, Y., Iwamoto, K., and Kitano, K. (1988) *Appl. Microbiol. Biotechnol.* 27, 533–537.
Skalak, R., Ozkaya, N., and Skalak, T.C. (1989) in *Annual Review of Fluid Mechanics* Vol. 21 (Lumley, J.L., van Dyke, M., and Reed, H.L., eds.), pp. 167–204, Annual Reviews, Palo Alto, CA.
Spier, R.E. (1988) in *Animal Cell Biotechnology* Vol. 3 (Spier, R.E., and Griffiths, J.B., eds.), pp, 29–53, Academic Press, London.
Sprague, E.A., Steinbach, B.L., Nerem, R.M., and Schwartz, C.J. (1987) *Circulation* 76, 648–656.
Stahl, P., and Schwartz, A.L. (1986) *J. Clin. Invest.* 77, 657–662.
Stockbridge, L.L., and French, A.S. (1988) *Biophys. J.* 54, 187–190.
Stryer, L. (1988) *Biochemistry* 3rd ed., Freeman, New York.
Sung, K.-L.P., Schmid-Schonbeim, G.W., Schuessler, G.B., and Chien, S. (1986) *Biorheology* 23, 284 (Abstract).
Sweere, A.P.J., Luyben, K.Ch.A.M., and Kossen, N.W.F. (1987) *Enzyme Microbiol. Technol.* 9, 386–398.
Taylor, G.I. (1935) *Proc. R. Soc. London A.* 151, 494–564.
Tolbert, W.R., Lewis, Jr., C., White, P.J., and Feder, J. (1985) in *Large-Scale Mammalian Cell Culture* (Feder, J., and Tolbert, W.R., eds.), pp. 97–123, Academic Press, New York.
Valberg, P.A., and Albertini, D.F. (1985) *J. Cell Biol.* 101, 130–140.
van Oss, C.J. (1989) *Cell Biophys.* 14, 1–16.
Varani, J., Wass, J.A., and Rao, K.M.K. (1983) *J. Natl. Cancer Inst.* 70, 805–809.
Wallach, D.F.H. (1987) *Fundamentals of Receptor Molecular Biology*, Marcel Dekker, New York.
Waugh, R.E., and Hochmuth, R.M. (1987) in *Cytomechanics. The Mechanical Basis of Cell Form and Structure* (Bereiter-Hahn, J., Anderson, O.R., and Reif, W.-E., eds.), pp. 249–260, Springer, Berlin.
Welch, W.J., and Suhan J.P. (1986) *J. Cell Biol.* 103, 2035–2952.
Whyte, P., Buchkovich, K.J., Horowitz, J.M., et al. (1988) *Nature* 334, 124–129.
Yarden, Y., and Ullrich, A. (1988a) in *Annual Review of Biochemistry* Vol. 57 (Richardson, C.C., Boyer, P.D., Dawid, I.B., and Meistner, A., eds.), pp. 443–478, Annual Reviews, Palo Alto, CA.
Yarden, Y., and Ullrich, A. (1988b) *Biochemistry* 27, 3113–3119.
Zaner, K.S. (1986) *Biorheology* 23, 209 (Abstract).

Protein Glycosylation: Function and Factors that Regulate Oligosaccharide Structure

Joseph K. Welply

A common modification of polypeptides made in eukaryotic cells is the covalent attachment of oligosaccharides. Sometimes the oligosaccharides are very large and they frequently contribute more to the mass of the glycoprotein than does the polypeptide backbone. They also often comprise a significant portion of the protein surface. Protein-bound oligosaccharides perform a varied assortment of biological functions, many of which are only beginning to be appreciated. Among these functions are that the oligosaccharides influence the thermal stability and solubility of the protein; they protect polypeptides from proteolysis by restricting the accessibility of susceptible peptide bonds to proteases; they influence the half-life of proteins in serum because several carbohydrate-specific receptors on liver cells and macrophages recognize and eliminate glycoproteins with the appropriate terminal sugars from the circulation by receptor-mediated endocytosis; they may be antigenic as are the ABO[H] blood-group structures of humans or, conversely, they may mask immunogenic areas within the polypeptide; they can affect the enzyme and bioactivity of the protein to which they are attached; and they can direct proteins to intra- or extracellular sites.

For cells, the oligosaccharides of glyoproteins as well as those of the glycolipids that are lodged within the plasma membrane provide a coating that covers the cell. This layer of sugar prevents contact between cells and pathogens; it is involved in the communication between cells and the substratum, and it affects cell migration and metastasis. Further, many oligosaccharides, particularly glycolipids, are tumor antigens and may participate in certain forms of cancer.

Often, specific oligosaccharide structures are required for specific biological events. For instance, the infection of cells by influenza virus involves recognition and binding of the viral hemagluttin protein to sialic acid residues, which are linked in a specific manner to an adjacent monosaccharide and are contained within a subset of glycolipids on the surface of the infected cells. Different strains of the virus bind to nearly identical structures that differ only in the linkage of their sialic acid residues (Rogers et al. 1985). In the absence of the appropriate and exact sugar structure, the virus will not infect the cell. The role of oligosaccharides in the regulation of blood coagulation also depends upon a specific sugar structure. In this case, the important sugar is heparin, a proteoglycan type of oligosaccharide. When antithrombin III binds to a small tetrasaccharide of defined and unique structure within the large heparin oligosaccharide, the protein is activated and converted into a powerful inhibitor of thrombin, a pivotal enzyme in the blood coagulation cascade (Ofosu et al. 1989). From these and other examples it can be concluded that a knowledge of the chemical structure of oligosaccharides is essential for understanding how they operate. Recent technical advances for sequencing carbohydrates are providing opportunities to better accomplish this goal (Ginsburg and Robbins 1984; Green and Baenziger 1989).

Several proteins of therapeutic interest (i.e., erythropoietin, tissue plasminogen activator, interleukins, and interferons) are glycoproteins that are being produced in their recombinant form. These proteins contain oligosaccharides. It is essential to characterize the structure of the sugars on recombinant proteins to determine how these structures may affect the bioactivity of the protein, and to be aware of factors that might alter oligosaccharide structure. Several events influence cellular glycosylation. In fact, altered structures of oligosaccharides are one of the most common responses of cells to external cues that initiate developmental or disease-related processes. Changes in the structure of protein-bound oligosaccharides may arise by several mechanisms. Minor substitutions in the amino acid sequence of a glycoprotein or modifications in the interaction of the protein with other subunits frequently change the structure of its sugar. The cell type chosen for the expression of the glycoprotein is another crucial determinant because all cells contain their own unique complement of the glycosyltransferases for formation of the oligosaccharides and, therefore, each cell may glycosylate a protein in a different way. Most importantly, a variety of agents and cell growth conditions that modify the metabolism of the cell result in changes in the synthesis and structure of oligosaccharides. The material presented within this chapter is intended to be an introduction to protein-bound oligosaccharides and their structures and as an overview of the factors and conditions of cell culture that may alter the structures of the oligosaccharides.

3.1 STRUCTURE AND HETEROGENEITY OF THE PROTEIN-BOUND OLIGOSACCHARIDES

Protein-bound oligosaccharides are placed into general classifications on the basis of the amino acid to which the oligosaccharide is linked and by the monosaccharide that is directly attached to the polypeptide. The major classes are listed in Table 3–1. N-linked, mucin, and proteoglycan types of oligosaccharides are the most common forms. The collagen type of oligosaccharide is found virtually exclusively in collagen. The linkage of the oligosaccharide is to hydroxylysine, a modified form of lysine. The serine/ threonine linked GlcNAc and the glycolipid tail linked oligosaccharides have been discovered recently and are currently a topic of much research. Many

TABLE 3–1 Types of Oligosaccharide Linkage to Proteins from Animal Sources

Common Name	Linkage[1]	Representative Proteins
Mucin type	GalNAc to Ser/Thr[2]	Mucins
Collagen type	Gal to Hydroxylysine[3]	Collagen
Proteoglycan type	Xyl to Ser/Thr[4]	Proteoglycans
O-linked GlcNAc	to Ser/Thr[5]	Nuclear/cytoskeletal
N-linked	GlcNAc to Asparagine[6]	Membrane/secreted
Glycolipid tails	C-terminus[7]	Parasite cell surface
Glucosylation	Glc to amino groups[8]	Aging, diabetes

GalNac, N-acetylgalactosamine; Gal, galactose; Xyl, xylose; GlcNAc, N-acetylglucosamine; Glc, glucose.

[1]Only the monosaccharide directly attached to the protein is shown.

[2]Mucin-type oligosaccharides are generally small and contain high levels of sialic acid and galactose, N-acetylglucosamine, and fucose residues.

[3]Only found in collagen. The galactose is often capped with glucose.

[4]The oligosaccharides of proteoglycans are called glycosaminoglycans. Glycosaminoglycans are very large, highly sulfated, and contain multiple repeating disaccharides made up of an amino sugar and an acidic sugar like glucuronic acid. Some of the better known proteoglycans are heparin and chondroitin sulfate. These differ from each other by size, composition, and sulfation.

[5]A single N-acetylglucosamine residue is attached to the polypeptide without additional sugar residues.

[6]Asparagine is part of an Asn-X-Ser/Thr sequence. These oligosaccharides generally contain 10 to 20 monosaccharide residues. Every N-linked oligosaccharide has a common core comprised of two N-acetylglucosamine and three mannose residues.

[7]Oligosaccharides with 8 to 10 residues are attached to the c-terminus of the protein via an ethanolamine linking arm. At the other end of the sugar is a covalently attached phosphatidylinositol that anchors the protein and the sugar to the membrane.

[8]Glucose addition to lysine side chains is nonenzymatic. It is found in proteins of aged red blood cells and in serum proteins of diabetics.

of the glycoproteins in the nucleus have single GlcNAc residues linked to threonine or serine as do a few plasma membrane and cytoskeletal proteins. The O-linked GlcNAc residue is highly immunogenic and is contained within some of the well-known transcriptional activators. A recent study showed that *Drosophila* chromatin is covered with proteins that contain O-linked GlcNAc, suggesting that this modification may have a role in the interaction of protein with DNA or in transferring proteins from the cytoplasm to the nucleus. The glycolipid tail oligosaccharides are located at the carboxyl terminus of a set of plasma membrane proteins, which are anchored to the membrane via the glycolipid. Upon treatment of cells with a specific phosphoinositol specific phospholipase, the bond between the lipid and sugar is broken and proteins that contain this form of oligosaccharide are released from the surface. Parasite coat proteins are among those that contain a glycolipid tail modification. The coat proteins of the parasite are released from the membrane during a process known as antigenic variation, the mechanism that parasites utilize to replace their coat proteins and thereby escape the immune system of the host. The last form of glycosylation commonly found on animal proteins is glucose addition to lysine residues. This is a nonenzymatic form of glycosylation that is found on proteins from aged cells and in serum of people with diabetes. The function of this form of glycosylation is unknown. The remainder of this review will discuss the N, mucin, and proteoglycan types since these are found in many of the proteins currently being developed for therapeutic use.

A representative structure for each of the N-linked, mucin, and proteoglycan type oligosaccharides is presented in Figure 3–1. Each class contains a huge number of structures. Several features of oligosaccharide contribute to its structure, including the composition and sequential ordering of monosaccharides. Another feature, perhaps the most important and different from most other macromolecules, is that their monosaccharide units are not linked uniformly to each another. Rather, each monosaccharide within an oligosaccharide is linked to an adjacent residue through any of three or four of its carbon atoms. Therefore, while two oligosaccharides may have an identical composition and order of monosaccharides, their structure and therein their function may be different because the linkages between the monosaccharides are not the same. Individual monosaccharides may also be linked to more than a single neighboring monosaccharide, which results in branching and adds another dimension to the complexity of oligosaccharides. Further, each linking bond has an alpha or beta configuration and each monosaccharide a preferred conformation. Finally, many sugars undergo additional modifications such as phosphorylation, sulfation, and acetylation. For instance, proteoglycan sugars contain multiple N- and O-linked sulfate. N-linked oligosaccharides of lysosomal proteins contain phosphates. Also, a variety of modified forms of acetylated sialic acid residues exist on N-linked and mucin type oligosaccharides.

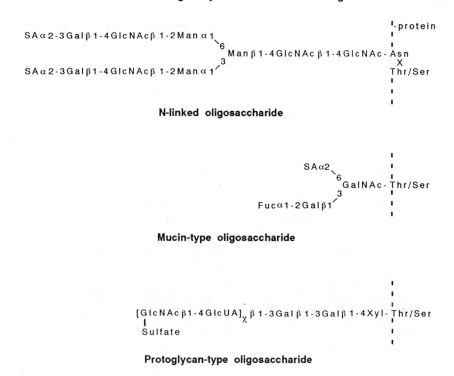

SAα2-3Galβ1-4GlcNAcβ 1-2Man α 1

Manβ 1-4GlcNAc β 1-4GlcNAc- Asn

SAα2-3Galβ1-4GlcNAcβ 1-2Man α 1

N-linked oligosaccharide

SAα2

GalNAc- Thr/Ser

Fucα1-2Galβ1

Mucin-type oligosaccharide

[GlcNAcβ1-4GlcUA]$_\chi$ β 1-3Gal β 1-3Gal β 1-4Xyl- Thr/Ser

Sulfate

Protoglycan-type oligosaccharide

FIGURE 3–1 Representative structures of the classes of protein-bound oligosaccharides.

The cellular or extracellular location of a glycoprotein as well as the type of oligosaccharide it contains often suggests functions for the oligosaccharide itself. Glycoproteins of the proteoglycan class are often found within the extracellular matrix between layers of different cell types. The structure of the sugars of proteoglycans, the glycosaminoglycans, are cell and tissue specific and are developmentally regulated. The extremely high content of negative charge on the glycosaminoglycans of proteoglycans causes water in some tissues (e.g., cartilage) to be highly organized and thereby contributes to the functional properties of the tissue. Glycosaminoglycans also interact with a variety of other proteins found in the extracellular matrix, such as fibronectin and laminin, thereby directly affecting the activity of these molecules and indirectly affecting several biological events. A variety of other proteins, such as several growth factors, bind to proteoglycans. Sometimes this binding permits the attachment of secreted enzymes to the surface of cells (e.g., cholesterol esterase binding to intestinal epithelial cells) and thereby promotes the biological activity of the enzyme.

Mucin type oligosaccharides cover all exposed epithelial cell surfaces in the body. The number of structurally unique oligosaccharides within the mucin family is enormous. Proteins that contain numerous mucin-like chains are very important as lubricants for epithelial cell surfaces. These proteins also protect the cells from infection by binding to infective organisms, many of which invade cells by attachment to sugars on the epithelial cell surface.

N-linked oligosaccharides are present in numerous secreted membrane and intracellular proteins (e.g., lysosomal proteins). Much is known about the synthesis and structure of the N-linked sugars. The addition of the N-linked oligosaccharides to the polypeptide can be predicted by the existence of Asn-X-Thr/Ser sequences, although these sites are not always glycosylated. The initial transfer of sugar to asparagine occurs in the endoplasmic reticulum, in contrast to mucin and proteoglycan synthesis which proceeds entirely within the Golgi apparatus. The N-linked structure later undergoes a series of modifications within the Golgi. A protein may contain more than one type of oligosaccharide. Some of the best examples of multisugar-type containing proteins are cell surface receptors such as the transferrin receptor and the class II antigen invariant chains.

An unusual and potentially important property of protein-bound oligosaccharides is heterogeneity. Mixtures of structurally related oligosaccharides are often present at individual glycosylation sites of the protein. This heterogeneity is referred to as oligosaccharide microheterogeneity and much may occur due to the cellular mechanism for the synthesis of oligosaccharides, which is not template driven and involves competition among similar glycosyltransferases that add monosaccharides sequentially to the protein-bound oligosaccharide as the protein migrates through different compartments of the cell. Heterogeneity is found not only within the terminal sugars but is also present at internal branches, and it is frequently different at every site within a protein that has multiple glycosylated sites. The structures of the sugars at each site are often closely related to one another. The function of the heterogeneity is unknown; however, under constant and stable conditions in vivo, the heterogeneity has been shown to be constant. Therefore, the precise heterogeneity at each site may be of biological significance. In effect, oligosaccharide heterogeneity produces a family of chemically defined proteins that differ from one another with respect to sugar but are identical in their polypeptide. Thus, glycosylation may be nature's way of diversifying proteins by a mechanism that requires only a few glycosyltransferases, is available to all glycoproteins, and is responsive to external physiological and developmental cues. Some recent evidence regarding tissue plasminogen activator suggests that individual glycoforms or classes of glycoforms of a protein have enzyme activities that are not identical (Hansen et al. 1988). It is too early to conclude whether oligosaccharide heterogeneity will universally result in significant differences in the bioactivity of proteins. But, if it is true, and there are already some examples among therapeutic proteins (e.g., erythropoietin and t-PA) (Hansen et al. 1988; Goto et al. 1988), then the challenge for those attempting to produce therapeutic proteins will be to understand the biological significance of the microheterogeneity and to establish conditions such that the desired oligosaccharide structural heterogeneity is attained. There are various ways that one might conceive that this could be accomplished, such as choosing the correct cell type for expressing the glycoprotein and adjusting conditions of cell growth [see below].

3.2   FACTORS THAT INFLUENCE THE STRUCTURE OF PROTEIN-BOUND OLIGOSACCHARIDES

3.2.1   Sequence of the Protein

Several lines of evidence suggest that the primary, tertiary,and quaternary structure of a protein play significant roles in determining the chemical nature of oligosaccharides at individual glycosylation sites. The importance of primary structure of the polypeptide was demonstrated initially when it was shown that influenza hemmagluttins with minor substitutions in amino acid sequence contain nonidentical oligosaccharide structures (Schwartz and Klenk 1981). This was also known to be true for other proteins, such as the homologous proteases, renin and cathepsin D (Faust et al. 1987). Although related in amino acid sequence, cathepsin D acquires phosphorylated mannose oligosaccharides and is shuttled to the lysosome whereas renin oligosaccharides are not phosphorylated and the protein is secreted by the cell.

The role that the tertiary conformation of a protein can have on the structure of the oligosaccharide is evident by the finding that denatured proteins that contain high mannose oligosaccharides are often susceptible to digestion with alpha mannosidase whereas the native protein is not. Accessibility is a function of the location of the oligosaccharide within the protein and of the particular glycosidase or glycosyltransferase that is acting upon the sugar. For glycosidases and glycosyltransferases that modify N-linked oligosaccharides, several studies suggest that alpha mannosidases and GlcNAc transferases are sensitive to the tertiary structure of the protein whereas galactosyl and siayltransferases are less affected.

The quaternary structure of glycoproteins is another important factor that influences oligosaccharide structure. It has been shown that the glycosylation of the common subunit of related dimeric proteins is controlled by their association with noncommon subunits. Thus, the glycosylation of the alpha subunit of follicle stimulating hormone is different from the glycosylation of the identical alpha subunit when in combination with the beta subunit of lutenizing hormone (Green and Baenziger 1988). Likewise the beta chain of beta macroglobulin is glycosylated differently when associated with its alpha chain than is the identical beta chain when associated with the alpha chain of lymphocyte function-associated antigen 1, even when both proteins are made simultaneously in the same cell (Dahms and Hart 1986).

Thus, the primary, tertiary, and quaternary structures of a protein are all important for determining the structure of its protein-bound oligosaccharides. Mutations that result in a change in the structure of the protein will influence the oligosaccharides. Amino acid alterations within the glycosylation site will prohibit glycosylation, and this loss may result in altered biological properties such as increased or decreased serum half-life or susceptibility to proteolysis. Likewise, alterations in other parts of the protein may create new glycosylation sites. The additional sugar may result in a protein that is resistant to neutralizing antibodies, as was observed for a strain of influenza virus (Skehel et al. 1984). Finally, an alteration in primary structure may lead to alterations in carbohydrate structure at distant glycosylation sites, as is the case for a variant of tissue plasminogen activator that has a single amino acid substitution at position 73, which results in a switch from a high mannose to complex N-linked oligosaccharide at position 117 (Welpy et al. 1989).

3.2.2 Cell Type

The cell in which a protein is produced has a dramatic effect upon the structure of protein-bound oligosaccharides. Oligosaccharide structures are created based upon the complement of transferases that the cell produces and upon the organization of the cell's secretory apparatus. While the primary and tertiary structure of the protein affect or govern the type of oligosaccharide at a particular site, N-linked versus mucin type, the cell fine tunes the structure of the oligosaccharide (i.e., high mannose versus complex). This tuning is subject to the metabolic and developmental state of the cell (see below). On the basis of exhaustive comparisons with recombinant tissue plasminogen activator and erythropoietin, where these proteins have been produced in several cell types, it can be concluded that virtually every cell will glycosylate the same protein differently (Parekh et al. 1989; Tsuda et al. 1988). Thus, the oligosaccharides that a protein inherits through expression in a particular cell type may not be those that will endow the protein with the full range of activities that it might contain were it produced within its natural cell type and environment. This issue is a difficult one for biotechnologists. Thus far, recombinant proteins have generally been expressed in cells that are most easily manipulated, or for which the most experience has been acquired (i.e., CHO, C127, or baby hamster kidney (BHK) cells), or cells that naturally produce large quantities of the protein (i.e., tumor-derived cells).

Another consideration is that cells from different species have been shown to synthesize oligosaccharides that are structurally dissimiliar. For instance, GlcNAcβ1-3Man is characteristic of serum glycoproteins of bovine and rat species, but not of humans. Likewise, Galα 1-3 Gal is produced on the N-linked oligosaccharides of New World monkeys and nonprimate mammals, including C127 cells, but not by Old World monkeys and humans. Additionally, humans and Old World monkeys have a naturally occurring antibody that will recognize this structure. Thus, caution must be exercised when deciding which cell line to use to produce proteins for therapeutic purposes in humans. For example, in humans, C127-produced proteins may be quickly recognized and cleared by virtue of the binding of Galα 1-3 Gal to naturally occurring anti-α Gal (Galili et al. 1987).

Tissue specific glycosylation is another feature of oligosaccharide synthesis. Several studies have shown that the glycosylation of the same protein will vary depending upon the tissue in which it is made. Thus, Thy 1 from rat brain contains different oligosaccharides than does Thy 1 from rat thymus (Parekh et al. 1987). These differences may modulate or regulate the biological activity of Thy 1 within the environment of the organ in which it is produced.

Different cell types such as yeast, insect, and plant, are being considered for the production of recombinant proteins. The yeast and insect cell lines do not synthesize the complex N-linked oligosaccharides, rather, they make primarily high mannose types. This may turn out to be a serious drawback for glycoproteins intended for therapeutic use in humans since high mannose chains are frequently recognized by a receptor on macrophages and, therefore, the proteins may have very short circulating half-lives. Further, yeast synthesizes unusual mannose-containing mucin type oligosaccharides. These sugars may be immunogenic. The insect mucin-type structures lack sialic acid, a very common monosaccharide of mammalian mucin-type chains. Plant N-linked oligosaccharides contain xylose and an altered fucose linkage and have been shown to be very highly immunogenic in mice and rabbits.

3.2.3 Cell Metabolism

Metabolic changes that accompany normal tissue development and differentiation or that occur upon aging or during abnormal development, such as in tumors, lead to dramatic alterations in oligosaccharide structures in vivo (Feizi and Childs 1987).

Structural alterations in glycosylation under tissue culture conditions may be difficult to control. Some of the conditions and reagents that have effects upon the structure and synthesis of oligosaccharides and glycoconjugates are listed in Table 3–2. Among these are the metabolic inhibitors of glycosylation, such as tunicamycin or castanospermine, etc., which cause dramatic changes in structure. Likewise conditions, such as glucose starvation, produce major effects. Often alterations in oligosaccharide structure due to the age of the cells, the growth rate of the cells, the substratum underlying the cells, growth factors, spontaneous mutations, autocrine released factors, and cell differentiation may be more subtle to detect and control. What is certain is that multiple external cues are capable of altering the structure of the oligosaccharides. These alterations may be biologically

TABLE 3-2 Events, Culture Conditions, and Reagents that Affect Oligosaccharide Structure

	Cell Types	Oligosaccharides Affected	References
Events:			
Viral transformation	BHK fibroblasts	N-linked	Yamashita et al. 1985
			Hunt et al. 1981
Oncogenic transformation	Uroepithelial NIH 3T3	N-linked	Debray et al. 1986
			Bolscher et al. 1981
Cell differentiation	Muscle T cells	N-, O-, and PG	Senechal et al. 1983
	Numerous others		Piller et al. 1988
Inflammation	Liver parotid	N-linked	Kaplan et al. 1983
Somatic mutations	CHO	N-, O-, and PG	Haspel et al. 1988
	Tumor lines		Dennis and Kerbel 1981
Growth rate	Fibroblasts	N-, and PG	Vischer et al. 1985
Cell age	RBC, muscle	Glucosylation	Vlassara et al. 1987
		PG	Ciba Foundation 1986
Conditions:			
Glucose deprivation	BHK	N-linked	Turco and Prickard 1982
	CHO		Rearick et al. 1981
	Rat Hepatoma		Baumann and Jahreis 1983
Sulfate deprivation	Sea Urchin	PG	Lane and Solursh 1988
Conditioned media	Muscle	PG	Berrou et al. 1988
Substratum	Muscle	PG	Ciba Foundation 1986

Reagents:

Phorbol esters	Leukemia cell	N-linked	Delia et al. 1982
Retinoic acid	Carcinoma	N-linked	Lotan et al. 1987
DMSO	Adenocarcinoma	Glycolipids	Siddiqui and Kim 1984
Orotic acid	Rat liver	N-linked	Narasimhan et al. 1988
Isoproterenol	Parotid	N-linked	Humphrey-Beher 1984
TRH	Rat pituitary	N-linked	Ronin et al. 1985
Thyroxine	Liver	N-linked	Naval et al. 1986
Insulin	Fibroblasts	PG	Moscatelli and Rubin 1976
Glucocorticoids	Macrophages	N-linked	Bar-Shavit et al. 1984
Interleukin 1	Synovium	PG	Hammerman and Wood 1984
Xylosides		PG	Datema et al. 1987
Monensin		N-, O-, PG	Datema et al. 1987
Metabolic inhibitors	Numerous	N-linked	Datema et al. 1987
Tunicamycin			
Deoxynorjirimycin			
Castanospermine			
Bromoconduritol			
Deoxymannojirimycin			
Swainsonine			

significant in vivo, and may create havoc for those trying to attain and maintain a constant oligosaccharide heterogeneity in therapeutic proteins. Undoubtedly, many other heretofore untested procedures will be found to affect the glycosylation process. These factors will only be discovered by a close monitoring of the oligosaccharides structure of recombinant proteins produced over extended times in the bioreactor. Technological advances in analytical methods for determining oligosaccharide structures are continuing to make this endeavor more feasible. Likewise, separation or elimination of undesirable glycoforms may be possible by treatments with exoglycosidases or by separation by lectin affinity chromatography. At the current time there is no formula for predicting what the alterations in oligosaccharide structure will be. Similarly, several conditions unique to growth in a bioreactor (i.e., shearing forces, high glucose media) may drastically affect the glycosylation process. Some cell types and proteins may be far more sensitive than others.

3.3 CONCLUSION

The structures of protein-bound oligosaccharides is influenced by the structure of the protein, the cell type in which it is produced, and the metabolic condition of the cell. Of these factors, cell metabolism may be the most difficult to control, particularly in tissue culture. Alterations in glycosylation may be magnified during scale-up procedures typical of those necessary for the production of recombinant proteins for therapeutic applications. Factors that might lead to altered glycosylation in bioreactors are the clonal selection of particular cell subtypes over the time of culture, alterations due to cell age or density, and changes due to shearing forces, oxygen tension, or other conditions unique to the environment within the bioreactor. A careful monitoring of cell morphology and oligosaccharide structures over an extended period of time will be necessary before it is known whether oligosaccharide structures and synthesis can be maintained when cells are grown in bioreactors. The alterations in oligosaccharides could have a profound effect upon properties of proteins intended for therapeutic use. In terms of making a safe and effacious product for human treatment, it is imperative to define and monitor the structures of the oligosaccharides on recombinant proteins and to be aware of the properties that these oligosaccharides may possess.

REFERENCES

Bar-Shavit, Z., Kahn, A.J., Pegg, L.E., Stone, K.R., and Teitlebaum, S.L. (1984) *J. Clin. Invest.* 73, 1277–1283.
Baumann, H., and Jahreis, G.P. (1983) *J. Biol. Chem* 258, 3942–3249.
Berrou, E., Breton, M., Deudon, E., and Picard, J. (1988) *J. Cell. Physiol.* 137, 430–438.

Bolscher, J.G.M., van der Bijl, M.M.W., Neffjes, J.J., et al. (1988) *EMBO J.* 11, 3361–3368.

Carter, S.R., Slomiany, A., Gwordzinski, K., Yun, H.L., and Slomiany, B.L. (1988) *J. Biol. Chem.* 263, 11977–11984.

Ciba Foundation, ed. (1986) *Functions of the Proteoglycans*, pp. 248. Wiley, New York.

Dahms, N.M., and Hart, G.W. (1986) *J. Biol. Chem.* 261, 13186–13196.

Datema, R., Olofsson, S., and Romero, P.A. (1987) *Pharamol. Therapeutics* 33, 221–286.

Debray, H., DeLannoy, P., Montreuil, J., et al. (1986) *Int. J. Cancer* 37, 607–611.

Delia, D., Graves, M.F., Newman, R.A., et al. (1982) *Int. J. Cancer* 29, 23–31.

Dennis, J.W., and Kerbel, R.S. (1981) *Cancer Res.* 41, 98–104.

Faust, P.L., Chirgwin, J.M., and Kornfeld, S. (1987) *J. Cell Biol.* 105, 1947–1954.

Feizi, T., and Childs, R.A. (1987) *Biochem. J.* 245, 1–11.

Galili, U., Clark, M.R. Shonet, S.B. Buehler, J., and Macher, B.A. (1987) *Proc. Natl. Acad. Sci. USA* 84, 1369–1373.

Ginsburg, V., and Robbins, P.W., eds. (1984) *Biology of Carbohydrates*, pp. 87–163, Wiley, New York.

Goto, M., Akai, K., Murakami, A., et al. (1988) *Biol/technology* 6, 67–70.

Green, E.D., and Baenziger, J.U. (1988) *J. Biol. Chem.* 263, 25–35.

Green, E.D., and Baenziger, J.U. (1989) *Trends Bio. Sci.* 14, 168–172.

Hammerman, D., and Wood, D.D. (1984) *Proc. Soc. Exp. Biol. Med.* 177, 205–210.

Hansen, L., Blue, Y., Barone, K. Collen, D., and Larsen, G.R. (1988) *J. Biol. Chem.* 263, 15713–15719.

Haspel, H.C., Revillame, J., and Rosen, O.M. (1988) *J. Cell. Physiol.* 136, 361–366.

Humphrey-Beher, M.G. (1984) *J. Biol. Chem.* 259, 5797–5802.

Hunt, L.A., Lamph, W., and Wright, S.E. (1981) *J. Virol.* 37, 207–215.

Kaplan, H.A., Woloski, B.M.R.N.J., Hellman, M., and Jamesion, J.C. (1983) *J. Biol. Chem.* 258, 11505–11509.

Lane, M.C., and Solursh, M. (1988) *Develop. Biol.* 127, 78–87.

Lotan, R., Sacks, P.G. Lotan, D., and Hong, W.K. (1987) *Int. J. Cancer* 40, 224–229.

Moscatelli, D., and Rubin, H. (1976) *J. Cell Physiol.* 91, 79–88.

Narasimhan, S., Schachter, H., and Rajalakshmi, S. (1988) *J. Biol. Chem.* 263, 1273–1281.

Naval, J., Calvo, M., Lampreave, F., and Pineiro, A. (1986) *Int. J. Biochem.* 18, 115–122.

Ofosu, F.A., Danishefsky, I., and Hirsh, J., eds. (1989) Heparin and Related Polysaccharides, pp. 81–94, *Annals of the New York Academy of Science* Vol. 556, New York.

Parekh, R., Parekh, R.B., Dwek, R.A., et al. (1989) *Biochemistry* 28, 7662–7669.

Parekh, R.B., Tse, A.G.D., Dwek, R.A., and Rademacher, T.W. (1987) *EMBO J.* 6, 1233–1244.

Piller, F., Piller, V., Fox, R.I., and Fukuda, M. (1988) *J. Biol. Chem.* 263, 15146–15250.

Rearick, J.L., Chapman, A., and Kornfeld, S. (1981) 6255–6261.

Rogers, G.H., Daniels, R.S., Skehel, J.J., et al. (1985) *J. Biol. Chem.* 260, 7362–7370.

Ronin, C., Stannard, B.S., and Weintraub, B.D. (1985) *Biochemistry* 24, 5626–5631.

Schwartz, R., and Klenk, H-D. (1981) *Virology* 113, 584–593.

Senechal, H., Delain, D., Schapira, G., and Wahrmann, J.P. (1983) *Exp. Cell Res.* 147, 341–350.

Siddiqui, B., and Kim, Y.S. (1984) *Cancer Res.* 44, 1648–1652.

Skehel, J.J., Stevens, D.J., Daniels, R.S., Douglas, A.R., and Knossow, M. (1984) *Proc. Natl. Acad. Sci. USA* 81, 1779–1783.

Tsuda, E., Goto, M., Murkami, A., et al. (1988) *Biochemistry* 27, 5646–5654.

Turco, S.J., and Pickard, J.L. (1982) *J. Biol. Chem.* 257, 8674–8679.

Vischer, P., and Buddecke, E. (1985) *Exp. Cell Res.* 156, 15–28.

Vlassara, H., Valinsky, J., Brownlee, M., et al. (1987) *J. Exp. Med.* 116, 539–549.

Welply, J.K., Pegg, L.E., Reitz, B.A., and Warren, T. (1989) *J. Cell. Biochem.* Suppl. 13a,108.

Yamashita, K., Tachibana, Y., Takashi, T., and Kobata, A. (1985) *J. Biol. Chem.* 260, 3963–3969.

Serum-Free Media

Kazuaki Kitano

Serum has been used most commonly to stimulate the growth of mammalian cells ever since cell culture systems were established in vitro. Serum, which can support the survival and growth of a wide variety of primary and established cells, is a complex mixture whose components are poorly characterized. Thus, to better understand the physiology of mammalian cells in detail, the need to cultivate cells under defined conditions has increased.

In 1946, White reported the earliest attempt to develop a chemically defined culture medium. He employed a nutrient mixture consisting of amino acids, vitamins, a carbohydrate, inorganic ions, and hormones such as insulin and thyroxine to maintain chick embryo skeletal cells and chick heart cells. Eagle and co-workers investigated the minimal essential nutritions required to grow several species of mammalian cell lines in the presence of dialyzed serum (Eagle 1955a, 1955b, 1956, and 1959; Eagle et al. 1958). They showed that 13 amino acids, eight vitamins, six inorganic ions, and glucose or some other carbohydrates were required to grow mammalian cells and established a minimal essential medium (MEM). However, dialyzed serum was an essential component in their medium.

In the 1960s, Ham and his group investigated the nutritional requirements of various cells in detail and established many chemically defined media that consisted of 50 to 70 ingredients to eliminate or reduce the amount of serum required (Ham and McKeehan 1978).

In the 1970s, Sato and co-workers constructed the hypothesis that the function of the serum was to supply hormones and growth factors required by the cells. They succeeded in cultivating a large number of cell lines in serum-free media supplemented with these factors (Barnes and Sato 1980). Up to now, the growth requirements of more than 200 kinds of established and primary cells have been investigated, and these cells have been cultivated under serum-free conditions.

As a result of recent progress in biotechnology, the need to cultivate mammalian cells on a large scale has been increased to produce medically important substances. A medium containing serum has several disadvantages for industrial applications. First, serum is very expensive; second, the quality of the serum is not necessarily uniform from lot to lot; and third, serum contains many unknown compounds that may disturb the downstream purification process.

In this chapter, we will review the current status of developing serum-free media and the use of such media to produce biologically active substances on a large scale.

4.1 CHEMICALLY DEFINED SERUM-FREE MEDIA

Mammalian cells have generally been cultivated in vitro in a nutrient mixture (basal medium) supplemented with 5 to 20% of serum, usually fetal calf serum (FCS). The contribution of serum involves a multitude of component substances. By replacing some of these with chemically defined compounds, many cells can grow without serum. Thus, a chemically defined serum-free medium consists of the basal medium and chemically defined supplements, which are chosen from hormones, growth factors, transport proteins, nutrients, attachment factors, and others.

4.1.1 Basal Media

The basal medium consists of amino acids, vitamins, inorganic salts, carbon sources, and others. Many basal media have been established for serum-supplemented cultures and are commercially available. Some others were developed for serum-free purposes. The basal media that have been used for serum-free cultures are listed in Table 4–1. Since many of these were developed for serum-supplemented cultures, they sometimes have to be modified for serum-free cultures.

Eagle's MEM and Dulbecco's modified Eagle's medium (DMEM) are often used as the base to establish serum-free media. For example, Yamane et al. (1981a) established the serum-free medium RITC80-7 for human diploid fibroblasts using MEM as the basal medium. They also used DMEM as the basal medium for RITC55-9, which is useful for human lymphoid cells (Sato et al. 1982). Iscove's modified Dulbecco's medium (IMDM) was

TABLE 4-1   Basal Media Used for Serum-Free Cultures

Medium	Developer	Example of Use
199	Morgan et al. 1950	Maciag et al. 1981
		Dodge and Sharma 1985
BME	Eagle 1955a	Bottenstein et al. 1980
		Fischer 1984
NCTC109	Evans et al. 1956	Bryant et al. 1961
		Franc et al. 1984
MEM	Eagle 1959	Yamane et al. 1981a
		Bertolero et al. 1984
MB752/1	Waymouth 1959	Morrison and Jenkin 1972
		Brannon et al. 1985
DMEM	Dulbecco and Freeman 1959	Giguere et al. 1982
		Smith et al. 1986
F10	Ham 1963	Chessebeuf and Padieu 1984
		Okabe et al. 1984
F12	Ham 1965	Hayashi and Sato 1976
		Ethier 1986
MAB87/3	Gorham and Waymouth 1965	McGowan et al. 1981
RPMI1640	Moore et al. 1967	Buhl and Regan 1972
		Brower et al. 1986
—	Higuchi and Robinson 1973	Bauer et al. 1976
MCDB104	McKeehan et al. 1977	Phillips and Cristofalo 1981
		Kaji 1985
IMDM	Iscove and Melchers 1978	Iscove et al. 1980
		Flesch and Ferber 1986
MCDB402	Shipley and Ham 1981	Shipley and Ham 1983a and
		1983b
		Shipley et al. 1984
MCDB152	Tsao et al. 1982	Kirk et al. 1985a and 1985b

developed for the serum-free culture of lymphoid cells and has been used widely. RPMI 1640 is also used quite often for lymphoid cell lines and Ham's F12 medium is favored for epithelial cells.

Combinations of commercially available basal media have been effectively employed for serum-free cultures (Table 4–2). Such use is an easy way to complement the ingredients required in a serum-free culture. The most commonly used combination is an equal volume mixture of DMEM and Ham's F12 (Barnes and Sato 1980). Mixtures of three basal media are also valuable. We use a mixture (1:1:2, vol/vol) of IMDM, F12, and L15 supplemented with 4.5 g/l N(2-hydroxyethyl)piperazine-N'-2-ethanesulfonic acid (HEPES) and 1 g/l sodium bicarbonate (TL-2 medium) for the serum-free culture of hybridomas (Kitano et al. 1986; Shintani et al. 1988 and 1989). A mixture (2:1:1, vol/vol) of RPMI 1640, DMEM, and F12, supple-

TABLE 4–2   Mixed Basal Media Used for Serum-Free Cultures

Medium	Mix Ratio	Example of Use
DMEM/F12	1:1	Mather and Sato 1979a and 1979b
		Orly and Sato 1979
		Barnes and Sato 1979
DMEM/F12	3:1	McClure 1983
		Chen et al. 1986
DMEM/F12	9:1	Kato and Gospodarowicz 1984
DMEM/F10	1:1	Barano and Hammond 1985
		Lino et al. 1985
DMEM/MCDB104	1:1	Heath and Deller 1983
DMEM/MCDB104	3:1	Allen et al. 1985
DMEM/199	1:1	Clark et al. 1982
IMDM/RPMI1640	1:1	Bjare and Rabb 1985
IMDM/F12	1:1	Mosien 1981
		Cleveland et al. 1983
IMDM/F12/L15 (TL-2)	1:1:2	Kitano et al. 1986
		Shintani et al. 1988 and 1989
MEM/MAB87–3	3:1	Darlington et al. 1987
MEM/F12	1:1	Harrison et al. 1985
RPMI1640/F12	1:1	Hiragun et al. 1983
RPMI1640/DMEM/F12 (RDF)	2:1:1	Kawamoto et al. 1983
		Hagiwara et al. 1985

mented with 2 mM glutamine, 0.01% sodium pyruvate, 15 mM HEPES, 2.2 g/l sodium bicarbonate and 2.0 g/l glucose (RDF medium) is also often used for lymphoid cell lines (Kawamoto et al. 1983) and eRDF medium was further developed from RDF medium by enriching the concentration of amino acids and some other components for high-density hybridoma cultures (Murakami et al. 1984).

4.1.2   Correction of the Ingredients of Basal Media

Since most basal media have been developed to be used with serum, corrections of the ingredients are sometimes necessary for serum-free cultures.

Cells in a serum-free medium are generally more sensitive to change in osmotic pressure and pH than those in a serum-supplemented medium. Murakami (1984) found that cellular growth in a serum-free culture of lymphoid cells was remarkably inhibited when the basal medium was even slightly diluted. They solved the problem by preparing the basal medium at a concentration 5% higher than that of the conventional formula. Osmolarity can also be adjusted by adding such inorganic salts as sodium chloride. To control the pH of the medium within an appropriate range,

HEPES is often added at a concentration of 10 to 15 mM. Sodium β-glycerophosphate is also used as a buffering agent (Ling et al. 1968; Sato et al. 1982). The concentration of sodium bicarbonate is adjusted to control both pH and osmotic pressure.

All basal media contain mixtures of amino acids, but their number, from 13 to 21, and proportions vary widely. Eagle (1959) showed that 13 amino acids (i.e., arginine, cystine, glutamine, histidine, leucine, isoleucine, lysine, methionine, phenylalanine, threonine, tryptophan, tyrosine, and valine) are essential for mammalian cell growth; they further added serine for some cell lines (Eagle and Piez 1962). For serum-free culture, glutamine is often enriched (Allegra and Lippman 1978; Kawamoto et al. 1983; Kovar and Franek 1984) and nonessential amino acids are also added (Chang et al. 1980; Amorosa et al. 1984; Reznikoff et al. 1987). Proline is required for the proliferation of rat parenchymal hepatocytes in a serum-free medium supplemented with insulin and epidermal growth factor (EGF) (Houck and Michalopoulos 1985). The consumption rate of amino acids during the cultivation must be analyzed and the limiting ones have to be supplemented on a case-by-case basis.

Eagle (1959) showed that eight vitamins (i.e., thiamine, riboflavin, pantothenic acid, folic acid, pyridoxal, nicotinamide, inositol, and choline) are necessary for cell growth. Folic acid and pyridoxal are sometimes replaced by folinic acid and pyridoxine, respectively. In addition to these, biotin and vitamin B_{12} are included in most basal media; biotin is sometimes also added for serum-free cultures (Gaillard et al. 1984; Amorosa et al. 1984; Griffin et al. 1985). Because the basal media containing ascorbic acid are limited, sodium ascorbate is also often added (Murakami and Masui 1980; Simonian et al. 1982; Libby and O'Brien 1983; Kovar and Franek 1987). Vitamin E (α-tocopherol) is required by some cell lines (Bettger et al. 1981; Simonian et al. 1982; Kirk and Alvarez 1986; Bodeker et al. 1987). It is also useful for the growth of human diploid fibroblasts at low inoculum size (Kan and Yamane 1983). Vitamin A or its analogue, retinoic acid, sometimes has important effects on cell differentiation, proliferation, and function, and has been used as a supplement (Mather and Sato 1979b, Bradshaw and Dubes 1983; Hiragun et al. 1983; Oka et al. 1984; Kirk and Alvarez 1986).

Purines and pyrimidine bases and nucleosides have been included in many basal media, but most cells appear to be able to synthesize nucleotides de novo. However, Peehl and Ham (1980) found that epidermal keratinocytes required a high concentration (0.18 mM) of adenine in MCDB151 medium. For serum-free cultures, bases and nucleosides are sometimes added (Kan and Yamane 1982; Eliason and Odartchenko 1985).

Inoganic ions commonly included in a basal medium are calcium, magnesium, potassium, sodium, chloride, and phosphate (Eagle 1956); sulfate is also included. The importance of trace metals was established by Shooter and Gey (1952). Zinc and iron have been added in most basal media since

1963. Copper and cobalt are also included in modern basal media. Selenium was found to be essential for the growth of the WI-38 diploid human fibroblasts (McKeehan et al. 1976) and is included in IMDM. Since the basal medium containing this ion is, however, very limited, it is generally added to the serum-free medium at a concentration range of 10^{-9} to 10^{-7} M, mostly 10^{-8} or 10^{-7} M. Hamilton and Ham (1977) achieved the clonal growth of Chinese hamster cell lines in a protein-free medium supplemented with a mixture of 23 trace elements. This complex mixture is also used as a component of a protein-free medium for hybridomas (Cleveland et al. 1983). The growth of WI-38 cells in a serum-free, growth factor-free medium is markedly stimulated by elevated concentration (5 mM) of calcium chloride (Praeger and Cristofalo 1986).

Eagle et al. (1958) showed that glucose, mannose, fructose, and galactose are generally utilized by cells, but glucose has been the principal energy source in most serum-free media. Fructose and galactose produce less lactic acid than glucose (Eagle et al. 1958) and thus assist in maintaining a stable pH. Galactose is used in L15 medium (Leibovitz 1963) and fructose is a carbon source for skin fibroblasts (Wolfrom et al. 1983). Glutamine is another major carbon and energy source (Reitzer et al. 1979). Pyruvate, which is included in many basal media, can also serve as a carbon source. It is often added to serum-free media (Richman et al. 1976; Flesch et al. 1984; Schonherr et al. 1987).

4.1.3 Media Supplements

Very few cell lines can grow in basal media and some others can grow after adaptation (Waymouth 1959; Katsuta and Takaoka 1973; Okabe et al. 1984). However, most cell lines require supplements selected from the following ingredients.

4.1.3.1 Insulin and Other Hormonal Supplements

Since Sato (1975) proposed the idea that the major role of serum in cell culture is to provide hormones, the hormonal requirements of various cell lines have been investigated. The hormones used as supplements in serum-free media are listed in Table 4–3.

The growth-promoting action of insulin was discovered by Gey and Thalhimer (1924), and partial substitution of serum by addition of insulin to a serum-free medium was demonstrated by Lieberman and Ove (1959). It has been added to almost all serum-free media at concentrations of 0.1 to 100 μg/ml, and in many cases, 5 to 10 μg/ml.

Glucocorticoids were known to suppress the growth of some cell lines, but their growth-promoting action on other cells was found by the early studies of Castor (1962) and Macieira-Coelho (1966). Hydrocortisone and

TABLE 4-3 Hormones Added to Serum-Free Media

Hormone	Concentration	Example of Addition to the Media
Insulin	$0.1–100\ \mu g/ml$ $(5–10\ \mu g/ml)^1$	Almost all cell lines
Hydrocortisone	$5 \times 10^{-9}–7 \times 10^{-5}\ M$ $(10^{-8}–10^{-6}\ M)$	Epithelial cells (Taub et al. 1979)
		Keratinocytes (Maciag et al. 1981)
		Astrocytes (Morrison and DeVellis 1981)
		Chondrocytes (Kato and Gospodarowicz 1984)
		Fibroblasts (Amorosa et al. 1984)
Dexamethasone	$10^{-8}–10^{-6}\ M$ $(10^{-7}–10^{-6}\ M)$	Breast cancer cells (Allegra and Lippman 1978)
		Muscle cells (Florini and Roberts 1979)
		Fibroblasts (Bettger et al. 1981)
		Pancreatic acinar cells (Brannon et al. 1985)
		Hepatomas (Darlington et al. 1987)
Triiodothyronine (T_3)	$10^{-12}–10^{-6}\ M$ $(10^{-9}–10^{-8}\ M)$	Pituitary cells (Hayashi and Sato 1976)
		Breast cancer cells (Allegra and Lippman 1978)
		Colon carcinomas (Murakami and Masui 1980)
		Fibroblasts (Yamane et al. 1981a)
		Epithelial cells (Taub et al. 1979)
Thyroxin (T_4)	$10^{-11}–10^{-10}\ M$	Heart cells (Mohamed et al. 1983)
Prostaglandins (PGE$_1$, PGE$_{2\alpha}$)	$10^{-8}–10^{-6}\ M$	Epithelial cells (Taub et al. 1979)
		Fibroblasts (Bettger et al. 1981)
		Mammary tumor cells (Barnes and Sato 1979)
		Astrocytes (Morrison and DeVellis 1981)
		Metanephric organ (Avner et al. 1985)
Estradiol (E_2)	$10^{-12}–10^{-8}\ M$ $(10^{-9}–10^{-8}\ M)$	Breast cancer cells (Allegra and Lippman 1978)
		Small cell carcinomas (Simms et al. 1980)
		Hypathalamus cells (Binoux et al. 1985)

(continued)

TABLE 4-3 (continued)

Hormone	Concentration	Example of Addition to the Media
Testosterone	10^{-9}–10^{-8} M	Melanomas (Mather and Sato 1979a)
		Lymphomas (Darfler et al. 1980)
		Heart cells (Libby 1984)
Glucagon	10^{-7}–10^{-6} M	Hepatocytes (Richman et al. 1976)
		Colon carcinoma cells (Murakami and Masui 1980)
Growth hormone	10–100 μg/ml	Sertoli cells (Mather and Sato 1979b)
		Hepatocytes (Enat et al. 1984)
Progesterone	10^{-9}–10^{-4} M (10^{-9}–10^{-8} M)	Neuroblastomas (Bottenstein et al. 1980)
		Lymphocytes (Mosien 1981)
		Keratinocytes (Tsao et al. 1982)
Prolactin	1–10 μg/ml	Epithelial cells (Hammond et al. 1984)
		Hepatocytes (Enat et al. 1984)
FSH	10–100 ng/ml	Melanomas (Mather and Sato 1979a)
		Sertoli cells (Mather and Sato 1979b)
		Epithelial cells (Oka et al. 1984)
LH-RH	0.1–1 ng/ml	Melanomas (Mather and Sato 1979a)
		Plasmacytomas (Murakami et al. 1981)
		Pituitary cells (Brunner 1982)
PTH	0.1–1 ng/ml	Pituitary cells (Hayashi and Sato 1976)
Somatostatin	1–100 ng/ml	Epithelial cells (Ambesi-Impiombato et al. 1980)
TRH	0.1–1 ng/ml	Pituitary cells (Hayashi and Sato 1976)

FSH, Follicle-stimulating hormone; LH-RH, luteinizing hormone-releasing hormone; PTH, parathyroid hormone; TRH, thyrotropin-releasing hormone.

*( ), Concentration generally used.

a synthetic glucocorticoid, dexamethasone, are widely used as a component of serum-free media.

Thyroid hormones, thyroxin (T_4) and triiodothyronine (T_3), activate cell metabolism and thus have been used in many serum-free media. T_3 is generally more active than T_4.

The hormonal requirements of the cells sometimes differ depending on the basal medium, the kind of attachment matrix (Gatmaitan et al. 1983), and combinations of hormones and growth factors.

4.1.3.2 Growth Factors To differentiate and proliferate, cells often require a small amount of proteinous factors, called growth factors. In mammalian tissues, these factors are supplied by the cells themselves (autocrine), by neighboring cells (paracrine), or by other cells via blood (endocrine) (Sporn and Todaro 1980). In mammalian cell culture in vitro, these factors are generally supplied from serum or by the cells themselves, but some factors must be added even in the serum-supplemented medium. Thus, it is essential to investigate growth factor requirements of the cells and to add these factors to the serum-free medium. Growth factors often used in serum-free cultures are listed in Table 4–4.

The first growth factor so named is nerve growth factor (NGF) (Levi-Montalcini et al. 1954), but its action is limited in a narrow spectrum of cells. On the other hand, EGF (Cohen 1962) promotes the growth of most normal and malignant cells except lymphoid cells. Platelet-derived growth factor (PDGF) (Kohler and Lipton 1974) stimulates the growth of fibroblasts, smooth muscle cells, glia cells, and 3T3 cells, but does not stimulate the growth of endothelial or lymphoid cells.

Fibroblast growth factor (FGF) was first found in the bovine pituitary by Gospodarowicz (1974) as a growth factor for fibroblasts. FGF is now classified into two types, basic FGF and acidic FGF. Both promote the growth of both fibroblasts and almost all mesodermal cells. However, the growth-promoting action of basic FGF is 30 to 100 times stronger than that of acidic FGF (Gospodarowicz et al. 1984; Bohlen et al. 1985). Various growth factors having different names such as eye-derived growth factor-I, hepatoma-derived growth factor, chondrosarcoma-derived growth factor, and cartilage-derived growth factor are identical to basic FGF. On the other hand, endothelial cell growth factor and eye-derived growth factor-II are identical to acidic FGF.

Somatomedins or insulin-like growth factors (IGF) can often replace the action of insulin at a lower concentration.

A great number of factors, other than those listed in Table 4–4, have been discovered. These factors are useful for the serum-free culture of targeted cells.

TABLE 4-4 Growth Factors Added to Serum-Free Media

Factor	Concentration	Example of Addition to the Media
EGF	5–5000 ng/ml (5–50 ng/ml)[1]	HeLa cells (Hutchings and Sato 1978)
		Epithelial cells (Dell'Aquila and Gaffney 1982)
		Fibroblasts (Yamane et al. 1981a)
		Hepatocytes (McGowan et al. 1981)
		Endothelial cells (Hoshi and McKeehan 1984)
		Keratinocytes (Maciag et al. 1981)
		Brain cells (Eccleston et al. 1985)
FGF	1–10000 ng/ml (10–100 ng/ml)	Astrocytes (Morrison and DeVellis 1981)
		Endothelial cells (Giguere et al. 1982)
		Fibroblasts (Darmon et al. 1981)
		Adrenocortical cells (Simonian et al. 1982)
		Kidney cells (Bradshaw et al. 1983)
IGF (Somatomedins)	1–100 ng/ml	Pituitary cells (Hayashi and Sato 1976; Hayashi et al. 1978)
		Epithelial cells (Reddan and Dziedzic 1982)
		Chondrocytes (Adolphe et al. 1984)
NGF	0.02–3 ng/ml	Melanoma cells (Mather and Sato 1979a)
		Neurons (Bottenstein et al. 1980)
		Plasmacytomas (Murakami et al. 1981)
PDGF	1–3 μg/ml	Glia cells (Heldin et al. 1980)
		Fibroblasts (Phillips and Cristofalo 1981)
		Lung cells (Malewicz et al. 1985)
EPO	1–10 U/ml	Bone marrow cells (Stewart et al. 1984)
		Hemopoietic progenitor cells (Eliason and Odartchenko 1985)
CSF	1–500 U/ml	Macrophages (Flesh and Ferber 1986)
		Hemopoietic progenitor cells (Sonoda et al. 1988)
TGF-beta	500 ng/ml	Nontransformed cells (Rizzino and Ruff 1986)

EGF, Epidermal growth factor; FGF, fibroblast growth factor; IGF, insulin-like growth factor; NGF, nerve growth factor; PDGF, platelet-derived growth factor; EPO, erythropoietin; CSF, colony-stimulating factor; TGF, transforming growth factor.
[1] Concentration generally used.

4.1.3.3 Transport Proteins Transferrin is one of the most abundant proteins in plasma and has been added to almost all hormone-supplemented serum-free media since its growth-promoting action in serum-free culture was discovered by Hayashi and Sato (1976). It is an iron-binding glycoprotein and promotes cell growth by means of its iron delivery function (Mather and Sato 1979a; Perez-Infante and Mather 1982).

Human transferrin is generally added to serum-free media at concentrations of 0.1 to 600 μg/ml, mostly 5 to 50 μg/ml. Human lactoferrin was reported to stimulate the growth of human lymphoid cell lines better than human transferrin (Hashizume et al. 1983). Horse serum-transferrin stimulates the growth of human myeloid leukemia cells at a low concentration (Yoshinari et al. 1989).

The function of transferrin can be replaced by a synthetic lipophilic iron chelator, pyridoxal isonicotinoyl hydrazone (PIH) (Landschultz et al. 1984). Two highly water soluble iron salts, ferric ammonium citrate and ferric ammonium sulfate, can completely replace transferrin to support the growth of human leukemic cell lines (Titeux et al. 1984). Kovar and Franek (1987) succeeded in establishing a chemically defined protein-free medium for mouse hybridomas by replacing transferrin with high concentration of ferric citrate. Iron choline citrate also promotes the growth of human-human or mouse-human hybridomas (Ill et al. 1988).

Bovine or human serum albumin (BSA or HSA) is often added to serum-free media at relatively high concentrations of 1 to 10 mg/ml. One of its major functions is to supply fatty acids such as oleic acid, linoleic acid, and arachidonic acid (Spector 1968; Yamane et al. 1975). Fatty acid-free albumin is sometimes used to separate the function of albumin and fatty acids (Iscove and Melchers 1978; Rockwell et al. 1980; Malan-Shibley and Lype, 1983). It is also used as a carrier of cholesterol (Sato et al. 1984). α-Cyclodextrin (Yamane et al. 1981b) or β-cyclodextrin (Ohmori and Yamatoma 1987) can substitute for BSA as a carrier of unsaturated fatty acids. Lipids can also be effectively introduced into cells in the form of liposomes (Poste et al. 1976; Bettger and Ham 1982).

Lipoproteins, low density lipoprotein (LDL) and high density lipoprotein (HDL), stimulate the growth of a number of cell lines (Gospodarowicz 1984). Murakami et al. (1988) isolated a very low density lipoprotein (VLDL) from egg yolk that promotes the growth of a wide variety of mammalian cell lines, including plasmacytomas and epithelial cell lines. We isolated SSGF-1, a LDL from swine serum, that is more active than human LDL for hybridomas (Shintani et al. 1989). These lipoproteins seem to function as carriers of cholesterol and phospholipids, which are difficult to dissolve in water (Sato et al. 1984). In some cases, however, apolipoproteins can substitute entirely for HDL (Chen et al. 1986).

4.1.3.4 Nutrients Nutrients not included in basal media are sometimes required in serum-free media.

The growth promoting action of ethanolamine and phosphoethanolamine was first demonstrated for a rat mammary carcinoma cell line, 62–64, by Kano-Sueoka et al. (1979; Kano-Sueoka and Errick 1981), and then for mouse hybridomas (Murakami et al. 1982). Mammalian cells in culture can be classified into ethanolamine-responsive and etanolamine-nonresponsive (Kano-Sueoka 1984). Many types of epithelial cells, including carcinomas and keratinocytes, and lymphoid cells, including hybridomas, are now known to require ethanolamine. Ethanolamine is a structural component of phosphatidylethanolamine, which is one of the most abundant phospholipids in mammalian cell membranes. Also, ethanolamine-responsive cells can synthesize phosphatidylethanolamine only when 10^{-6} to 10^{-4} M (generally 1 to 5×10^{-5} M) of ethanolamine is in the medium. Phosphoethanolamine is as effective as ethanolamine for rat mammary carcinoma cells (Kano-Suekoa and Errick 1981) but it is 10-fold less effective for hybridomas (Murakami et al. 1982). Sometimes, both are used together in a serum-free medium (Minna et al. 1982; Tsao et al. 1982, Bertolero et al. 1984; Kirk et al. 1985a and 1985b).

The effect of phospholipids in serum-free cultures was first shown by Iscove and Melchers (1978). They reported the effect of soy bean lecitin on B lymphocytes. Honma et al. (1979) added phosphatidylcholine to a serum-free medium for mouse myeloid leukemia cells. It is also required by vascular endothelial, smooth muscle, and corneal endothelial cells (Fujii et al. 1983). Soy bean phospholipid is required for the stirred culture of myeloma MPC-11 cells in a serum-free medium, though it is not required for a standing culture in a dish (Murakami et al. 1983).

Fatty acids are generally not required, even in serum-free cultures, when serum albumin and other proteins are added to the medium. However, in the absence of serum proteins (Bauer et al. 1976) or in the presence of delipidated albumin (Iscove et al. 1980), unsaturated fatty acids are sometimes required.

Cholesterol is a major lipid component of plasma membranes and is required by several cell lines. Cholesterol was first found to be required for the clonal growth of HeLa cells in the presence of dialyzed serum (Sato et al. 1957). Primary human diploid fibroblasts (Holmes et al. 1969), a porcine kidney cell line (Higuchi 1970), and human kidney cells (Gonzalez et al. 1974) also require cholesterol for proliferation in a serum-free or lipid-deficient medium. It is also required by mouse myelomas NS-1 (Sato et al 1984), P3-X63-Ag8, X63-Ag8-653 (Sato et al. 1987), and the human monocyte/macrophage-like cell line U937 (Esfahani et al. 1986). But hybridomas of normal lymphocytes fused with these cells do not require cholesterol. High productivity subclones derived from a mouse·human-human heterohybridoma require cholesterol for their growth, though the parental heterohybridoma does not (Shintani et al. 1989). Cholesterol has to be added as a complex with the serum-albumin because of its low solubility in water.

LDL is also a good source of cholesterol (Kawamoto et al. 1983; Shintani et al. 1989).

4.1.3.5 Attachment Factors Mammalian cells can be classified into two categories according to their growth pattern; anchorage-dependent and anchorage-independent. Most cells, other than lymphoid cells and some malignant cells, must be attached to a substrate in order to replicate.

Erhmann and Gey (1956) discovered that many cell lines and tissues can grow more rapidly on a collagen matrix than on a glass plate. Subsequently, a great number of cell lines have been cultivated on a collagen matrix (Kleinman et al. 1981). Collagen is also coated on microcarriers (Gebb et al. 1982) or used as a material for porous microcarriers (Van Brunt 1986). Collagen is classified into five types, type I to type V, and different cells sometimes favor different types of collagen for maximum growth. For example, chondrocytes adhere preferentially to type II collagen, and epithelial cells favor type IV collagen, but fibroblasts bind well to all types of collagen (Kleinman et al. 1981).

Fibronectin, also called cold-insoluble globulin (CIg), was originally discovered as a glycoprotein in serum (Morrison et al. 1948; Vaheri et al. 1978), and its growth-promoting action in a serum-free culture was reported by Orly and Sato (1979). It promotes the attachment and subsequent spreading of many cells by forming bridges between the cells and the collagen matrix or other substrata. It is added as a component of serum-free media at concentrations of 1 to 10 μg/ml (Rizzino and Crowley 1980; Kan and Yamane 1982; Shipley and Ham 1983a; Calvo et al. 1984; Chen et al. 1986). It is sometimes coated on the surface of dishes (Hoshi and McKeehan 1984; Epstein-Almog and Orly 1985; Simonian et al. 1987). Brower et al. (1986) showed that treating plates with both fibronectin and collagen is very effective for the serum-free culture of human non-small cell lung cancer.

Laminin, a glycoprotein found in basement membrane (Timpl et al. 1979), mediates the attachment of epithelial cells to type IV collagen, which is characteristic of the basement membrane (Vlodavsky and Gospodarowicz 1981). It is added to the medium (Pixley and Cotman 1986) or coated on dishes (Fujii and Gospodarowicz 1983).

Vitronectin, which was isolated from plasma (Haymann et al. 1983) is also known as α_1-protein (Barnes and Sato 1979), epibolin (Stenn 1981), and serum-spreading factor (Barnes and Silnutzer 1983). It can mediate attachment and spreading of a number of animal cell types in a serum-free culture (Silnutzer and Barnes 1984).

Extracellular matrix (ECM) is an organized complex of collagens, elastin, proteoglycans, and glycoproteins. ECM produced by bovine corneal endothelial cells or PF-HR-9 endodermal cells is sometimes used as an attachment matrix in a serum-free culture (Ill and Gospodarowicz 1982; Giguere et al. 1982; Kato and Gospodarowicz 1984; Bellot et al. 1985). Basic polymers such as polylysine is also used as an attachment factor (Brunner 1982; McKeehan et al. 1977).

4.1.3.6 Miscellaneous Supplements   The fact that various compounds other than those described above have been added to chemically defined serum-free media demonstrates the multiple functions of serum in mammalian cell culture.

Ham (1964) showed that polyamines such as putrescine, spermidine, and spermine are required for the clonal growth of a strain of Chinese hamster cells in a medium that contains serum albumin or linoleic acid. Putrescine has been added as a supplement to various serum-free media at concentrations of 10^{-7} to 10^{-4} M (generally 10^{-4} M) (Katsuta et al. 1975; Bottenstein and Sato 1979; Bottenstein et al. 1980; Morrison and DeVellis 1981; Delinassions 1983; Griffin et al. 1985; Binoux et al. 1985; Hendelman et al. 1985).

Protamine was first found to promote attachment of cells to glass in a serum-free medium (Lieberman and Ove 1958), but its growth-promoting action was found by Neuman and Tytell (1961) for chick embryo cells. Since then it has also been used as a component of some serum-free media (Higuchi 1963; Higuchi and Robinson 1973; Bauer et al. 1976).

Enhancement of antibody synthesis in vitro by 2-mercaptoethanol (2ME) was reported by Click et al. (1972). It was also found to stimulate the proliferation of mouse leukemia cells (Broome and Teng 1973) and is added to various serum-free media at concentrations of 10^{-7} to 10^{-4} M (mostly $1-5 \times 10^{-5}$ M) (Kawamoto et al. 1983; McClure 1983; Brown et al. 1983; Hagiwara et al. 1985).

Though 2ME noneffective cells can take up cystine smoothly from the medium, 2ME effective cells can not. When 2ME and cystine are in the medium, however, a mixed disulfide of 2ME and cysteine is formed and is rapidly taken up by the cells (Ishii et al. 1981). Other than 2ME, α-thioglycerol, cysteamine, and 3-mercaptopropionic acid are also effective (Broome and Teng 1973).

Cholera toxin increases cellular cyclic AMP concentration by activating adenylate cyclase via a cell surface receptor (Holmgren 1981). Its growth-promoting action was first reported by Green (1978) and is a component of serum-free media for various epithelial cells (Iguchi et al. 1983; Tomooka et al. 1983; Peehl and Stamey 1986), keratinocytes (Smith et al. 1986), Schwann cells (Needham et al. 1987), and so on.

Thrombin induces the cell proliferation of chicken embryonic fibroblasts (Chen and Buchanan 1975). It is also a component of serum-free media for some cell lines (Simonian et al. 1982; Harrison et al. 1985; Buck and Schomberg 1987).

A variety of nonprotein polymers have been tested in attempts to replace some serum factors. Methylcellulose was first found to protect cells in an agitated suspension culture (Kuchler et al. 1960). Since then, it has been employed as a component of serum-free media for mouse fibroblasts (Bryant et al. 1961; Birch and Pirt 1970), human leukemia cell lines (Buhl and Regan 1972), human lymphoblastoid cell lines (Birch 1980), and others. Alginic acid, dextran, carboxymethylcellulose, and hydroxyethylstarch (Katsuta et al. 1959a; Mizrahi and Moore 1970) also promote cell growth.

Synthetic polymers can also be a good substitute for serum. Polyvinyl-pyrolidones (PVP), which have been used as a plasma substitute, can replace most of the function of serum for rat ascites hepatoma cells, mouse L-cells, and human HeLa cells (Katsuta et al. 1959a, 1959b, and 1960). Pluronic polyols are known to promote cell growth in suspension culture (Swim and Parker 1960; Mizrahi 1975). Tween 20 or 80 is a component of some basal media (Morgan et al. 1950; Evans et al. 1956; Neuman Tytell 1960).

We recently found that polyethyleneglycols (PEG) and polyvinylalcohols (PVA), both known as fusogens for mammalian cells and microbial or plant protoplasts (Kao and Michayluk 1974; Nagata 1978), markedly stimulate the growth of mouse·human-human heterohybridomas at concentrations of less than one-hundredth of that used for cell fusion (Figure 4–1). A serum-free medium, PEG-86-1, was established by adding 4.5 g of HEPES, 1 g of sodium bicarbonate, 2 mg of insulin, 2 mg of transferrin, 2×10^{-6} M ethanolamine, 2.5×10^{-8} M sodium selenite, and 1 g of PEG20,000 to 1 l of a mixture (1:1:2, vol/vol) of IMDM, F12, and L15 (Shintani et al. 1988). Though the growth-promoting mechanism of PEG is unknown, cell growth was further stimulated in the presence of a small amount of FCS. PEG-containing serum-free medium is useful for a wide variety of cell lines (Figure 4–2).

4.2 SERUM-FREE MEDIA-CONTAINING SERUM-SUBSTITUTES

The establishment of a chemically defined serum-free medium is important to understanding the physiology of cells. However, some cells still cannot survive and proliferate in chemically defined media because their nutritional requirements are not completely defined. Thus, extracts from various tissues or conditioned media of other cell lines have been used to support the growth of these cell lines (Table 4–5).

Though a chemically defined serum-free medium is ideal for industrial use, it has some problems: it is generally specific for a cell line, the growth factors required are sometimes very expensive, and cell growth is not necessarily very stable. Supplements that can overcome these disadvantages are, therefore, very important for industrial purposes.

Some bacterial media ingredients have been investigated for use in tissue culture media. Waymouth (1956) found that Bacto peptone has a tremendous effect on L-cell growth in ML192/2 medium. Pumper et al. (1965)

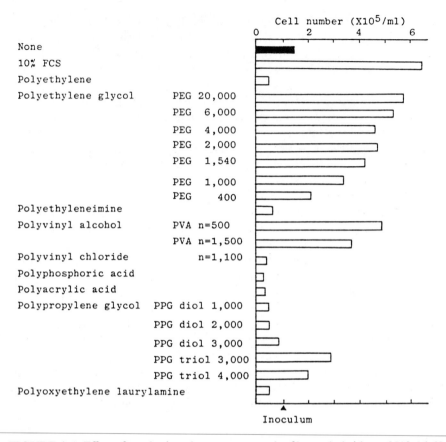

FIGURE 4–1 Effect of synthetic polymers on growth of heterohybridoma N12-16.63 in a serum-free medium. The cells were suspended at 1×10^5/ml in TL-2 medium supplemented with ITES and 0.2% (wt/vol) of each polymer, seeded in a volume of 1 ml in the well of 24-well multiplates and incubated at 37°C for four days. Reproduced with permission from Shintani et al. (1988).

reported autoclaved peptone or its dialyzate is also useful as a serum substitute. Peptone has been used for various cell lines, including mouse lung cells (Pumper 1958), Chang liver cells (Holmes 1959), rabbit myocardium cells (Taylor et al. 1972), mammary tumor cells (Lasfargues et al. 1973), BHK cells (Keay 1975), and human diploid fibroblasts (Delinassions 1983). Tryptose phosphate broth (Ginsberg et al. 1955; Reuveny et al. 1980), yeast extract (Katsuta et al. 1960), and casein (Darfler et al. 1980) have also been used.

Lactoalbumin hydrolyzate was a good amino acid and nutrient source of serum-free media at an early stage of the research (Newman and Tytell 1960; Takaoka et al. 1960). Primatone PL, a water soluble peptic digest of

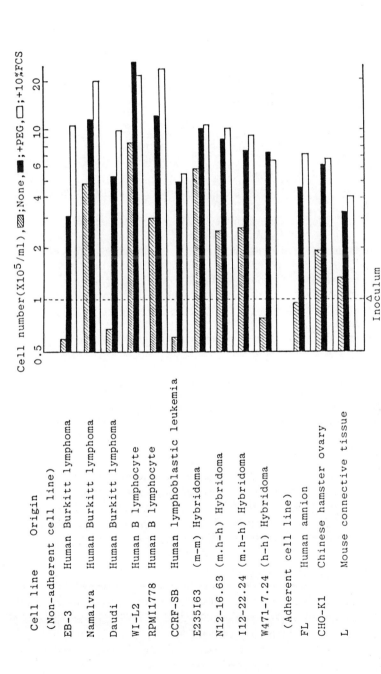

FIGURE 4-2 Effect of PEG on growth of various cell lines in a serum-free medium. TL-2 medium supplemented with ITES was used and 0.1% (wt/vol) PEG 20,000 was added. FCS of 1% (vol/vol) and 2% (vol/vol) was added for CHO-K1 and L cell lines, respectively. The cells of each cell line were inoculated at 1×10^5/ml and incubated at 37°C for five days. Reproduced with permission from Shintani et al. (1988).

TABLE 4–5   Tissue Extracts and Conditioned Media Used As a Serum Substitute

Extraction or Conditioned Medium	Example Where Used As a Serum Substitute
Tissue Extract	
Bovine brain extract	Endothelial cells (Hoshi and McKeehan 1984)
	Keratinocytes (Maciag et al. 1981)
Bovine hypothalamas extract	Endothelial cells (Hoshi and McKeehan 1984)
	Epithelial cells (Chaproniere and McKeehan 1986)
	Tracheal cells (Wu et al. 1985)
Bovine pituitary extract	Keratinocytes (Bertolero et al. 1984; Woodley et al. 1985)
	Epithelial cells (Hammond et al. 1984; Thompson et al. 1985)
	Endothelial cells (Hoshi and McKeehan 1984)
	Urothelial cells (Kirk et al. 1985a and 1985b)
Pregnant mouse-uterus extract	Hemopoietic colony forming cells (Baines et al. 1986)
Human platelet extract	Granulosa cells (Lino et al. 1985)
Rat submaxillary-gland extract	Preadipocyte cell lines (Gaillard et al. 1984)
Conditioned Medium	
Human hepatoma cells	Endothelial cells (Hoshi and McKeehan 1984)
Mouse spleen cells	Hemopoietic colony forming cells (Baines et al. 1986)
Human placental cells	Hemopoietic colony forming cells (Baines et al. 1986)
Mouse L cells	Macrophages (Flesch et al. 1984)

animal tissue, is also useful for large-scale production (Mizrahi 1977). Chicken egg yolk (Fujii and Gospodarowicz 1983) and a protein fraction from chicken egg yolk (Martis and Schwarz 1986) are also used as growth stimulants. An extract of a thermophilic blue-green alga, *Synechococcus elongatus* var. promotes the growth of human lymphoid cell lines (Shinohara et al. 1986).

Various growth factors have been isolated from serum, but the serum is still full of unknown compounds. Some interesting attempts have been made to fractionate serum proteins by alcohol precipitation (Sanford et al. 1955) or by electrophoresis (Holmes and Wolfe 1961). PEG-treated serum is also used instead of FCS (Inglot et al. 1975).

An interesting work along this line was carried out recently by Sasai et al. (1987) of our group. Though serum from adult cattle is very cheap and easily obtainable, it is not an appropriate supplement for mammalian cell cultures because it is often cytotoxic. Sasai and co-workers tried to remove the cytotoxic substances from the serum and succeeded in obtaining a growth factor fraction (named GFS) by collecting proteins salted out by ammonium sulfate at concentrations from 55 to 70% saturation. They also succeeded in sterilizing the GFS with ethylene oxide to remove contamination by mycoplasma and viruses without any substantial drop in growth-promoting activity. The GFS prepared from different serum lots are quite uniform in their growth-promoting activity, though their original serum often shows strong cytotoxicity, as shown in Figure 4–3. GFS is a complex consisting of several proteins, including serum albumin. However, when each component is separated and further purified, the strong growth-promoting action of GFS is not reproduced by combinating these compounds. Thus, a serum-free medium that contains an equal volume mixture of IMDM and F12, a mixture of insulin, transferrin, ethanolamine and sodium selenite (ITES), and GFS (GIT-medium) was established. This medium is now commercially available from Wako Pure Chemical Industries, Ltd. (Osaka, Japan). It is applicable to a wide variety of cells, including mouse

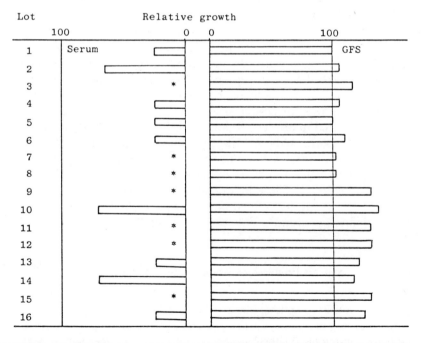

FIGURE 4–3 Comparison of growth-promoting activity of various lots of GFS and its original serum. Cells of mouse myeloma MPC-11 were passed two times each in serum- or GFS-containing medium before use. The cells were inoculated at 1×10^5/ml and incubated at 37°C for four days. Increase of cells in a 10% FCS-containing medium was taken as 100. *Cell death due to toxicity of the serum.

and human myelomas, various kinds of hybridomas, and most adherent cell lines. A great improvement of fusion efficiency in developing hybridomas was reported by use of this medium (Kudo et al. 1987 and 1988). Medium that contains GFS is also useful to produce human monoclonal antibodies by mouse·human-human heterohybridomas (Kitano et al. 1986). Since bovine immunoglobulins is mostly removed in GFS, it is especially advantageous for producing monoclonal antibodies.

4.3 PRODUCTION OF BIOLOGICALLY ACTIVE SUBSTANCES BY SERUM-FREE CULTURES

The biologically active substances listed in Table 4–6 are the subjects of large-scale mammalian cell cultures. This process involves the cultivation of primary cells, cell strains, and cell lines, including normal, transformed, and hybridoma cell lines. To optimize the economics of the process, the necessity for a serum-free culture has increased each year.

4.3.1 Vaccine Production

Vaccine production is one of the oldest and biggest areas of mammalian cell culture. Cells are generally proliferated in a serum-containing medium, then transferred to a serum-free medium, and infected with a virus (White et al. 1971; Keay and Schlesinger 1974). The yield of the virus depends greatly on the presence of glucose, but much less so on the presence of amino acids, vitamins, or serum (Zwartouw and Algar 1968). However, glutamine is required to produce Sendai virus in BHK cells (Ito et al. 1974).

Serum-free culture in both cell-proliferation and virus-infection processes is sometimes very effective for vaccine production. Foot-and-mouse disease virus (FMDV) is produced by the serum-free culture of BHK cells (Tomei and Issel 1975). These investigators used BHK-S cells, which had been adapted to chemically defined media, and a glutamine-free serum-free chemically defined medium (GFAD) (Nagle and Brown 1971). Poliovirus

TABLE 4–6 **Biologicals Produced by Mammalian Cell Culture**

Viral vaccines
Interferons and other lymphokines
Monoclonal antibodies
Nonantibody immunomodulators
Hormones
Polypeptide growth factors
Enzymes
Animal cells

and measles virus are produced by the serum-free culture of lymphoblastoid cells using IMDM or a mixture (1:1) of IMDM and RPMI1640 supplemented with HSA, Intralipid[R] (Kabi Pharmaceuticals) and human transferrin (Bjare and Rabb 1985). Poliovirus is also produced by African green monkey kidney cells (Vero), which can be cultivated in an equal volume mixture of DMEM and Medium 199 supplemented with insulin, transferrin, putrescine, fetuin, EGF, BSA, and fibronectin on Cytodex-1 microcarriers (Clark et al. 1982).

The human hepatoma cell line PLC/PRF/5 has several copies of hepatitis B virus DNA integrated into host DNA and produces hepatitis B virus surface antigen particle in the culture medium. This cell line is cultivated in RPMI1640 supplemented with 15 mM HEPES, 10^{-8} M sodium selenite, 10^{-10} M hydrocortisone, and 0.5 μg/ml insulin (Bagnarelli et al. 1985).

4.3.2 Interferons

Alpha-type interferons have been produced by cultured lymphoma cells. Namalva cells are cultivated in a medium that contains serum substitute (Reuveny et al. 1980). RPMI1640 medium supplemented with 3.5% (vol/vol) heat inactivated and PEG-treated bovine serum, 0.35% (wt/vol) Primatone RL, and 0.2% Pluronic polyol F-68 is used for large scale propagation. Cells are grown by a semicontinuous culture in a 130-L fermentor, harvested every three to four days, spun down, and resuspended in a serum-free RPMI1640 medium. After being primed by a small amount of interferon, Sendai virus induces the interferon in large amounts. The interferon of 2.5 to 5 $\times$ 10^4 units/10^7 cells/ml is accumulated. Namalva cells can also be cultivated in RPMI1640 supplemented with BSA and Pluronic polyol F-68 for large-scale production (Lazar et al. 1982). IMDM or an equal-volume mixture of IMDM and RPMI1640 supplemented with HSA, Intralipid[R], and human transferrin is also applied, and sodium butyrate stimulates the production (Bjare and Rabb 1985). Very recently, Namalva cell lines, which can grow continuously in a glutamine-free chemically defined medium were reported by Hosoi et al. (1988). These cells are expected to be useful for producing interferon.

Interferon-beta is produced by various cells, including normal human diploid fibroblasts (Havell and Vilcek 1972), MG-63 cells (Billiau et al. 1977), and C-10 cells (Tan 1981). However, human diploid fibroblasts have been adopted for industrial production because of safety concerns. Human diploid foreskin cells (FS-4) are grown to confluency on microcarrier in DMEM supplemented with 10% FCS. The interferon is then induced by polyI·polyC in chemically defined DMEM and produced by incubating the cells in DMEM supplemented with 0.5% Plasmanate (Giard et al. 1982).

4.3.3   Monoclonal Antibodies

Large-scale production of monoclonal antibodies (MoAb) is the most exciting area in recent mammalian cell cultures.

Iscove and Melchers (1978) developed a serum-free medium for new bacterial lipopolysaccharide (LPS)-stimulated murine B lymphocytes. They have developed a formula for the basal medium (IMDM) and added delipidated BSA, transferrin, and soy bean lipid. Chang et al. (1980) succeeded in cultivating several mouse hybridomas in MEM (or RPMI1640) supplemented with insulin and transferrin. Murakami et al. (1982) found that ethanolamine and selenite are also required and established a serum-free medium that contained insulin, transferrin, ethanolamine, and selenite (ITES). 2ME is also required by mouse myeloma NS-1 and NS-1 hybridomas (Kawamoto et al. 1983). Though NS-1 requires cholesterol for growth, NS-1 hybridomas does not. Thus, a serum-free medium, consisting of RDF medium, ITES, 2ME, and fatty acid-free bovine serum albumin is effectively used to selectively isolate hybridomas.

Cleveland et al. (1983) succeeded in producing mouse MoAb under protein-free conditions using a medium supplemented with 22 trace elements. An effective basal medium, eRDF, was established for the high-density culture of hybridomas (Murakami et al. 1984). Another serum-free medium, SFH, containing ITES, fatty acid-free BSA-linoleic acid complex, ascorbic acid, hydrocortisone, and 12 trace elements was established for mouse hybridomas (Kovar and Franek 1984).

Murine MoAb can be produced by large-scale serum-free culture using mechanical agitators (Martin et al. 1987), air-lift fermentors (Lambert et al. 1987), a hollow-fiber cell culture unit (Klerx et al. 1988), and porous microcarriers (Dean et al. 1987).

The production of human MoAb is the newest target for serum-free cultures. Mouse·human-human heterohybridomas are effectively cultivated in TL-2 medium supplemented with ITES and 3 mg/ml GFS (GITL-2 medium) using a perfusion culture system equipping a cell sedimentation column (Kitano et al. 1986). Maximum cell density reaches 1.2×10^7 cells/ml and the human MoAb is produced constantly for long periods (Figure 4–4). A high-productivity subclone of a mouse·human-human heterohybridoma (Kitano et al. 1988a) shows high titer of human MoAb in a chemically defined serum-free medium, PEG-86-3, which consists of TL-2 medium, ITES, 0.1% PEG20,000 and 10 μg/ml swine LDL (Shintani et al. 1989).

Mouse-human heterohybridomas are also cultivated in eRDF medium supplemented with ITES (or BSA and ITES) (Takazawa et al. 1988). These investigators used a culture vessel with a cell settling zone for perfusion and an ultrafiltration unit to recycle high molecular weight components.

A serum-free medium, RITC57-1 was developed to culture human lymphoid cell lines (Minamoto and Mitsugi 1985). Human-human hybridomas are effectively cultivated in PEG-86-1 medium using a ceramic-ma-

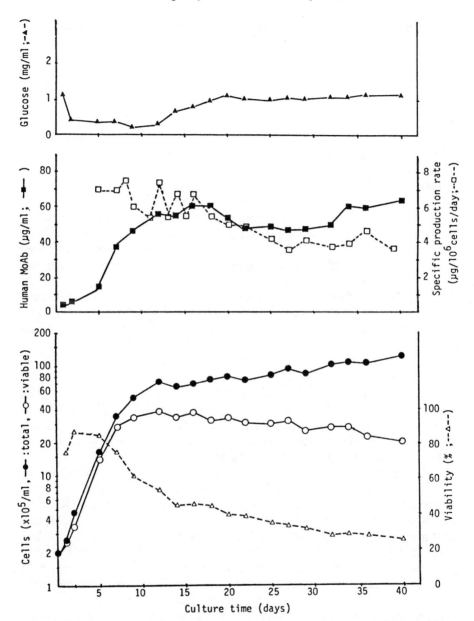

FIGURE 4–4 Production of human MoAb by perfusion culture of N12-16.63 in a 2-l jar fermentor. Cells (2 × 10⁵/ml) were inoculated into 1 l of GITL-2 medium in a 2-l round-bottomed jar fermentor equipped with a cell sedimentation column. Cultivation was carried out at 37°C at an agitation speed of 25 rpm. The medium supplemented with 1.576 g/l glucose was perfused at a rate of 0.69 volume per day from the third day of the cultivation. Dissolved oxygen was controlled at around 1 ppm. Reproduced with permission from Kitano et al. (1986).

trix reactor (Kitano et al. 1988b) or in a serum-free medium containing ITES, α-lipoic acid, linoleic acid, and BSA using a porous collagen matrix (Hayman and Tung 1988).

4.3.4 Enzymes

Plasminogen activators catalyze the conversion of plasminogen to active plasmin, which plays an important role in the balance between fibrinolysis and coagulation. They are classified into two groups based on differences in their immunological and biochemical properties, urokinases, and tissue plasminogen activators (t-PA) (Rijken et al. 1981).

Urokinase has been purified from human urine to be used clinically in thrombolytic therapy, but its content in urine is very low. Human kidney cells produce the same urokinase that is found in urine (Barlow and Lazer 1972) and the production by mammalian cell cultures has been investigated using serum-supplemented medium (Lewis 1979). However, when human kidney cells are cultivated in a serum-free medium, another inactive compound is accumulated (Kasai et al. 1985). It is a single-chain prourokinase, converted to the active urokinase by a catalytic amount of plasmin. When murine 3T3 cells infected with murine sarcoma virus are maintained in serum-free DMEM, inactive prourokinase is also accumulated (Skriver et al. 1982).

t-PA is produced by Bowes melanoma cells (Rijken and Collen 1981), HeLa cells (Bernik et al. 1981), human vascular endotherial cells (Booyse et al. 1981), human fibroblasts (Kadouri and Bohak 1983), and human embryonic lung cells (Brouty-Boye et al. 1984). Guinea pig ear keratinocytes (GPK cell line) are also used (Griffiths et al. 1985).

Though the t-PA-producing cells are generally grown in serum-containing medium, they are transferred to the serum-free medium at the production stage to facilitate enzyme purification from the supernatant (Kadouri and Bohak 1983; Kruithof et al. 1985; Griffiths et al. 1985). Bowes melanoma cells are grown to confluency in DMEM supplemented with FCS, washed with phosphate-buffered saline (PBS), and then incubated in DMEM supplemented with insulin, transferrin, progesterone, cortisol, aprotinin, and a mixture of trace elements (Kruithof et al. 1985). The same production rate can be maintained for over eight months by treating the culture with serum every six to eight weeks. This cell line is also maintained in an equal volume mixture of DMDM and F12 supplemented with insulin, transferrin, cholesterol, vitamin E, vitamin E acetate, oleic acid, phosphoethanolamine, BSA, and glutamine on a ceramic core reactor to effectively produce t-PA (Bodeker et al. 1987). Normal human embryonic lung cells are also maintained in a serum-free medium consisting of a mixture (3:1, vol/vol) of DMEM and F12, insulin, transferrin, selenite, concanavalin A, and aprotinin for the same purpose (Brouty-Boye et al. 1984).

4.3.5 Recombinant Products

Recombinant DNA technology has been used to successfully produce many biologically active substances by microorganisms. However, the microbial products do not necessarily satisfy requirements; the products are sometimes inactive because of the wrong folding in the tertiary or quaternary structure, and are sometimes very unstable in vivo because of the nonglycosylation. Thus, the need to produce these substances by genetically engineered mammalian cells has increased rapidly. However, the development processes are mostly industrial secrets and published work is scarce.

BHK cells harboring the human interleukin-2 (IL-2) gene are cultivated in a serum-free medium on microcarriers (Wagner and Lehmann 1988). A continuous perfusion vessel (1-L) equipped with a bubble-free aeration system is used. The cell density increases up to 8×10^6 cells/ml with a production of 700 μg/ml of IL-2 per day.

Rhodes and Birch (1988) established two systems for recombinant t-PA production by serum-free mammalian cell culture. The first system used the mouse C127 cell line transfected with bovine papilloma virus vector for gene expression. A microcarrier culture process was scaled up to 40 l in a spin-filter stirred perfusion reactor. Cells were grown in low serum medium and a protein-free medium was then continuously perfused. The t-PA concentrations reached up to 55 mg/l over a period of more than one month. Another system uses the rat myeloma cell line Yo with an expression vector employing the Rous sarcoma virus long terminal repeat sequence. A serum-free batch culture process using an air-lift fermentor gives more than 50 mg/l of t-PA after nine days of cultivation.

4.4 CONCLUSION

During the past 10 years, the growth requirements of a great number of cells or cell lines have been investigated and serum-free media for these cells have been established. This has led to a deeper understanding of the cells and the discovery of many biologically active substances related to the differentiation or proliferation of the targeted cells.

The industrial application of mammalian cells, including normal, transformed, hybrid, and recombinant, has increased rapidly. However, one of the disadvantages of the mammalian cell culture process is that it is far more costly than the microbial process. To free the culture medium of serum is expected to greatly reduce costs.

Though serum-free media containing a serum substitute have a somewhat long history in vaccine production, the application of chemically defined media has just started. To choose between a medium that contains serum-substitute and a chemically defined medium will depend on the cell line, the culture method, and the product. The defined medium is ideal, but the cultivation is sometimes not sufficiently stable. Thus, the choice of appropriate serum-substitutes is an important subject. Adaptation of cells to the medium may be required by some cell lines. Since a number of serum-free media or serum substitutes are now commercially available, they may be beneficial in some cases.

This has been a brief overview of the present status of serum-free media. But, a medium should always be considered in relation to the culture methods and conditions. If associated problems can be overcome, the serum-free culture is expected to become the mainstream method in industrial mammalian cell culture.

REFERENCES

Adolphe, M., Froger, B., Ronot, X., Corvol, M.T., and Forest, N. (1984) *Exp. Cell Res.* 155, 527–536.

Allegra, J.C., and Lippman, M.E. (1978) *Cancer Res.* 38, 3823–3829.

Allen, R.E., Dodson, M.V., Luiten, L.S., and Boxhorn, L.K. (1985) *In Vitro Cell. Develop. Biol.* 21, 636–640.

Ambesi-Impiombato, F.S., Parks, L.A.M., and Coon, H.G. (1980) *Proc. Natl. Acad. Sci. USA* 77, 3455–3459.

Amorosa, L.F., Khachadurian, A.K., Harris, J.N., Schneider, S.H., and Fung, C.H. (1984) *Biochim. Biophys. Acta* 792, 192–198.

Avner, E.D., Sweeney, W.E., Jr., Piesco, N.P., and Ellis, D. (1985) *In Vitro Cell. Develop. Biol.* 21, 297–304.

Bagnarelli, P., Brugia, M., and Clementi, M. (1985) *Develop. Biol. Stand.* 60, 45–53.

Baines, P., Cormier, F., Lucien, N., and Boffa, G.A. (1986) *Int. J. Cell Cloning* 4, 103–114.

Barano, J.L.S., and Hammond, J.M. (1985) *Endocrinology* 116, 51-58.

Barlow, G.H., and Lazer, L.V. (1972) *Thromb. Res.* 1, 201–208.

Barnes, D., and Sato, G. (1979) *Nature* 281, 388–389.

Barnes, D., and Sato, G. (1980) *Anal. Biochem.* 102, 255–270.

Barnes, D.W., and Silnutzer, J. (1983) *J. Biol. Chem.* 258, 12548–12552.

Bauer, R.F., Arthur, L.O., and Fine, D.L. (1976) *In Vitro* 12, 558–563.

Bellot, F., Luis, J., El Battari, A., et al. (1985) *Int. J. Cancer* 36, 609–615.

Bernik, M.B., Wijngaards, G., and Rijken, D.C. (1981) *Ann. N.Y. Acad. Sci.* 370, 592–608.

Bertolero, F., Kaighn, M.E., Gonda, M.A., and Saffiotti, U. (1984) *Exp. Cell Res.* 155, 64–80.

Bettger, W.J., Boyce, S.T., Walthal, B.J., and Ham, R.G. (1981) *Proc. Natl. Acad. Sci. USA* 78, 5588–5592.

Bettger, W.J., and Ham, R.G. (1982) in *Growth of Cells in Hormonally Defined Media Book A* (Sato, G.H., Pardee, A.B., and Sirbasku, D.A., eds.), pp. 61–64, Cold Spring Harbor, New York.

Billiau, A., Edy, V.G., Heremans, H., et al. (1977) *Antimicrob. Agents Chemother.* 12, 11–15.

Binoux, M., Faivre-Bauman, A., Lassarre, C., Barret, A., and Tixier-Vidal, A. (1985) *Develop. Brain Res.* 21, 319–321.

Birch, J.R. (1980) *Develop. Biol. Stand.* 46, 21–27.

Birch, J.R., and Pirt, S.J. (1970) *J. Cell Sci.* 7, 661–670.

Bjare, V., and Räbb, I. (1985) *Develop. Biol. Stand.* 60, 349–354.

Bödeker, B.G.D., Klimetzek, V., Klein, U., Hewlett, G., and Schlumberger, H.D. (1987) *Develop. Biol. Stand.* 66, 291–297.

Böhlen, P., Esch, F., Baird, A., Gospodarowicz, D. (1985) *EMBO J.* 4, 1951–1956.

Booyse, F.M., Scheinbuks, J., Radek, J., et al. (1981) *Thromb. Res.* 24, 495–504.

Bottenstein, J.E., and Sato, G.H. (1979) *Proc. Natl. Acad. Sci. USA* 76, 514–517.

Bottenstein, J.E., Skaper, S.D., Varon, S.S., and Sato, G.H. (1980) *Exp. Cell Res.* 125, 183–190.

Bradshaw, G.L., and Dubes, G.R. (1983) *In Vitro* 19, 735–742.

Bradshaw, G.L., Sato, G.H., McClure, D.B., and Dubes, G.R. (1983) *J. Cell. Physiol.* 114, 215–221.

Brannon, P.M., Orrison, B.M., and Kretchmer, N. (1985) *In Vitro Cell. Develop. Biol.* 21, 6–14.

Broome, J.D., and Teng, M.W. (1973) *J. Exp. Med.* 138, 574–592.

Brouty-Boye, G.C., Maman, M., Marian, J.C., and Choay, P. (1984) *Biotechnology* 2, 1058–1062.

Brower, M., Carney, D.N., Die, H.K., Gazdar, A.F., and Minna, J.D. (1986) *Cancer Res.* 46, 798–806.

Brown, R.L., Griffith, R.L., Neubauer, R.H., and Rabin, H. (1983) *J. Cell.Physiol.* 115, 191–198.

Brunner, G. (1982) *Cell Tissue Res.* 224, 553–561.

Bryant, J.C., Evans, V.J., Schilling, E.L., and Earle, W.R. (1961) *J. Natl. Cancer Inst.* 26, 239–252.

Buck, P.A., and Schomberg, D.W. (1987) *Biol. Reprod.* 36, 167–174.

Buhl, S.N., and Regan, J.D. (1972) *Proc. Sci. Exp. Biol. Med.* 140, 1224–1227.

Calvo, F., Brower, M., and Carnoy, D.N. (1984) *Cancer Res.* 44, 4553–4559.

Castor, C.W. (1962) *J. Lab. Clin. Med.* 60, 788–798.

Chang, J.H., Steplewski, Z., and Koprowski, H. (1980) *J. Immunol. Methods* 39, 369–375.

Chaproniere, D.M., and Mckeehan, W.L. (1986) *Cancer Res.* 46, 819–824.

Chen, L.B., and Buchanan, J.M. (1975) *Proc. Natl. Acad. Sci. USA* 72, 131–135.

Chen, J.-K., LaBrake-Farmer, S., and McClure, D.B. (1986) *J. Cell. Physiol.* 128, 413–420.

Chessebeuf, M., and Padieu, P. (1984) *In Vitro* 20, 780–795.

Clark, J.M., Gebb, C., and Hirtenstein, M.D. (1982) *Develop. Biol. Stand.* 50, 81–91.

Cleveland, W.L., Wood, I., and Erlanger, B.F. (1983) *J. Immunol. Methods* 56, 221–234.

Click, R.E., Benck, L., and Alter, B.J. (1972) *Cell. Immunol.* 3, 155–160.

Cohen, S. (1962) *J. Biol. Chem.* 237, 1555–1562.

Darfler, F.J., Murakami, H., and Insel, P.S. (1980) *Proc. Natl. Acad. Sci. USA* 77, 5993–5997.

Darlington, G.J., Kelly, J.M., and Buffone, G.J. (1987) *In Vitro Cell. Develop. Biol.* 23, 349–354.

Darmon, M., Serrero, G., Rizzino, A., and Sato, G. (1981) *Exp. Cell Res.* 132, 313–327.

Dean, R.C., Karkare, S.B., Ray, N.G., Runstadler, P.W., and Venkatasubramanian, K. (1987) *Ann. N.Y. Acad. Sci.* 506, 129–146.

Delinassions, J.G. (1983) *Exp. Cell Biol.* 51, 315–321.

Dell'Aquila, M.L., and Gaffney, E.V. (1982) *Exp. Cell Res.* 137, 441–446.

Dodge, W.H., and Sharma, S. (1985) *J. Cell. Physiol.* 123, 264–268.

Dulbecco, R., and Freeman, G. (1959) *Virology* 8, 396–397.

Eagle, H. (1955a) *Proc. Soc. Exp. Biol. Med.* 89, 362–364.

Eagle, H. (1955b) *Science* 122, 501–504.

Eagle, H. (1956) *Arch. Biochem. Biophys.* 61, 356–366.

Eagle, H. (1959) *Science* 130, 432–437.

Eagle, H., Barban, S., Levy, M., and Schulze, H.O. (1958) *J. Biol. Chem.* 233, 551–558.

Eagle, H., and Piez, K. (1962) *J. Exp. Med.* 116, 29–43.

Eccleston, P.A., Gunton, D.J., and Silberberg, D.H. (1985) *Develop. Neurosci.* 7, 308–322.

Eliason, J.F., and Odartchenko, N. (1985) *Proc. Natl. Acad. Sci. USA* 82, 775–779.

Enat, R., Jefferson, D.M., Ruiz-Opazo, N., et al. (1984) *Proc. Natl. Acad. Sci. USA* 81, 1411–1415.

Epstein-Almog, R., and Orly, J. (1985) *Endocrinology* 116, 2103–2112.

Erhmann, R.L., and Gey, G.O. (1956) *J. Natl. Cancer Inst.* 16, 1375–1403.

Esfahani, M., Scerbo, L., Lund-Katz, S., et al. (1986) *Biochim. Biophys. Acta* 889, 287–300 (1986).

Ethier, S.P. (1986) *In Vitro Cell. Develop. Biol.* 22, 485–490.

Evans, V.J., Bryant, J.C., Fioramonti, M.C., et al. (1956) *Cancer Res.* 16, 77–86.

Fischer, G. (1984) *J. Neurosci. Res.* 12, 543–552.

Flesch, I., Ketelsen, U.-P., and Ferber, E. (1984) *Agents Actions* 15, 33–34.

Flesch, I., and Ferber, E. (1986) *Immunobiology* 171, 14–26.

Florini, J.R., and Roberts, S.B. (1979) *In Vitro* 15, 983–992.

Franc, J.L., Hovsepian, S., Fayet, G., Bouchilloux, S. (1984) *Mol. Cell Endocrinol.* 37, 233–239.

Fujii, D.K., Cheng, J., and Gospodarowicz, D. (1983) *J. Cell. Physiol.* 114, 267–278.

Fujii, D.K., and Gospodarowicz, D. (1983) *In Vitro* 19, 811–817.

Gaillard, D., Negrel, R., Serrero-Dave, G., Cermolacce, C., and Ailhaud, G. (1984) *In Vitro* 20, 79–88.

Gatmaitan, Z., Jefferson, D.M., Ruiz-Opazo, N., et al. (1983) *J. Cell Biol.* 97, 1179–1190.

Gebb, C., Clark, J.M., Hirtenstein, M.D., et al. (1982) *Develop. Biol. Stand.* 50, 93–102.

Gey, G.O., and Thalhimer, W. (1924) *J. Am. Med. Assoc.* 82, 1609.

Giard, D.J., Fleischaker, R.J., and Sinskey, A.J. (1982) *J. Interferon Res.* 2, 471–477.

Giguere, L., Cheng, J., and Gospodarowicz, D. (1982) *J. Cell. Physiol.* 110, 72–80.

Ginsberg, H.S., Gold, E., and Jordan, W.S. (1955) *Proc. Sci. Exp. Biol. Med.* 89, 66–71.

Gonzalez, R., Dempsey, M.E., Elliot, A.Y., and Fraley, E.E. (1974) *Exp. Cell Res.* 87, 152–158.

Gorham, L.W., and Waymouth, C. (1965) *Proc. Soc. Exp. Biol. Med.* 119, 287–290.

Gospodarowicz, D. (1974) *Nature* 249, 123–127.

Gospodarowicz, D. (1984) in *Cell Culture Methods for Molecular and Cell Biology* Vol. 1 (Barnes, D.W., Sirbasku, D.A., and Sato, G.H., eds.), pp. 69–86, Alan R. Liss, Inc., New York.

Gospodarowicz, D., Cheng, J., Lui, G.-M, Baird, A., and Böhlen, P. (1984) *Proc. Natl. Acad. Sci. USA* 81, 6963–6967.

Green, H. (1978). *Cell* 15, 801–811.

Griffin, M., Law, P.Y., and Loh, H. (1985) *Brain Res.* 360, 370–373.

Griffiths, J.B., McEntee, I.D., Electricwala, A., et al. (1985) *Develop. Biol. Stand.* 60, 439–446.

Hagiwara, H., Ohtake, H., Yuasa, H., et al. (1985) in *Growth and Differentiation of Cells in Defined Environment* (Murakami, H., Yamane, I., Barnes, D.W., et al., eds.), pp. 117–122, Kodansha/Springer-Verlag, Tokyo/Berlin.

Ham, R.G. (1963) *Exp. Cell Res.* 29, 515–526.

Ham, R.G. (1964) *Biochem. Biophys. Res. Commun.* 14, 34–38.

Ham, R.G. (1965) *Proc. Natl. Acad. Sci. USA* 53, 288–293.

Ham, R.G., and McKeehan, W.L. (1978) in *Nutritional Requirements of Cultured Cells* (Katsuta, H., ed.), pp. 63–115, Japan Scientific Societies Press, Tokyo.

Hamilton, W.G., and Ham, R.G. (1977) *In Vitro* 13, 537–547.

Hammond, S.L., Ham, R.G., and Stampfer, M.R. (1984) *Proc. Natl. Acad. Sci. USA* 81, 5435–5439.

Harrison, J.J., Soudry, E., and Sager, R. (1985) *J. Cell Biol.* 100, 429–434.

Hashizume, S., Kuroda, K., and Murakami, H. (1983) *Biochim. Biophys. Acta* 763, 377–382.

Havell, E.A., and Vilcek, J. (1972) *Antimicrob. Agents Chemother.* 2, 476–484.

Hayashi, I., Larner, J., and Sato, G. (1978) *In Vitro* 14, 23–30.

Hayashi, I., and Sato, G.H. (1976) *Nature* 259, 132–134 (1976).

Hayman, E.G., Pierschbacher, M.D., Öhgren, Y., and Ruoslahti, E. (1983) *Proc. Natl. Acad. Sci. USA* 80, 4003–4007.

Hayman, E.G., and Tung, A.S. (1988) in *Cell Culture Engineering*, Engineering Foundation, New York (Abstract M8).

Heath, J.K., and Deller, M.J. (1983) *J. Cell. Physiol.* 115, 225–230.

Heldin, C.H., Wasteson, A., and Westermark, B. (1980) *Proc. Natl. Acad. Sci. USA* 77, 6611–6615.

Hendelman, W.J., de Savigny, N., and Marshall, K.C. (1985) *In Vitro Cell. Develop. Biol.* 21, 129–134.

Higuchi, K. (1963) *J. Infec. Dis.* 112, 213–220.

Higuchi, K. (1970) *In Vitro* 6, 239.

Higuchi, K., and Robinson, R.C. (1973) *In Vitro* 9, 114–121.

Hiragun, A., Sato, M., and Mitsui, H. (1983) *Exp. Cell Res.* 145, 71–78.

Holmes, R. (1959) *J. Biophys. Biochem. Cytol.* 6, 535–536.

Holmes, R., Helms, J., and Mercer, G. (1969) *J. Cell Biol.* 42, 262–271.

Holmes, R., and Wolfe, S.W. (1961) *J. Biophys. Biochem. Cytol.* 10, 389–401.

Holmgren, J. (1981) *Nature* 292, 413–417.

Honma, Y., Katsuta, T., Okabe, J., and Hozumi, M. (1979) *Exp. Cell Res.* 124, 421–428.

Hoshi, H., and McKeehan, W.L. (1984) *Proc. Natl. Acad. Sci. USA* 81, 6413–6417.

Hosoi, S., Mioh, H., Anzai, C., Sato, S., and Fujiyoshi, N. (1988) *Cytotechnology* 1, 151–158.

Houck, K.A., and Michalopoulos (1985) *In Vitro Cell. Develop. Biol.* 21, 121–124.

Hutchings, S.E., and Sato, G.H. (1978) *Proc. Natl. Acad. Sci. USA* 75, 901–904.

Iguchi, T., Uchima, F-D.A., Ostrander, P.L., and Bern, H.A. (1983) *Proc. Natl. Acad. Sci. USA* 80, 3743–3747.

Ill, C.R., Brehm, T., Lydersen, B.K., Hernandez, R., and Burnett, K.G. (1988) *In Vitro Cell. Develop. Biol.* 24, 413–419.

Ill, C.R., and Gospodarowicz, D. (1982) *J.Cell. Physiol.* 113, 373–384.

Inglot, A.D., Godzinska, H., and Chndzio, T. (1975) *Acta Virol.* 19, 250–254.

Iscove, N.N., and Melchers, F. (1978) *J. Exp. Med.* 147, 923–933.

Iscove, N.N., Guilbert, L.J., and Weyman, C. (1980) *Exp. Cell Res.* 126, 121–126.

Ishii, T., Bannai, S., and Sugita, Y. (1981) *J. Biol. Chem.* 256, 12387–12392.

Ito, Y., Kumura, Y., and Kunii, A. (1974) *J. Virol.* 13, 557–566.

Kadouri, A., and Bohak, Z. (1983) *Biotechnology* 1, 354–358.

Kaji, K. (1985) in *Growth and Differentiation of Cells in Defined Environment* (Murakami, H., Yamane, I., Barnes, D.W., et al., eds.), pp. 265–270, Kodansha/ Springer-Verlag, Tokyo/Berlin.

Kan, M., and Yamane, I. (1982) *J. Cell. Physiol.* 111, 155–162.

Kan, M., and Yamane, I. (1983) *Tohoku J. Exp. Med.* 139, 389–398.

Kano-Sueoka, T. (1984) in *Cell Culture Methods for Molecular and Cell Biology* Vol. 2. (Barnes, D.W., Sirbasku, D.A., and Sato, G.H., eds.), pp. 89–104, Alan R. Liss, Inc., New York.

Kano-Sueoka, T., Cohen, D.M., Yamaizumi, Z., et al. (1979) *Proc. Natl. Acad. Sci. USA* 76, 5741–5744.

Kano-Sueoka, T., and Errick, J.E. (1981) *Exp. Cell Res.* 136, 137–145.

Kao, K.N., and Michayluk, M.R. (1974) *Planta* 115, 355–367.

Kasai, S., Arimura, H., Nishida, M., and Suyama, T. (1985) *J. Biol. Chem.* 260, 12377–12381.

Kato, Y., and Gospodarowicz, D. (1984) *J. Cell. Physiol.* 120, 354–363.

Katsuta, H., and Takaoka, T. (1973) *Methods Cell Biol.* 6, 1–42.

Katsuta, H., Takoaka, T., Furukawa, T., and Kawana, M. (1960) *Jpn. J. Exp. Med.* 30, 147–157.

Katsuta, H., Takoaka, T., Hosaka, S., et al. (1959a) *Jpn. J. Exp. Med.* 29, 45–70.

Katsuta, H., Takaoka, T., Mitamura, K., et al. (1959b) *Jpn. J. Exp. Med.* 29, 191–201.

Katsuta, H., Takaoka, T., Nose, K., and Nagai, Y. (1975) *Jpn. J. Exp. Med.* 45, 345–354.

Kawamoto, T., Sato, J.D., Le, A., McClure, D.B., and Sato, G.H. (1983) *Anal. Biochem.* 130, 445–453.

Keay, L. (1975) *Biotechnol. Bioeng.* 17, 745–764.

Keay, L., and Schlesinger, S. (1974) *Biotechnol. Bioeng.* 16, 1025–1044.

Kirk, D., Kagawa, S., Narayan, S.K., and Ohnuki, Y. (1985a) *Exp. Cell Res.* 160, 221–229.

Kirk, D., Kagawa, S., and Vener, G. (1985b) *In Vitro Cell. Develop. Biol.* 21, 165–171.

Kirk, D., and Alvarez, R.B. (1986) *In Vitro Cell. Develop. Biol.* 22, 604–614.

Kitano, K., Iwamoto, K., Shintani, Y., and Akiyama, S. (1988a) *J. Immunol. Methods* 109, 9–16.

Kitano, K., Shintani, Y., Ichimori,Y., et al. (1986) *Appl. Microbiol. Biotechnol.* 24, 282–286.

Kitano, K., Shintani, Y., Iwamoto, K., and Ichimori, Y. (1988b) in *Cell Culture Engineering*, Engineering Foundation, New York (Abstract T2).

Kleinman, H.K., Klebe, R.J., and Martin, G.R. (1981) *J. Cell Biol.* 88, 473–485.

Klerx, J.P.A.M., Verplanke, C.J., Blonk, C.G., and Twaalfhoven, L.C. (1988) *J. Immunol. Methods* 111, 179–188.

Kohler, N., and Lipton, A. (1974) *Exp. Cell Res.* 87, 297–301.

Kovar, J., and Franek, F. (1984) *Immunol. Lett.* 7, 339–345.

Kovar, J., and Franek, F. (1987) *Biotechnol. Lett.* 9, 259–264.

Kruithof, E.K.O., Schleuning, W-D., and Bachmann, F. (1985) *Biochem. J.* 226, 631–636.

Kuchler, R.J., Marlowe, M.L., and Merchant, D.J. (1960) *Exp. Cell Res.* 20, 428–437.

Kudo, T., Asao, A., and Tachibana, T. (1988) *Tohoku J. Exp. Med.* 154, 345–355.

Kudo, T., Morishita, R., Suzuki, R., and Tachibana, T. (1987) *Tohoku J. Exp. Med.* 153, 55–66.

Lambert, K.J., Boraston, R., Thompson, P.W., and Birch, J.R. (1987) *Develop. Ind. Microbiol.* 27, 101–106.

Landschultz, W., Thesleff, I., and Ekblom, P. (1984) *J.Cell Biol.* 98, 596–601.

Lasfargues, E.Y., Coutinho, W.G., Lasfargues, J.C., and Moore, D.H. (1973) *In Vitro* 8, 494–500.

Lazar, A., Reuveny, S., Traub, A., et al. (1982) *Develop. Biol. Stand.* 50, 167–171.

Leibovitz, A. (1963) *Am. J. Hyg.* 78, 173–180.

Levi-Montalcini, R., Meyer, H., and Hamburger, V. (1954) *Cancer Res.* 14, 49–57.

Lewis, L.J. (1979) *Thrombos. Haemostas.* 42, 895–900.

Libby, P. (1984) *J. Mol. Cell. Cardiol.* 16, 803–811.

Libby, P., and O'Brien, K.V. (1983) *J. Cell. Physiol.* 115, 217–223.

Lieberman, I., and Ove, P. (1958) *J. Biol. Chem.* 233, 637–642.

Lieberman, I., and Ove, P. (1959) *J.Biol. Chem.* 234, 2754–2758.

Ling, C.T., Gey, G.O., and Richters, V. (1968) *Exp. Cell Res.* 52, 469–489.

Lino, J., Baranao, S., and Hammond, J.M. (1985) *Endocrynology* 116, 2143–2151.

Maciag, T., Nemore, R.E., Weinstein, R., and Gilchrest, B.A. (1981) *Science* 211, 1452–1454.

Macieira-Coelho, A. (1966) *Experientia* 22, 390–391.

Malan-Shibley, L., and Lype, P.T. (1983) *In Vitro* 19, 749–758.

Malewicz, B., Anderson, L.E., Crilly, K., and Jenkin, H.M. (1985) *In Vitro Cell. Develop. Biol.* 21, 470–476.

Martin, N., Brennon, A., Denome, L., and Shaevitz, J. (1987) *Biotechnology* 5, 837–840.

Martis, M.J., and Schwarz, R.I. (1986) *In Vitro Cell. Develop. Biol.* 22, 241–246.

Mather, J.P., and Sato, G.H. (1979a) *Exp. Cell Res.* 120, 191–200.

Mather, J.P., and Sato, G.H. (1979b) *Exp. Cell Res.* 124, 215–221.

McClure, D.B. (1983) *Cell* 32, 999–1006.

McGowan, J.A., Strain, A.J., and Bucher, N.L.R. (1981) *J. Cell. Physiol.* 108, 353–363.

McKeehan, W.L., Hamilton, W.G., and Ham, R.G. (1976) *Proc. Natl. Acad. Sci. USA* 73, 2023–2027.

McKeehan, W.L., McKeehan, K.A., Hammond, S.L., and Ham, R.G. (1977) *In Vitro* 13, 399–414.

Minamoto, Y., and Mitsugi, K. (1985) in *Growth and Differentiation of Cells in Defined Environment* (Murakami, H., Yamane, I., Barnes, D.W., et al., eds.), pp. 127–130, Kodansha/Springer-Verlag, Tokyo/Berlin.

Minna, J.D., Carney, D.N., Oie, H., Bunn, P.A., and Gazdar, A.F. (1982) *Cold Spring Harbor Conf. Cell Prolif.* 9, 627–639.

Mizrahi, A. (1975) *J. Clin. Microbiol.* 2, 11–13.

Mizrahi, A. (1977) *Biotechnol. Bioeng.* 19, 1557–1561.

Mizrahi, A., and Moore, G.E. (1970) *Appl. Microbiol.* 19, 906–910.

Mohamed, S.N.W., Holmes, R., and Hartzell, C.R. (1983) *In Vitro* 19, 471–478.

Moore, G.E., Gerner, R.E., and Franklin, H.A. (1967) *J. Am. Med. Assoc.* 199, 519–524.

Morgan, J.F., Morton, H.J., and Parker, R.C. (1950) *Proc. Sci. Exp. Biol. Med.* 73, 1–8.

Morrison, P.R., Edsall, J.T., and Miller, S.G. (1948) *J. Am. Chem. Soc.* 70, 3103–3108.

Morrison, R.S., and DeVellis, J. (1981) *Proc. Natl. Acad. Sci. USA* 78, 7205–7209.

Morrison, S.J., and Jenkin, H.M. (1972) *In Vitro* 8, 94–100.

Mosien, D.E. (1981) *J. Immunol.* 127, 1490–1493.

Murakami, H. (1984) *Cell Culture Methods for Molecular and Cell Biology* Vol. 4 (Barnes, D.W., Sirbasku, D.A., and Sato, G.H., eds.), pp. 197–205, Alan R. Liss, Inc., New York.

Murakami, H., Edamoto, T., Shinohara, K., and Omura, H. (1983) *Agric. Biol. Chem.* 47, 1835–1840.

Murakami, H., and Masui, H. (1980) *Proc.Natl. Acad. Sci. USA* 77, 3464–3468.

Murakami, H., Masui, H., Sato, G., and Raschke, W.C. (1981) *Anal. Biochem.* 114, 422–428.

Murakami, H., Masui, H., Sato, G.H., et al. (1982) *Proc. Natl. Acad. Sci. USA* 79, 1158–1162.

Murakami, H., Okazaki, Y., Yamada, K., and Omura, H. (1988) *Cytotechnology* 1, 159–169.

Murakami, H., Shimomura, T., Nakamura, T., Ohashi, H., Sinohara, K., and Omura, H. (1984) *J. Agric. Biol. Chem.* 58, 575–583.

Nagata, T. (1978) *Naturwissenschaften* 65, 263–264.

Nagle, S.C., and Brown, B.L. (1971) *J. Cell. Physiol.* 77, 259–263.

Needham, L.K., Tennekoon, G.L., and McKhann, G.M. (1987) *J. Neurosci* 7, 1–9.

Neumann, R.E., and Tytell, A.A. (1960) *Proc. Soc. Exp. Biol. Med.* 104, 252–256.

Neumann, R.E., and Tytell, A.A. (1961) *Proc. Soc. Exp. Biol. Med.* 106, 857–862.

Ohmori, H., and Yamatoma, I. (1987) *Eur. J. Immunol.* 17, 79–83.

Oka, M.S., Landers, R.A., and Bridge, C.D.B. (1984) *Exp. Cell Res.* 154, 537–547.

Okabe, T., Fijisawa, M., and Takaku, F. (1984) *Proc. Natl. Acad. Sci. USA* 81, 453–455.

Orly, J., and Sato, G.H. (1979) *Cell* 17, 295–305.

Peehl, D.M., and Ham, R.G. (1980) *In Vitro* 16, 526–538.

Peehl, D.M., and Stamey, T.A. (1986) *In Vitro Cell. Develop. Biol.* 22, 82–90.

Perez-Infante, V., and Mather, J.P. (1982) *Exp. Cell Res.* 142, 325–332.

Phillips, P.D., and Cristofalo, V.J. (1981) *Exp. Cell Res.* 134, 297–302.

Pixley, S.K.R., and Cotman, C.W. (1986) *J. Neurosci. Res.* 15, 1–17.

Poste, G., Papahadjopoulos, D., and Vail, W.J. (1976) *Methods Cell Biol.* 14, 33–71.

Praeger, F.C., and Cristofalo, V.J. (1986) *In Vitro Cell. Develop. Biol.* 22, 355–359.

Pumper, R.W. (1958) *Science* 128, 363–364.

Pumper, R.W., Yamashiroya, H.M., and Molander, L.T. (1965) *Nature* 207, 662–663.

Reddan, J.R., and Dziedzic, D.C. (1982) *Exp. Cell Res.* 142, 293–300.

Reitzer, L.Z., Wice, M.B., and Kennell, D. (1979) *J. Biol. Chem.* 256, 2669–2676.

Reuveny, S., Bino, T., Kosenberg, H., Traub, A., and Mizrahi, A. (1980) *Develop. Biol. Stand.* 46, 281–288.

Reznikoff, C.A., Loretz, L.J., Pesciotta, D.M., Oberley, T.D., and Ignjatovic, M.M. (1987) *J. Cell. Physiol.* 131, 285–301.

Rhodes, M., and Birch, J. (1988) *Biotechnology* 6, 518–523.

Richman, R.A., Claus, T.H., Pilkis, S.J., and Friedman, D.L. (1976) *Proc.Natl. Acad. Sci. USA* 73, 3589–3593.

Rijken, D.C., and Collen, D. (1981) *J. Biol. Chem.* 256, 7035–7041.

Rijken, D.C., Wijngaards, G., and Welbergen, J. (1981) *J. Lab. Clin. Med.* 97, 477–486.

Rizzino, A., and Crowley, C. (1980) *Proc. Natl. Acad. Sci. USA* 77, 457–461.

Rizzino, A., and Ruff, E. (1986) *In Vitro Cell. Develop. Biol.* 22, 749–755.

Rockwell, G.A., Sato, G.H., and McClure, D.B. (1980) *J. Cell. Physiol.* 103, 323–331.

Sanford, K.K., Westfall, B.B., Fioramonti, M.C., et al. (1955) *J. Natl. Cancer Inst.* 16, 789–802.

Sasai, S., Fujimoto, T., and Tsukamoto, K. (1987) U.S. Patent 4,654,304.

Sato, G.H. (1975) in *Biochemical Actions of Hormones* Vol. III (Litwak, G., ed.), pp. 391–396, Academic Press, New York.

Sato, G.H., Fisher, W., and Puck, T.T. (1957) *Science* 126, 951–964.

Sato, J.D., Kawamoto, T., McClure, D.B., and Sato, G.H. (1984) *Mol. Biol. Med.* 2, 121–134.

Sato, J.D., Kawamoto, T., and Okamoto, T. (1987) *J. Exp. Med.* 165, 1761–1766.

Sato, T., Minamoto, Y., Yamane, I., Kudo, T., and Tachibana, T. (1982) *Exp. Cell Res.* 138, 127–134.

Schönherr, O.T., van Gelder, P.T.J.A., van Hees, P.J., van Os, A.M.J.M., and Roelofs, H.W.M. (1987) *Develop. Biol. Stand.* 66, 210–220.

Shinohara, K., Okura, Y., Koyano, T., et al. (1986) *Agric. Biol. Chem.* 50, 2225–2230.

Shintani, Y., Iwamoto, K., and Kitano, K. (1988) *Appl. Microbiol. Biotechnol.* 27, 533–537.

Shintani, Y., Iwamoto, K., and Kitano, K. (1989) *Cytotechnology* 2, 9–17.

Shipley, G.D., Childs, C.B., Volkenant, M.E., and Moses, H.L. (1984) *Cancer Res.* 44, 710–716.

Shipley, G.D., and Ham, R.G. (1981) *In Vitro* 17, 656–670.

Shipley, G.D., and Ham, R.G. (1983a) *Exp. Cell Res.* 146, 249–260.

Shipley, G.D., and Ham, R.G. (1983b) *Exp. Cell Res.* 146, 261–270.

Shooter, R.A., and Gey, G.O. (1952) *Br. J. Exp. Pathol.* 33, 98–103.

Silnutzer, J., and Barnes, D.W. (1984) in *Cell Culture Methods for Molecular and Cell Biology* Vol. 1 (Barnes, D.W., Sirbasku, D.A., and Sato, G.H., eds.), pp. 245–268, Alan R. Liss, Inc., New York.

Simms, E., Gazdar, A.F., Abrams, P.G., and Minna, J.P. (1980) *Cancer Res.* 40, 4356–4363.

Simonian, M.H., White, M.L., and Foggia, D.A. (1987) *In Vitro Cell. Develop. Biol.* 23, 247–252.

Simonian, M.H., White, M.L., and Gill, G.N. (1982) *Endocrynology* 111, 919–927.

Skriver, L., Nielsen, L.S., Stephens, R., and Dano, K. (1982) *Eur. J. Biochem.* 124, 409–414.

Smith, E.L., Walworth, N.C., and Holick, M.F. (1986) *J. Invest. Dermatol.* 86, 709–714.

Sonoda, Y., Yang, Y-C., Wong, G.G., Clark, S.C., and Ogawa, M. (1988) *Proc. Natl. Acad. Sci. USA* 85, 4360–4364.

Spector, A.A. (1968) *Ann. N.Y. Acad. Sci.* 149, 768–783.

Sporn, M.B., and Todaro, G.J. (1980) *New Eng. J. Med.* 303, 878–880.

Stenn, K.S. (1981) *Proc. Natl. Acad. Sci. USA* 78, 6907–6911.

Stewart, S., Zhu, B., and Axelrad, A. (1984) *Exp. Hematol.* 12, 309–318.

Swim, H.E., and Parker, R.F. (1960) *Proc. Soc. Exp. Biol. Med.* 103, 252–254.

Takaoka, T., Katsuta, H., Kaneko, K., Kawana, M., and Furukawa, T. (1960) *Jpn. J. Exp. Med.* 30, 391–408.

Takazawa, Y., Tokashiki, M., Hamamoto, K., and Murakami, H. (1988) *Cytotechnology* 1, 171–178.

Tan, Y.H. (1981) *Methods in Ezymol.* 78, 120–125.

Taub, M., Chuman, L., Saier, M.H., and Sato, G. (1979) *Proc. Natl. Acad. Sci. USA* 76, 3338–3342.

Taylor, W.G., Taylor, M.J., Lewis, N.J., and Pumper, R.W. (1972) *Proc. Soc. Exp. Biol. Med.* 139, 96–99.

Thompson, A.A., Dilworth, S., and Hay, R.J. (1985) *J. Tissue Cult. Methods* 9, 117–122.

Timpl, R., Rode, H., Robey, G.P., et al. (1979) *J. Biol. Chem.* 254, 9933–9937.

Titeux, M., Testa, V., Louache, F., et al. (1984) *J. Cell. Physiol.* 121, 251–256.

Tomei, L.D., and Issel, C.J. (1975) *Biotechnol. Bioeng.* 17, 765–778.

Tomooka, Y., Imagawa, W., Nandi, S., and Bern, H.A. (1983) *J. Cell. Physiol.* 117, 290–296.

Tsao, M., Walthal, B.J., and Ham, R.G. (1982) *J. Cell. Physiol.* 110, 219–229.

Vaheri, A., Alitalo, K., Hedman, K., et al. (1978) *Ann. N.Y. Acad. Sci.* 312, 343–353.

van Brunt, J. (1986) *Biotechnology* 4, 505–510.

Vlodavsky, I., and Gospodarowicz, D. (1981) *Nature* 289, 304–306.

Wagner, R., and Lehmann, J. (1988) *Trends Biotechnol.* 6, 101–104.

Waymouth, C. (1956) *J. Natl. Cancer Inst.* 17, 315–325.

Waymouth, C. (1959) *J. Natl. Cancer Inst.* 22, 1003–1016.

White, P.R. (1946) *Growth* 10, 281–289.

White, A., Berman, S., and Lowenthal, J.P (1971) *Appl. Microbiol.* 22, 909–913.

Wolfrom, C., Loriette, C., Polini, G., et al. (1983) *Exp. Cell Res.* 149, 535–546.

Woodley, D.T., Briggaman, R.A., Gammon, W.R., and O'Keefe, E.J. (1985) *Biochem. Biophys. Res. Commun.* 130, 1267–1272.

Wu, R., Nolan, E., and Turner, C. (1985) *J. Cell. Physiol.* 125, 167–181.

Yamane, I., Kan, M., Hoshi, H., and Minamoto, Y. (1981a) *Exp. Cell Res.* 134, 470–474.

Yamane, I., Kan, M., Minamoto, Y., and Amatsuji, Y. (1981b) *Proc. Jpn. Acad. Ser. B.* 57, 385–389.

Yamane, I., Murakami, O., and Kato, M. (1975) *Proc. Soc. Exp. Biol. Med.* 149, 439–442.

Yoshinari, K., Yuasa, K., Iga, F., and Mimura, A. (1989) *Biochim. Biophys. Acta* 1010, 28–34.

Zwartouw, H.T., and Algar, D.J. (1968) *J. Gen. Virol.* 2, 243–250.

Nuclear Magnetic Resonance Spectroscopy of Dense Cell Populations for Metabolic Studies and Bioreactor Engineering: A Synergistic Partnership

Bruce E. Dale
Robert J. Gillies

5.1 PHENOMENA THAT CAN AND CANNOT BE MEASURED BY NMR

A number of important physiological and metabolic parameters can be measured by in vivo nuclear magnetic resonance (NMR). Table 5–1 summarizes some of these parameters. Since NMR is a noninvasive/nondestructive technique, and since the energy levels of NMR are low, the sample receives minimum perturbation. In fact, Boltzmann distribution differences are about 1 in 100,000, i.e., approximately one molecule in 100,000 is perturbed by the magnetic field (Gillies et al. 1989). This lack of sensitivity is both an advantage (minimum perturbation) and a disadvantage of the technique. Because of the lack of sensitivity, the basic requirement for effective NMR is large and/or concentrated samples. More specifically, some of the

TABLE 5-1   Metabolic Parameters Observable by Whole-Cell NMR

Nucleus	Sensitivity[1]	Application
^{1}H	1.00	Water structure
		Intermediates: lactate, pyruvate, creatine, glutathione, etc.
^{19}F	0.94–0.83	Intracellular Ca^{2+}, Zn^{2+}, pH
^{31}P	0.41–0.066	Phosphorylated intermediates: ATP, AMP, PCr, PArg, Pi, ADP
		Intracellular pH, Mg
		Enzyme kinetics (by saturation transfer)
^{23}Na	0.26–0.093	Intracellular Na^+
^{13}C	0.25–0.015	Glycolytic intermediates
		Gluconeogenic intermediates
		Transamination intermediates
^{15}N	$0.101–10^{-5}$	Intracellular pH
		Protein turnover
^{35}Cl	$0.098–10^{-3}$	Intracellular Cl^-
^{14}N	$0.072–10^{-3}$	Urea, NH_3, TMA, amino acids
^{39}K	$0.046–10^{-4}$	Intracellular K^+

[1]Relative to the ^{1}H nucleus.

TABLE 5-2   Important Biological Phenomena Not Visible by NMR In Vivo[1]

Macromolecular Metabolism
  DNA replication
  Gene transcription
  Protein synthesis
Macromolecular Dynamics
  Protein/protein interaction
  Protein/DNA interactions
Low Abundance Compounds
  For ^{1}H <0.01 mM intracellular
  For ^{13}C <1 mM (labeled)
  For ^{31}P <0.15 mM (e.g., cAMP)
  For ^{19}F < 0.01 mM

[1]General limitations in vivo are molecular weight (e.g., correlation times) <1500 and concentrations (of homogeneous signal) $> \sim 0.1$ mM.

limitations of NMR for biological studies are summarized in Table 5–2. The general limitations for in vivo NMR are molecular weights (correlation times) less than 1,500 and concentrations of homogeneous signal greater than about 0.1 mM.

5.2 SOME REPRESENTATIVE NMR SPECTRA

The two most popular isotopes for biological NMR studies are ^{13}C and ^{31}P. Figure 5–1 is a ^{13}C spectrum of Erlich ascites tumor (EAT) cells grown in spinner culture. This spectrum was obtained from samples concentrated by a factor of 100 prior to the experiment and then resuspended in an NMR tube (Gillies et al. 1989). The sample was fed with ^{13}C-labeled glucose, labeled at the C-1 position. Fifteen minutes after feeding the label, the spectrum was acquired for 5 min at 90 MHz. The distribution of the label in various intracellular compounds is shown. In this way, both kinetic and mechanistic studies can be performed. As this spectrum demonstrates, the information density of NMR is potentially very high.

A similar ^{31}P spectrum for EAT cells grown and harvested in the same way is given in Figure 5–2. Since this nucleus is 100% naturally abundant, it is usually not necessary to enrich it with labeled compound as is done for ^{13}C. Different inorganic phosphate (Pi) peaks represent cellular compartments at different pH. Differences between intracellular and extracellular pH are an important means of assessing cellular health and metabolic state (Huang et al. 1986). Cellular energization, also an important indicator of viability and metabolic state, is reflected in ATP/ADP ratios, which can be determined in such experiments.

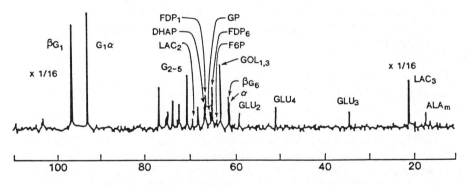

FIGURE 5–1 ^{13}C spectrum for Erlich ascites tumor cells fed ^{13}C-1-labeled glucose. G, glucose; GLU, glutamate and glutamine; LAC, lactate; ALA, alanine; GOL, glycerol; GP, glycerol phosphate. α and β refer to glucose anomers; numbers refer to carbon number in compound.

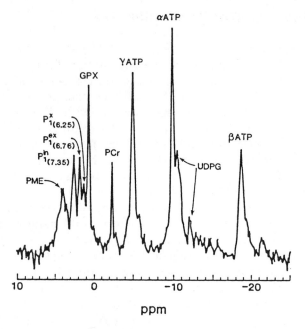

FIGURE 5–2 ³¹P spectrum for Erlich ascites tumor cells (22% cell volume). GPX, glycerol phosphoryl choline; PME, phosphomonesters; γ, α, and β are terminal, primary, and middle phosphate groups of ATP, respectively.

5.3 CURRENT LIMITATIONS ON WHOLE CELL NMR: SENSITIVITY AND THE NEED FOR WELL-DEFINED BIOREACTORS

For the most part, however, studies such as these are not routinely performed on cultured cells, primarily due to problems with NMR sensitivity and/or cell stability. Sensitivity is quantitatively expressed as the signal-to-noise (S/N) ratio for a particular chemical species or nucleus. Although a complete discussion of the factors affecting S/N is beyond the scope of this paper, it is important to recognize for our purposes that the signal is directly proportional to cell density times the NMR-visible sample volume (i.e., proportional to the number of nuclei present) while sample resistance is the main determinant of noise. Sample resistance is proportional to sample conductivity whereas conductivity is inversely proportional to cell density (Gillies et al. 1986). Therefore, noise should decrease with increasing cell density until coil resistance becomes the dominant factor. S/N is also proportional to the acquisition time to the one-half power. This mathematical relationship between S/N and other parameters is given below.

$$S/N = (AT)^{1/2} \, K \, D$$

where AT is the total acquisition time (seconds), K is a constant derived from theory, which includes the relationships for field strength, coil efficiency, nucleus, and linewidths, and D is the cell density in the coil volume (Gillies et al. 1986).

A graphical representation of the relationship between cell density and the acquisition time required to obtain a given S/N for the β phosphate of ATP is presented in Figure 5–3. We have followed current practice in Figure 5–3 by reporting cell density as cells per unit volume. This practice, however, should be changed. Since cell sizes vary, it is important to report results not as cells per unit volume but rather as cell volume fraction. Alternatively, if cell volume is directly proportional to protein content, results can be reported as protein concentration in the sample volume. Only in this way can different experimental systems be rigorously compared.

The problem of sensitivity in NMR studies of cell culture is generally addressed, therefore, by attempting to increase the number of cells in the sample. This can be approached in two different ways. It has been common practice to grow cells outside the magnet, spin down a relatively dense cell population, and then subject this concentrated cell suspension to NMR analysis. In spite of the valuable information that has been obtained this way, the disadvantages of this technique are numerous; "on-line" studies cannot be done and phenomena with time frames shorter than the sample processing time are obscured. Probably the most important drawback is that the sample environment changes with unknown and uncontrollable effects on cellular physiology.

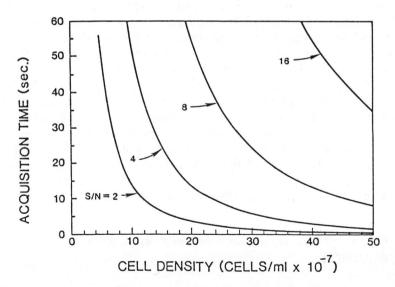

FIGURE 5–3 A graphical representation of the relationship between cell density and the acquisition time required to obtain a given signal-to-noise ratio for the β phosphate of ATP.

The second approach is to grow cells to sufficiently high density within a vessel compatible with NMR geometry and conditions. For instance, cells have been immobilized in agarose threads (Knop et al. 1984), on microcarrier beads (Ugurbil et al. 1981), within hollow fiber systems (Gonzalez-Mendez et al. 1982), or simply suspended in a gas-sparged NMR tube. Although growing cells to high density is technically difficult, it has a number of advantages over the alternative of high-density inoculation. The principal advantage, other than those listed above, is the stability imparted to the culture by a bioreactor system capable of supporting growth (Gillies et al. 1986). Therefore, bioreactor systems are required that are capable of growing cells to extremely high densities.

5.4 BIOREACTOR ENGINEERING CONSIDERATIONS

Thus, whole-cell NMR analyses require cultures that are both extremely dense and very stable. These goals are shared by biochemical engineers interested in the products obtainable from such cultures. Immobilized cell culture systems represent a particular class of reactors known to chemical engineers as heterogeneous catalytic systems. In such systems, reactants and products of reaction must diffuse between phases (e.g., between the flowing liquid phase in the lumena of hollow fiber bioreactors and the cell phase immobilized between the fibers) before the reaction(s) can occur. Since diffusion occurs at a finite rate, it is possible that the maximum rate of diffusion is less than the maximum possible rate of reaction (for instance, see Thiele 1939 and Zeldowich 1939). If this occurs, the system is said to be diffusion limited and the kinetic data do not represent the intrinsic or true kinetics of the system. Instead, the "kinetic" data represent, to a greater or lesser extent, the rates of diffusion rather than those of reaction.

In the case of dense cell culture devices for use in on-line NMR studies of cell metabolism, this is a crucial consideration. Metabolic data obtained from a badly diffusion-limited bioreactor may be useless for understanding the fundamentals of cellular physiology. If incorrect decisions are based on such data, the effects of poor data will be compounded. Unfortunately, an appreciation of the effects of diffusion and reaction and the degree to which kinetic data can be compromised by these effects seems to be almost entirely missing from the literature on NMR of dense cell cultures.

5.5 QUANTITATIVE MEASURES OF DIFFUSION AND REACTION

A number of quantitative tools have been developed for determining the relative importance of diffusion and reaction in heterogeneous catalytic systems. Probably the most useful of these tools for experimental work is the

Weisz-Prater criterion, which is based only on observable parameters (Weisz and Prater 1954). This modulus is the ratio of the observed reaction rate to the maximum rate of diffusion, e.g., for diffusion and reaction in a simple porous catalyst

$$\Phi = \frac{[dn/dt]}{\left[D_{eff} \dfrac{C_i}{L^2} \right]}$$

where

  dn/dt is the measured rate of reaction per unit volume of the catalyst;
  D_{eff} is the effective diffusivity within the porous catalyst of the species whose rate of reaction is being determined;
  L is the characteristic length of the diffusion path (e.g., R/3 for a sphere, R/2 for a cylinder); and
  C_i is the concentration of the species of interest in the phase from which diffusion must occur.

If the value of this observable modulus is less than about 0.3, the system is largely controlled by the rate of reaction rather than the rate of diffusion and the kinetic data (i.e., metabolic data) are reliable. If the value of the modulus is greater than about 3.0, the rate of reaction is controlled by the rate of diffusion and the "kinetic" data do not adequately represent actual reaction rate processes.

We have used such diffusion/reaction analyses to predict the oxygen concentration profiles in hollow fiber bioreactors (HFBR) (Chresand et al. 1988). Figure 5-4 is a representative plot. The ratio of oxygen concentration at any position to the inlet oxygen concentration (Co) is given for a desired cell density using per cell oxygen consumption rates obtained in batch culture and estimated values of oxygen diffusivity in dense cell aggregates. These data predict oxygen depletion before the end of the bioreactor, thus confirming the anecdotal information on nonuniform cell growth in HFBR. There is obvious practical and scientific value in being able to design and operate bioreactors to avoid diffusion limitations. How can we design these reactors and then verify that they are operating properly?

5.6 DESIGN PROCEDURES INVOLVING WEISZ'S MODULUS

We suggest the following protocol for bioreactor design:

1. Determine the maximum consumption rates for important and/or limiting substrates in batch culture on a per volume of cells basis (or per

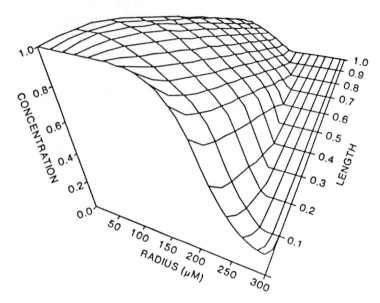

FIGURE 5-4 A representative plot of diffusion/reaction analyses to predict the oxygen concentration profiles in hollow fiber bioreactors.

unit mass of cells). For aerobic cultures, careful attention to oxygen requirements is especially important.

2. Determine the desired cell volume fraction in the bioreactor configuration of interest. For NMR studies this will be dictated largely by S/N considerations.

3. Estimate the characteristic diffusion path length for the substrates and products. For HBFR, a well-defined interfiber spacing is important so that this distance may be known with some accuracy.

4. Estimate the diffusivities of the species of interest in the phases through which they must diffuse. Such data are limited. Additional research to measure diffusivities of biomolecules, particularly in the dense cell aggregates characteristic of commercial bioreactors, is badly needed.

5. Calculate a value of the observable (or Weisz's) modulus for each of the species of interest using the batch kinetic data.

If the calculated value of the modulus is less than 0.3 for all reacting species, the kinetic data should be reliable and the system is probably not diffusion-limited. If the value of Φ is greater than 3.0, either the diffusion path length must be shortened (this factor has the most impact since it is squared), or the concentration of the reacting species must be increased, or the volume fraction of cells must be decreased (or all of the above). Given the saturation kinetics characteristic of biological catalytic systems, a reasonable design goal might be to keep the concentration of the species of interest in the zero-order kinetic (or saturation) region. Repeat the process once kinetic data are available in the actual experimental system to confirm reaction control.

A similar procedure might be followed for potentially inhibitory species. The rate of production of inhibitory compounds on a per volume of cells basis would be determined and the maximum permissible concentration of the species set using batch kinetic experiments. These and the other required parameters would be entered in the Weisz's modulus calculation. If Φ is less than about 0.3, the maximum rate of diffusion of the inhibitory compound out of the cell phase should be greater than its rate of production, and significant inhibition would not be anticipated.

The most general definition of the observable modulus is (Weisz and Prater 1954):

$$\Phi \equiv \emptyset^2 \eta$$

where $\emptyset$ is the Thiele modulus for the system of interest and η is the effectiveness factor. Our own research has emphasized HFBR for metabolic studies. If we design for zero-order kinetic behavior in HFBR, the relevant expressions for the Thiele modulus and the effectiveness factor have already been derived (Webster et al. 1979). For the zero-order kinetic limit they are:

$$\emptyset_0^2 = \left\{ \frac{1}{2} (\gamma^2 \ln\gamma) + \frac{1}{4} (1 - \gamma^2) + [\ln (\alpha) (1 - \gamma^2)]/2\Psi \right\}^{-1}$$

$$\eta = \left[\frac{2}{(\beta^2 - 1)\ln\gamma} \left(\frac{1}{\emptyset_0^2} - \frac{1 - \gamma^2}{4} - \frac{1}{2} \ln\gamma \right) \right.$$
$$\left. + \frac{\Psi}{(\beta^2 - 1) \ln (1/\alpha)} \left(\frac{2}{\emptyset_0^2} - \frac{1 - \gamma^2}{2} - \ln\gamma \right) \right]^{-1}$$

where α, β, and γ are geometrical parameters characteristic of the particular HFBR and $\Psi = K_p D_M/D_E$ where K_p is a substrate partition coefficient (unity in the absence of electrical interaction effects), D_M is the effective substrate diffusivity in the hollow fiber membrane wall, and D_E is the effective substrate diffusivity in the cell mass. From this analysis, the importance of good diffusivity data and well-defined interfiber spacings (since this determines β) is apparent.

5.7 EXPERIMENTAL CONFIRMATIONS OF REACTION RATE CONTROL: TOWARD GRADIENTLESS BIOREACTORS

A variety of experimental tests are also available for confirming reaction control. Since most systems of interest are immobilized cells with a flowing liquid stream containing the reactants and products, an obvious test is to sample the exit fluid from the reactor and compare the concentrations of the important reactants and products with desired levels. In particular, for the HFBR we would check reactor effluent concentrations to be certain they do not drop below or exceed desired limits (for limiting reactants and inhibitory products). Batch kinetic data supplemented by operating experience would provide these limits. A stirred vessel to which reactants are slowly added and products withdrawn at desired rates can serve as the reservoir to supply the immobilized cell system. This is the bioreactor analog of the differential packed bed recycle reactor used in many catalyst studies.

A classical test is to perform experiments with different characteristic diffusion path lengths (different interfiber spacings in the HFBR case). If the observed rates (per volume or mass of cells) are the same for different diffusion path lengths, then the kinetic data should be reliable. If the observed rates are not the same within experimental error, then the data are probably more or less diffusion limited. An additional *partial* test is to vary the fluid velocity in the system of interest. As long as the reaction rate varies with fluid velocity (except for shear-sensitive biocatalysts) the rate data are diffusion influenced. However, it is *not* true that the failure of the reaction rate to change with changing fluid velocity guarantees reaction rate control.

A further test is to calculate the activation energy for the reaction of interest if it is possible to vary the temperature of the system sufficiently without deactivating the biocatalyst. Activation energies of about 5–10 kcal/gmol or less are symptomatic of diffusion-influenced processes since in the region of strong diffusion resistance the observed activation energy is one-half the sum of the activation energies of diffusion and reaction (Froment and Bischoff 1979). The higher the observed activation energy, the more likely the kinetic data are to be reliable.

To the degree that diffusion limitations might limit cell growth, non-uniform cell growth in reactors would also be symptomatic of diffusion and not kinetic control. This might be detectable by microscopic examination of the immobilized cell aggregate. Since the conductivity of cell aggregates has been shown to vary as a function of cell concentration (Lovitt et al. 1986; Blute et al. 1988), simple conductivity electrodes might also be placed in the bioreactor to measure cell growth and cell density. For instance, the predicted variation in oxygen concentration with axial position in HFBR should lead to nonuniform cell growth, which might be detectable by in-place conductivity sensors for cell density placed at different axial locations in the HFBR.

5.8   USES OF NMR IN BIOREACTOR ANALYSIS AND DESIGN

As indicated at the beginning of this chapter, NMR has great potential to help us understand and, therefore, manipulate cellular metabolism. As the reactions are better understood, the reactors may be designed better. For instance, there is evidence that cellular behavior can change upon immobilization (Doran and Bailey 1986; Payne et al. 1988). If so, the batch kinetic data used in developing the initial estimate of the observable modulus (Φ) may not be adequate for high-density immobilized cell reactors. Such changes should be apparent in metabolic studies by NMR. NMR also might be used to measure mass transfer parameters and to assess possible diffusion limitations. Gradients in substrate and product concentrations may be visible by imaging NMR if the spatial resolution is consistent with bioreactor dimensions. For instance, axial gradients in cell concentrations in HFBR should be readily apparent with proton imaging NMR or phosphorus imaging NMR. Radial gradients would be more difficult to detect but may be possible given adequate resolution and well-defined interfiber spacings.

The paucity of diffusivity data for biomolecules, particularly in cell aggregates, has already been mentioned. These diffusivities might be measured using ^{13}C NMR of labeled compounds with an isotopic shift experiment. Cells would be perfused with a medium that contains metabolites enriched in ^{13}C. Once steady-state NMR signals are obtained, the circulating medium would be changed for one that contains substrate more enriched in ^{13}C. The increase in labeled compounds within the cell layer as a function of time would be used to estimate diffusivities by standard mathematical relationships (e.g., Crank 1975). Simultaneous information on metabolic intermediates formed from the labeled compounds would also be generated. In short, the ability to simultaneously generate kinetic and transport information of great value has yet to be properly exploited in whole-cell NMR.

5.9 SUMMARY

Commercial exploitation of the fruits of recombinant DNA and cell fusion technologies is significantly limited by the lack of fundamental metabolic information on the cell lines of interest, whether these are plant, animal, insect, or microbial cells. NMR can help to provide this information and thereby improve bioreactor design and operation. However, in the case of on-line NMR of dense cell culture devices for metabolic studies, these devices are inherently heterogeneous bioreactors. To ensure that the metabolic information generated is reliable, a number of precautions should be taken. These are the same precautions that should be taken to ensure that commercial bioreactors operate in a reaction-controlled regime. Therefore, reactor engineering methodologies, particularly diffusion and reaction analyses and reaction monitoring by whole-cell NMR must go hand in hand, each extending, complementing, and validating the other.

REFERENCES

Blute, T., Dale, B.E., and Gillies, R.J. (1988) *Biotech. Prog.* 4, 202–209.
Chresand, T.J., Gillies, R.J., and Dale, B.E. (1988) *Biotech. Bioeng.* 32, 983–992.
Crank, J. (1975) *The Mathematics of Diffusion*, Clarendon Press, Oxford.

Doran, P.M., and Bailey, J.E. (1986) *Biotech. Bioeng.* 28, 1814–1831.

Froment, G.F., and Bischoff, K.B. (1979) *Chemical Reactor Analysis and Design*, p. 186, John Wiley and Sons, New York.

Gillies, R.J., Chresand, T.J., Drury, D.D., and Dale, B.E. (1986) *Rev. Nuclear Mag. Res. Med.* 1, 155–179.

Gillies, R.J., MacKenzie, N.E., and Dale, B.E. (1989) *Bio/Technology* 7, 50–54.

Gonzalez-Mendez, R., Wemmer, D., Hahn, G., Wade-Jardetzky, N., and Jardetzky, O. (1982) *Biochim. Biophys. Acta* 720, 274–280

Huang, L., Forsberg, C.W., and Givving, L.N. (1986) *Appl. Environ. Microbiol.* 51, 1230–1234.

Knop, R.H., Chen, C-W., Mitchell, J.B., Russo, A., McPherson, S., and Cohen, J.S. (1984) *Biochim. Biophys. Acta* 804, 275–284.

Lovitt, R.W., Walter, R.P., Morris, J.G., and Kell, D.B. (1986) *Appl. Microbiol. Biotech.* 23, 168–173.

Payne, G.F., Payne, N.N., and Shuler, M.L. (1988) *Biotech. Bioeng.* 31, 905–912.

Thiele, E.W. (1939) *Ind. Eng. Chem.* 31, 916–921.

Ugurbil, K., Guernsey, D.L., Brown, T.R., et al. (1981) *Proc. Natl. Acad. Sci. USA* 78, 4843–4847.

Webster, I.A., Shuler, M.L., and Rony, P.R. (1979) *Biotech. Bioeng.* 21, 1725–1748.

Weisz, P., and Prater, C.D. (1954) in *Advances in Catalysis and Related Subjects* Vol. VI (Frankenburg, W.G., Komarewsky, V.I., and Rideal, E.K., eds.), pp. 144–195, Academic Press, New York.

Zeldowich, Ia.B. (1939) *Zhur. Fiz. Khim.* 13, 163–170.

Regulation of Animal Cell Metabolism in Bioreactors

William M. Miller
Harvey W. Blanch

The cultivation of mammalian cells for the production of a wide variety of therapeutic agents has expanded rapidly over the past several years. As the need for increasing amounts of these products has grown, new and larger-scale bioreactors have been examined for cultivation of mammalian cells. The requirements imposed by such large-scale cultivation of both hybridoma and anchorage-dependent cells often result in nutrient and cell concentrations far removed from those found in T-flask culture studies. Extended culture times, higher glucose and glutamine concentrations required to support higher cell concentrations, and accumulation of by-products are common to both suspension and attached-cell bioreactors. These factors have emphasized our lack of understanding of the behavior of these cells under conditions employed for industrial production of therapeutic products. While there is considerable information available that qualitatively describes cell metabolism, there is far less quantitative kinetic information available that can be employed in the design and analysis of bioreactors. The control of bioreactors requires a fundamental understanding of the dynamics of cellular metabolism, as well as the availability of models that describe metabolic responses to changes in the external environment of the cell.

The objectives of this chapter are thus to review the key factors involved in the metabolism of mammalian cells. We shall emphasize hybridoma cells because kinetic information collected under well-defined cultivation conditions has recently become available, and this provides a background for discussions of anchorage-dependent cells.

6.1 METABOLISM OF CULTURED CELLS

Most information available on cell metabolism has been obtained in tissue culture flasks or batch reactors. As discussed below, this is not conducive to quantitative analysis or metabolic modeling. However, the results obtained do provide insight into the metabolic differences between cells grown in vivo and in culture.

6.1.1 Patterns of Nutrient Metabolism

Glucose and glutamine are the major carbon and energy sources in most cell culture media. The major pathways for glucose and glutamine metabolism are shown in Figure 6–1. Both nutrients provide unique biosynthetic precursors. Glucose is required for nucleoside synthesis (Wice et al. 1981; Renner et al. 1972), as well as for the synthesis of glucosamine 6-phosphate and the precursor glyceraldehyde 3-phosphate (McKeehan 1986). Many cell types can use fructose, galactose, or other sugars in place of glucose (Morgan and Faik 1986); and can be grown without sugar if the medium is supplemented with uridine (Wice et al. 1981). Cells can also use maltose and starch as glucose sources, due to the presence of maltase and α-amylase in fetal bovine serum (Rheinwald and Green 1974). Glutamine is required for purine synthesis and for the formation of guanine nucleotides (Raivio and Seegmiller 1973). It is also the primary amino group donor in the synthesis of pyrimidines, amino sugars, and asparagine. Although some cell lines can use glutamate directly (Griffiths and Pirt 1967), most cells require glutamine for optimal growth. This may be partially explained by the observation that exchangeability between intracellular and extracellular glutamine is much greater than that for glutamate in cultured human fibroblasts (Darmaun et al. 1988). Glutamine may also be supplied in a more stable form by the dipeptides glycyl-glutamine and alanyl-glutamine (Brand et al. 1987; Roth et al. 1988).

Glucose and glutamine are complementary nutrients for the production of other metabolites such as aspartate (Zielke et al. 1980 and 1981) and for energy production. The extent to which each of the pathways shown in Figure 6–1 is employed depends on the cell line as well as the growth conditions (Eigenbrodt et al. 1985; McKeehan 1986). The available evidence suggests that either glucose or oxidative nutrients (primarily glutamine) can provide most cellular energy demands. Reitzer et al. (1979) found that cells

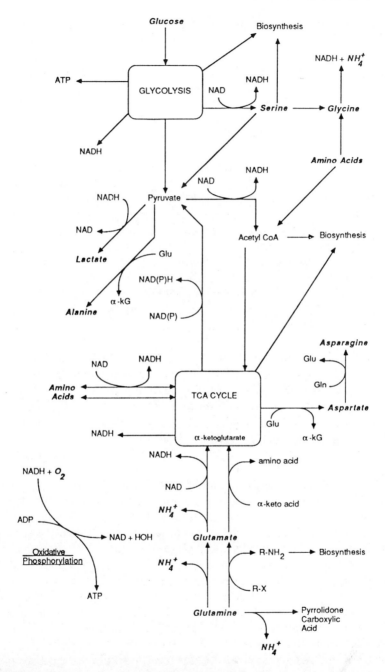

FIGURE 6–1 Schematic diagram of animal cell metabolic pathways. Metabolites shown in bold face italic type were measured by Miller et al. (1988b, 1988c, 1989a, and 1989b). Glu, glutamate; Gln, glutamine; and α-kG, α-ketoglutarate. Reproduced with permission from Miller et al. (1988b).

grew as well on fructose as on glucose; but showed negligible rates of glycolysis, and derived very little energy from carbohydrate metabolism. Glycolytic rates were also dramatically reduced for cells grown at glucose concentrations of less than 100 μM (Zielke et al. 1978; Rheinwald and Green 1974). At the other extreme, respiration-deficient Chinese hamster fibroblasts were shown to derive essentially all their ATP from glycolysis (Donnelly and Scheffler 1976).

6.1.1.1   Glucose Metabolism in Quiescent Normal Cells   Glucose is converted to pyruvate via glycolysis in all mammalian cells. Most normal (nontumor) cells are under growth regulation in vivo and divide slowly. In the presence of excess oxygen, these quiescent normal cells convert pyruvate to acetyl CoA for oxidation in the tricarboxylic acid (TCA) cycle. NADH from the TCA cycle and glycolysis is regenerated to NAD$^+$ inside the mitochondria in a process coupled to ATP production via oxidative phosphorylation. Glycolysis is regulated, at the level needed to provide sufficient acetyl CoA for energy production, via inhibition of phosphofructokinase (PFK) by ATP (Figure 6–2). Glucose 6-phosphate (glucose 6-P), which accumulates as a result of the inhibition of PFK, inhibits its own synthesis via hexokinase.

At low oxygen concentrations, mitochondrial ATP production is reduced and PFK is deinhibited. Increased fructose 1,6-diphosphate (fructose 1,6-P) concentrations further stimulate PFK and pyruvate kinase activity (Eigenbrodt et al. 1985). The resulting depletion of glucose 6-P enhances hexokinase activity, and large amounts of pyruvate are produced. Due to the low rate of oxidative phosphorylation, pyruvate is converted to lactate to reoxidize NADH generated by glycolysis.

6.1.1.2   Glucose Metabolism in Tumors and Cultured Cells   Cultured cells, tumors, and other proliferating cells exhibit high rates of aerobic glycolysis that differentiate them from quiescent normal cells (Morgan and Faik 1986; Eigenbrodt et al. 1985). The high rates of lactate production are similar to those observed for normal cells under oxygen limitation. Many transformed cells have increased concentrations of glycolytic enzymes and exhibit reduced depletion of glycolytic intermediates via reverse or side reactions (Pedersen 1978). Several glycolytic isozymes also exhibit different kinetic and regulatory properties. For example, hexokinase is bound to mitochondria, and shows reduced inhibition by glucose 6-P (Pedersen 1978). Cells that exhibit high rates of aerobic glycolysis also contain a type of pyruvate kinase (PK) that has reduced affinity for phosphoenolpyruvate (Eigenbrodt et al. 1985). This allows the concentration of fructose 1,6-P to increase until it overcomes inhibition of PFK and PK by ATP. The fully activated hexokinase and deinhibited PFK and PK lead to greatly increased levels of glycolysis (Eigenbrodt et al. 1985).

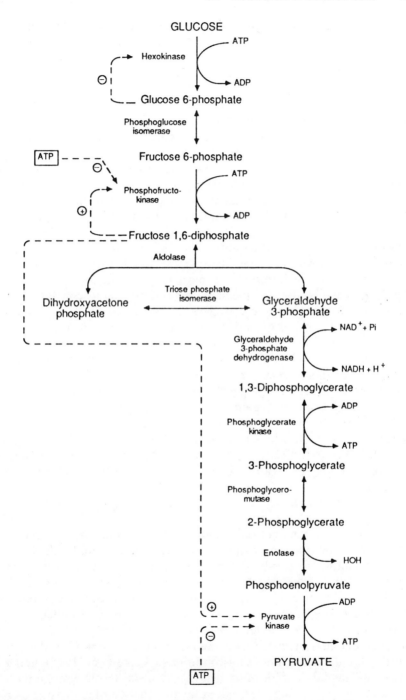

FIGURE 6-2 Schematic diagram of the glycolytic pathway and regulatory mechanisms. As discussed in the text, hexokinase in tumor cells is much less sensitive to regulation by glucose 6-phosphate. Reproduced with permission from Miller (1987).

High rates of glycolytic ATP production may increase the cytosolic ATP/ADP ratio and decrease the rate of mitochondrial ATP production due to reduced [ADP]. This would cause the mitochondrial transmembrane potential to increase, which would in turn reduce the rate of oxidative phosphorylation and increase the mitochondrial NADH/NAD$^+$ ratio (Williamson and Cooper 1980). A high mitochondrial NADH/NAD$^+$ ratio, combined with the smaller number of mitochondria in glycolytic tumors (Pedersen 1978) and possible defects in the malate/aspartate shuttle (Eigenbrodt et al. 1985) would be expected to dramatically decrease the amount of cytosolic NADH oxidized in mitochondria. As a result, pyruvate must be converted to lactate in order to oxidize NADH generated by glycolysis. It should be noted, however, that studies with intact hearts indicate that oxygen consumption is regulated by NADH, rather than by ADP (Balaban and Heineman 1989; Katz et al. 1987).

Cellular metabolic patterns may also be influenced by the extent of association with other cells. For example, the specific oxygen consumption rate of hybridoma cells immobilized in calcium alginate increased (from an initial level equal to that for the same hybridomas in suspension culture) by a factor of three during colony growth (Shirai et al. 1988). Increased oxygen consumption was associated with decreased glycolysis, as indicated by decreasing yield of lactate from glucose and specific glucose consumption rate; and may be due to cell-cell interactions similar to those encountered in vivo. Part of this response may be explained by the associated decrease in specific growth rate (see below), but the metabolic changes appeared before the growth rate decreased.

Proliferating normal cells in vivo share many metabolic characteristics with tumor cells, but to a lesser extent (Eigenbrodt et al. 1985). Proliferating normal cells also differ from tumor cells in that changes to the enzymes are reversible and that pyruvate oxidation is not reduced at higher glucose consumption rates.

6.1.1.3 Glutamine Metabolism in Cultured Cells

Many normal cells produce glutamine in vivo. In contrast, tumor cells and normal cells that proliferate in culture generally exhibit high glutamine consumption rates (McKeehan 1986). Most of the glutamine carbon skeleton enters the TCA cycle as α-ketoglutarate. The important reactions involved in glutaminolysis have been discussed by McKeehan (1986) and Moreadith and Lehninger (1984) and are described in detail by Glacken in chapter 15 of this volume. More than one-half of the CO_2 generated by normal diploid fibroblasts is derived from glutamine under all conditions (Zielke et al. 1978), and glutamine accounts for 70–80% of oxygen uptake by 6C3HED lymphoma cells (Lavietes et al. 1974). Reported values for the fraction of glutamine carbon atoms converted to CO_2 range from 35% (Reitzer et al. 1979) to 55% (Stoner and Merchant 1972). Other major products include lactate, glutamate, citrate, alanine, and macromolecules (Reitzer et al. 1979; Zielke et al. 1978; Lanks and Li 1988) with the relative amounts, as well as the glutamine consumption rate, dependent on the growth environment. The extent of glucose and glutamine oxidation also depends on the cell line. Lanks and Li (1988) examined 12 cell lines (grown in medium containing 25 mM glucose and 4 mM glutamine) and observed CO_2 production rates ranging from 420–2,600 nmol (glucose plus glutamine)/mg cell protein/24 hours. In all cases, however, oxidation was incomplete, as evidenced by the excretion of large amounts of metabolic intermediates into the culture medium.

High rates of glutamine consumption in cultured cells may contribute to elevated lactate production, as evidenced by increased glycolytic lactate production in L929 cells after glutamine addition (Lanks 1986a). High rates of glutamine consumption would be expected to generate large amounts of NADH and $FADH_2$ via glutamate dehydrogenase and TCA cycle enzymes. This is consistent with the reduction of intramitochondrial $NAD(P)^+$ by added glutamine (Kovacevic and Morris 1972). Additional NADH generated from glutamine would increase the mitochondrial $NADH/NAD^+$ ratio and further inhibit regeneration of glycolytic NADH via oxidative phosphorylation.

6.1.1.4 Other Amino Acids Consumption rates of the other amino acids are much less than that for glutamine, and vary with the cell line and culture conditions. Thomas (1986) reviewed amino acid consumption rates for a variety of cell lines. Generally, elevated consumption rates are observed for the branched-chain amino acids leucine, isoleucine, and valine, as well as for arginine. Other amino acids consumed at relatively high rates by various cell lines include lysine, threonine, asparagine, and methionine. All the cells evaluated produced alanine, and a large number also produced glutamate, aspartate, glycine, and proline. Similar patterns of amino acid metabolism have been observed for several hydridoma cell lines (Miller et al. 1989b; Adamson et al. 1987) and for mouse adrenal cortex Y-1 cells, which also consumed large amounts of histidine (Dhainaut et al. 1987). Some amino acids are used for energy production as well as for protein synthesis. Increased consumption of branched-chain amino acids has been observed at low glutamine concentrations in continuous hybridoma culture (Miller et al. 1989b) and during glucose-limited culture of primary rat myoblasts (Pardridge et al. 1981). Use of branched-chain amino acids as an energy source is also consistent with the observation that leucine oxidation by Ehrlich-Lettré ascites cells is inhibited by glucose (Lazo 1981).

6.1.2 Environmental Parameters that Affect Cell Metabolism

The growth and metabolism of all mammalian cells are strongly influenced by the culture environment. Although there are differences between cell types, similar responses are observed following changes in many environmental variables.

6.1.2.1 Glucose and Glutamine Concentrations The specific glucose consumption rate increases dramatically in batch culture as the initial glucose concentration is increased from 10 μM to 5 mM (Renner et al. 1972; Zielke et al. 1978), but is generally unchanged by further increases in glucose concentration (Low and Harbour 1985; Miller et al. 1988a). The product distribution from glucose metabolism is strongly influenced by the specific glucose consumption rate. For initial glucose concentrations at or below 0.5 mM, at least half of the glucose used by rat hepatoma cells was incorporated into nucleotides; but at 5 mM, 90% of the much greater amount of glucose consumed was converted to lactate (Renner et al. 1972). Higher apparent yields of lactate from glucose at higher specific glucose consumption rates have also been observed for batch culture of human FS-4 cells (Glacken et al. 1986). Higher glucose consumption rates after glucose addition have been shown to decrease the specific oxygen consumption rate (Nakashima et al. 1984; Glacken et al. 1986; Frame and Hu 1985; Paul 1965; Miller et al. 1989a; Krömer and Katinger 1982) as well as the extent of glutamine oxidation (Zielke et al. 1978; Lazo 1981; Donnelly and Scheffler 1976; Brand et al. 1987). This is probably due to inhibition of oxidative phosphorylation by increased glycolytic ATP production, as evidenced by an increased yield of lactate from glucose (Miller et al. 1989a) and an increase in the adenylate energy charge (Krömer and Katinger 1982).

A number of specific glucose effects on cultured cells have been observed in addition to the metabolic effects described above. Glucose deprivation induces expression of a set of glucose-regulated proteins (GRPs; Lanks 1986b) and represses expression of most other cellular proteins. GRP induction is associated with a deficiency of ribose 5-phosphate (Gstraunthaler et al. 1987) and may be due to improper protein glycosylation (Lee 1987). Other examples of specific glucose effects are provided by the observation that tumor cells grown at low glucose concentrations are less sensitive to the cytotoxic drug misonidazole (Sutherland 1988), and that interferon production by Namalwa cells is reduced at high glucose concentrations (Krömer and Katinger 1982).

Less information is available on the metabolic effects of different glutamine concentrations. Glutamine has been shown to increase the oxidation rate of the glucose 1-carbon via the pentose phosphate pathway and to inhibit oxidation of the 6-carbon via the TCA cycle in human diploid fibroblasts (Zielke et al. 1978) and Ehrlich-Lettré ascites tumor cells (Lazo 1981). The overall effect on oxygen consumption is generally small (e.g., Nakashima et al. 1984), but a 75% increase in q_{O2} was observed for mesenteric rat lymphocytes after glutamine addition (Ardawi and Newsholme 1983). Glutamine addition may enhance (rat lymphocytes; Ardawi and Newsholme 1983) or inhibit (mouse macrophages; Newsholme et al. 1987) glucose consumption. Glutamine addition to L929 cells (Lanks 1986a) resulted in a large increase in lactate production with only a small change in glucose oxidation. As discussed above, these responses may be influenced more strongly by culture conditions than by the cell type. Higher initial glutamine concentrations (up to about 10 mM) in batch culture generally lead to increased maximal cell concentrations (Butler and Spier 1984; Radlett et al. 1971; Miller et al. 1988a) and increased specific glutamine consumption rates (Glacken et al. 1986; Miller et al. 1988a). Although the cell growth rate is not sensitive to glutamine concentrations between 1 to 8 mM (Glacken et al. 1988; Miller et al. 1988a), it drops off rapidly at lower concentrations (Glacken et al. 1988).

6.1.2.2   pH   Different cell types exhibit widely different pH optima (Eagle 1973). The optimum pH for cell growth may be sharp, but an optimum plateau about 0.5–1.0 pH units wide is more common. Many investigators have observed lower glucose consumption rates and/or lower apparent yields of lactate from glucose at low pH (Paul 1965; Barton 1971; Birch and Edwards 1980; Leist et al. 1986; Miller et al. 1988a). Inhibition of glycolysis by low pH is also consistent with the greatly increased death rate of anoxic CHO cells at pH 6.5 compared to pH 7.3–8.0 (Haveman and Hahn 1981). Akatov et al. (1985) found that the pH in medium adjacent to confluent cell monolayers may be more than 1 pH unit below that in the bulk medium, and proposed that inhibition by low pH may be responsible for "contact inhibition" of cell growth.

6.1.2.3   Oxygen Concentration   Oxygen is required for energy production via oxidative phosphorylation, and is also used in the synthesis of cellular components such as cholesterol and tyrosine. Oxygen is only slightly soluble in cell culture media, and the oxygen supply rate may limit the productivity of commercial cell culture reactors (Fleischaker and Sinskey 1981). The specific oxygen consumption rate varies with cell type (McLimans et al. 1968), but is generally between 0.05–0.5×10^{-9} mmol O_2 per cell per hour for cultured animal cells. Oxygen consumption in tissues shows even more variation. For example, consumption rates of 0.5–2.5×10^{-9} mmol O_2 per cell per hour have been reported for rat liver slices (McLimans et al. 1968), while the much lower rate of 0.02×10^{-9} mmol O_2 per cell per hour has been reported for rat bone marrow cells (Olander 1972). The optimal partial pressure (pO_2) for growth varies with cell type, but is generally near 80 mm Hg (50% saturation with air) for cultured animal cells in batch reactors (Thomas 1986). However, continuous culture experiments (Miller et al. 1987; Boraston et al. 1984) suggest much lower optimal pO_2. It should also be noted that the optimal pO_2 for cell growth is not generally the same as those for other cellular functions. For example, the optimal pO_2 for antibody production by hybridomas (Miller et al. 1987; Reuveny et al. 1986a; Mizrahi et al. 1972) is different than that for cell growth. pO_2 also has a pronounced effect on cell metabolism. Glycolysis is generally enhanced at low pO_2, as indicated by increased lactate production (Kilburn et al. 1969; Miller et al. 1987). However, the oxygen consumption rate may either increase (Kilburn et al. 1969; 2–320 mm Hg), decrease (Suleiman and Stevens 1987; 130–300 mm Hg); or remain constant (Kekonen et al. 1987; 3–100 mm Hg) with increasing pO_2.

Although oxygen is essential for growth, it is toxic at high concentrations. Oxidative damage has been shown to cause DNA degradation, lipid peroxidation, polysaccharide depolymerization, and hydroxylation of aromatic compounds, but the molecular mechanisms are not well understood (Fee 1981). Although cell death is only observed at pO_2 values above those normally experienced in vivo, there is ample evidence to suggest that oxygen can adversely affect cell viability at concentrations within the physiological range. For example, hybridoma cell viability continued to increase as pO_2 was decreased from 160 mm Hg to ~2 mm Hg (Miller et al. 1987). Hepatocyte viability was also higher at reduced pO_2 (Suleiman and Stevens 1987). The effects are even more pronounced for in vitro culture of normal cells. Packer and Fuher (1977) found that normal diploid WI-38 and IMR-90 cells had increased culture lifetimes (number of population doublings before senescence) when cultured under atmospheres containing 2% or 10% oxygen, compared to those containing 20% oxygen. Other experiments suggest that the extended lifetime may be due to decreased chromosomal damage. Mouse embryo cells grown with 1% O_2 in the gas phase had a lower number of abnormal chromosomes than those grown with 18% O_2, and maintained a higher frequency of diploid cells during the first 35 days in culture (Parshad et al. 1977). Genomic damage is probably due to excited oxygen species (Taylor 1983; Richter et al. 1988) and may vary for different types of cells. The protective effects of reduced oxygen concentrations are also illustrated by increased resistance of hypoxic tumor cells to radiation and cytotoxic drugs (Sutherland 1988).

Oxygen also regulates the functional activities of a variety of cells. Bone marrow from rats grown under hypoxic conditions showed increased levels of erythropoiesis and respiration compared to that from control rats (Olander 1972). Macrophages grown under gas atmospheres containing 2% oxygen became rounded, detached from the culture dish, and secreted active angiogenesis factor into the medium (Knighton et al. 1983). The cells reattached and ceased production of active angiogenesis factor when returned to culture under a gas phase containing 20% oxygen.

6.1.2.4 By-product Concentrations   Ammonia and lactate are the major metabolic by-products from glutamine and glucose metabolism (see Figure 6-1). The final ammonia concentration in batch culture is typically between 2 and 5 mM and generally increases with the feed glutamine concentration (Butler and Spier 1984; Glacken et al. 1986). The final lactate concentration depends on the glucose concentration used (Glacken et al. 1986; Zielke et al. 1978; Miller et al. 1988a) and may be as high as 35 mM.

The inhibitory ammonia concentration differs markedly between cell types. The growth of mouse L cells (Ryan and Cardin 1966), 3T3 cells (Visek et al. 1972), and BHK cells (Butler and Spier 1984) was inhibited by less than 1 mM added NH_4Cl. Two mouse hybridoma cell lines (Reuveny et al. 1986a; Glacken et al. 1988) were inhibited by 2–3 mM NH_4Cl, whereas two other hybridoma cell lines (Adamson et al. 1987; Miller et al. 1988c) and MDCK cells (Glacken et al. 1986) only showed significant inhibition at medium concentrations ≥ 4 mM. Transformation of 3T3 cells by SV-40 decreased the extent of inhibition caused by NH_4Cl (Visek et al. 1972). The inhibitory effects of ammonia are also influenced by other medium components. For example, the concentration of mouse myeloma cells grown in serum-free medium was reduced by 64% when 2.9 mM NH_4Cl was added, but the cell concentration was only reduced by 30% for cells grown with serum (Iio et al. 1985).

Less information is available on the effects of lactate. In systems without pH control, the low pH due to high lactate concentrations may inhibit cell growth. The inhibitory lactate concentration varies with cell type, but added lactate is generally much less inhibitory than ammonia for cells grown at constant pH. One mouse hybridoma cell line was inhibited by lactate concentrations above 4 mM (Thorpe et al. 1987), while human hybridoma cells were not affected by the addition of 4.9 mM lactate in batch culture (Iio et al. 1985). The growth of a second mouse hybridoma cell line was stimulated by added lactate up to 22 mM, but concentrations greater than 28 mM were found to be inhibitory (Reuveny et al. 1986a). Two other mouse hybridoma cell lines showed no inhibition of growth at 40 mM lactate (Miller et al. 1988c; Glacken et al. 1988), although one was inhibited at 70 mM (Glacken et al. 1988). Added lactate may stimulate (Reuveny et al. 1986a) or inhibit (Glacken et al. 1988) antibody synthesis by hybridomas.

Studies on two human cell lines (Kimura et al. 1987) indicate that inhibition of oxygen consumption by ammonia and lactate is much less extensive than the corresponding inhibition of cell growth. However, it should be noted that the extent of inhibition does vary with the cell line (Zimber and Topping 1970; Kimura et al. 1987).

6.1.2.5 Shear Effects   Animal cells are sensitive to shear due to the lack of a cell wall. The effects of shear on cell growth and metabolism are discussed in Chapters 2 and 9 of this volume.

6.2 METHODS FOR OBTAINING METABOLIC INFORMATION IN BIOREACTORS

6.2.1 Reactor Systems
Most of the metabolic effects discussed above were identified in tissue culture flasks or batch reactors without pH and dissolved oxygen control. In order to fully characterize the responses and the interrelations between cell

growth and metabolism, it is necessary to conduct experiments under controlled conditions. Several general reviews of cell culture technology have appeared during the last ten years (Spier 1982; Glacken et al. 1983; Hu and Dodge 1985; Randerson 1985). Many of the techniques proposed to increase reactor productivity employ various forms of cell immobilization. Immobilized-cell reactors may be required for optimal production, but due to metabolite concentration gradients and the difficulty of obtaining representative cell samples, they are not well suited for analyses of cell growth and metabolism. In contrast, the uniform cell and metabolite concentrations characteristic of suspension culture reactors facilitate analysis of cell growth, metabolism, and product formation. The status of suspension cell culture has been reviewed by Katinger and Scheirer (1982). Batch culture has been employed in the large majority of suspension culture investigations.

Continuous culture offers a number of advantages over batch experiments (see Tovey 1985 for an earlier review). Values for metabolic quotients can be obtained at steady state, whereas conditions are constantly changing during batch growth. The effect of specific growth rate on nutrient consumption and product formation can be obtained by operating at different cell residence times. Transient and steady-state responses of cell metabolism to changes in culture conditions are obtained at high cell concentrations. This eliminates the additional stress due to low initial cell concentrations in batch culture experiments. Another advantage of continuous culture is that it allows for adaptation by cells to inhibitory conditions. Adaptation by cells to low oxygen supply rates or high concentrations of inhibitory by-products has important implications for the operation and control of commercial cell culture reactors. Numerous investigators have used continuous culture with or without cell retention, as shown in Table 6–1. Important results obtained from a number of these studies will be discussed in section 6.3.

Continuous suspension culture facilitates the analysis of transient responses to step or pulse changes employed to probe cellular regulatory mechanisms, but relatively few such experiments have been conducted. Transient responses have been reported for glucose pulse during batch (Frame and Hu 1985) and continuous (Krömer and Katinger 1982) culture, for continuous culture step changes in glucose concentration (Graff et al. 1965; Moser and Vecchio 1967), and in glucose feed concentration (Tovey and Brouty-Boyé 1976; Pirt and Callow 1964; Ray et al. 1989). Limited metabolic data have been reported for most of these studies. More extensive metabolic data were obtained by Miller et al. (1988b, 1988c, 1989a, and 1989b) after a variety of metabolite pulse and step changes, but no intracellular measurements were made. Tovey et al. (1979) evaluated the transient responses in cyclic AMP and cyclic GMP levels in mouse leukemia cells after interferon addition in continuous culture.

TABLE 6-1 Continuous Animal Cell Suspension Culture Experiments

Cell Type	Parameters Controlled[1]	Cell Retention Method	Parameters Evaluated[2]	Transient Analysis	Extracellular Measurements	Intracellular Measurements	Reference(s)
Transformed embryo rabbit kidney	tc, pH[3]	None	Gas percent O2, flow control mechanism	No	tc	None	Cooper et al. 1959 and 1958
HeLa-S3	D, pH[3]	None	D	No	tc, cell vol., gluc, AA	Protein	Cohen and Eagle 1961
Mouse L and ERK	D, pH	None	D, pH, gluc	Yes (gluc)	gluc, tc, pH	None	Pirt and Callow 1964
Mouse ascites mast (P815Y)	D, pH	None	gluc	Yes	tc	frac DNA and RNA syn	Graff et al. 1965
Mouse ascites mast (P815Y)	D, pH[3]	None	D, gluc	Yes (gluc)	tc, gluc, lac	frac DNA syn	Moser and Vecchio 1967
Mouse LS	D, pH[3]	None	D, gln vs. glu	No	tc, gln, AA	Cell nitrogen	Griffiths and Pirt 1967
HeLa (subline Gey)	D, pH[3]	None	D, virus titer	No	tc, virus titer	None	Holmström 1968
HeLa	tc	None	tc	No	tc	None	Peraino et al. 1970
BHK cells infected w/ rubella virus	D, pH, DO[3]	None	D, limiting nutrient	No	pH, tc, virus titer, lac, gluc, dry wt.	RNA, DNA, virus	Kilburn and van Wezel 1970
Mouse leukemia (L1210)	Perfusion rate	Rotating filter	Amethopterin	Yes	tc, vc, amethopterin	Enzyme act, RNA syn. amethopterin, DNA syn	Thayer et al. 1970
Mouse LS	D	None	D, gluc, vc, IF inducer conc.	Yes	gluc, lac, keto acids, vc, interferon	None	Tovey et al. 1973
Mouse LS	D (semicont.), DO[3]	None	D	No	tc, gluc, lac, qO2	DNA, protein, enzyme act, RNA and syn rate	Sinclair 1974
Mouse leukemia (L1210)	D, pH[3]	None	D, interferon, prostaglandin	Yes	pH, vc, tc, gluc, lac	DNA, RNA, Ptn syn rates; cAMP, cGMP	Tovey et al. 1975 and 1979; Tovey and Brouty-Boyé 1979; Tovey 1980; Tovey and Rochette-Egly 1980
Mouse leukemia (L1210)	D, pH[3]	None	D, gluc	Yes (gluc)	pH, vc, tc, gluc, lac	DNA, RNA, Ptn and synthesis rates	Tovey and Brouty-Boyé 1976

(continued)

TABLE 6–1 (continued)

Cell Type	Parameters Controlled[1]	Cell Retention Method	Parameters Evaluated[2]	Transient Analysis	Extracellular Measurements	Intracellular Measurements	Reference(s)
Mouse leukemia (L1210)	D	None	Interferon, gluc	Yes	vc, tc	Thy and dAden incorp.; Thy, dAden, and dgluc uptake; DNA	Brouty-Boyé and Tovey 1978
Mouse leukemia (L1210)	D, pH[3]	None	D, gluc	No	pH, tc	DNA syn rate, cAMP, cGMP, protein	Tovey et al. 1980
Mouse LS	D, DO[3]	None	D	No	tc, gluc, lac, cell size	Cytosolic and mitochondrial enzyme act.	Sinclair 1980
Walker 256 rat tumor human hepatoma (SK-HEP-1)	D, DO, pH	Rotating filter	Retention, percent serum	No	tc, %v, gluc, lac	None	Tolbert et al. 1981; Feder and Tolbert 1985
Namalwa	D, pH, DO	None	gluc	Yes	qO2, gluc, prod, DO, tc	ATP, ADP, AMP	Krömer and Katinger 1982
Mouse hybridoma (N-527)	tc[3]	None	tc	No	tc, Ab	None	Fazekas de St. Groth 1983
Mouse hybridoma (NB1)	pH, DO, D	None	D, limiting nutrient	No	vc, tc, qO2, Ab	None	Birch et al. 1984; Boraston et al. 1984
Unspecified hybridoma	Unspecified	Rotating screen	Retention	No	vc, dc, Ab	None	van Wezel et al. 1985
Mouse hybridoma (VII H-8)	DO, D, pH[3]	Rotating screen	Retention, D	No	vc, dc, lac, gluc, NH₃, pH, Ab	None	Reuveny et al. 1986b; Velez et al. 1987
Mouse hybridoma (Ab2-143.2)	pH, D, DO	None	D, pH	Yes (pH)	vc, %v, gluc, lac, gln, NH₃, Ab	None	Miller et al. 1986 and 1988a
Mouse hybridoma (9.2.27)	D, recirculation rate, DO[3], pH[3]	Total cell recycle across filter	D, gln	No	DO, pH, gluc, gln, ala, lac, vc, tc, cell vol., Ab, Ab characterization	None	Flickinger et al. 1987
Mouse hybridoma (Ab2-143.2)	pH, D, DO	None	DO	Yes	vc, %v, gluc, lac, gln, IAA, NH₃, Ab, qO2	None	Miller et al. 1987 and 1988b

Cell type		Automatic control (in addition to temperature)[1]	Culture time		Measured[2]		Reference
Hybridoma	D	None		No	tc, Ab, metabolites, fraction Ab-producing	None	Hu et al. 1987
Human hematopoietic (RPMI 8226)	D	None	D	Washout	vc, gluc, lac, NH_3, gln	None	Mano et al. 1987
Unspecified recombinant	pH, DO, D	None	D, pH, DO	No	vc, %v, lac, gluc, gln, NH_3, prod (unspec.)	None	Tajiri et al. 1987
Hybridoma	D	None	D	No	vc, tc, gluc, lac, DO, qO_2, Ab	None	Low et al. 1987
Bowes melanoma	D, DO, pH, tc	Rotating filter	tc	No	tc, gluc, lac, product, qO_2, perfusion rate	None	Feder and Barker 1987
Mouse hybridoma (Ab2–143.2)	pH, D, DO	None	lac, NH_3	Yes	vc, %v, gluc, lac, gln, lAA, NH_3, Ab, qO_2	None	Miller et al. 1988c
Mouse hybridoma (Ab2–143.2)	pH, D, DO	None	gluc, gln	Yes	vc, %v, gluc, lac, gln, lAA, NH_3, Ab, qO_2	None	Miller et al. 1989a and 1989b
Mouse hybridomas (VX-7, VX-12)	D, pH, DO[3]	None	D, gluc	Yes (gluc)	tc, vc, lac, NH_3, gluc, Ab	None	Ray et al. 1989

[1]Automatic control (in addition to temperature).

[2]For effects explicitly evaluated; DO, dissolved oxygen concentration; D, dilution rate = 1/residence time; vc, viable cells; %v, fraction of viable cells; dc, dead cells; tc, total cells; lac, lactate; gluc, glucose; gln, glutamine; NH_3, ammonia; Ab, antibody; prod, product; frac DNA syn, fraction of cells labeled with ^{3}H-thymidine; lAA, primary amino acids; AA, amino acids; glu, glutamate; qO_2, oxygen consumption rate; Ptn, protein; Thy, ^{3}H-thymidine; dAden, ^{3}H-deoxyadenosine; dgluc, ^{3}H 2-deoxy-D-glucose.

[3]Parameter checked periodically; adjustments made if necessary to keep it within the desired range.

6.2.2 Characterization of Cell Metabolism

6.2.2.1 Measurement of Cell Extracts and Use of Radioactive Substrates

Measurement of intracellular metabolite concentrations and enzyme activities was instrumental in investigations evaluating the regulation of cell metabolism in culture (c.f. Reitzer et al. 1979; Renner et al 1972; Ardawi and Newsholme 1984) and the effects of oxygen on cell metabolism (Brosemer and Rutter 1961; Self et al. 1968). Radiolabeled glucose and glutamine have been used to determine the metabolic fate of these substrates under various environmental conditions (c.f. Zielke et al. 1984; Donnelly and Scheffler 1976; Lazo 1981; Darmaun et al. 1988). Radioactive precursors are also useful for quantifying protein or DNA synthesis rates. Limited measurements of intracellular metabolite concentrations, enzyme activities, and macromolecular synthesis rates were reported for the studies cited in Table 6–1. Sinclair (1980) correlated changes in glucose consumption and lactate production rates at different dilution rates with changes in the activities of cytosolic and mitochondrial enzymes. Krömer and Katinger (1982) measured intracellular concentrations of ATP, ADP, and AMP. Graff et al. (1965) measured the fractions of cells synthesizing DNA and RNA. Tovey and coworkers (Tovey 1980; Tovey and Brouty-Boyé 1979; Tovey and Rochette-Egly 1980; Tovey et al. 1975, 1979, and 1980) reported intracellular concentrations for DNA, RNA, protein, cyclic AMP, and cyclic GMP. They also measured the incorporation rates of ^{3}H-thymidine, ^{3}H-uridine, ^{14}C-labeled amino acids, and ^{3}H-2-deoxy-D-glucose.

6.2.2.2 Noninvasive Spectroscopic Techniques

In situ (^{31}P, ^{13}C, ^{1}H, and ^{19}F) NMR is a powerful tool for studying cellular metabolism (Fernandez and Clark 1987; Avison et al. 1986). High cell concentrations are required for adequate resolution, and the major limitation for widespread use of this technique is the ability to maintain viable high-density cell cultures for extended time periods in the absence of metabolite concentration gradients. The application of NMR to animal cell culture is reviewed by Dale in chapter 5 of this volume.

Another powerful spectroscopic technique is electron paramagnetic resonance, which can be used to follow changes in the environments of free radicals and paramagnetic metals such as iron, copper, cobalt, manganese, and molybdenum (Foster 1984). The use of lipid- and water-soluble stable organic free radicals to study membrane properties and cell responses to environmental changes and drugs has been reviewed by Dodd (1984).

6.2.2.3 Flow Cytometry

Flow cytometers can be used to quantify the distribution of an intracellular or surface-bound molecule in a cell population (Parks et al. 1986). Dyes are available for measuring DNA, RNA, and protein. Labeled antibodies and cDNA probes also allow detection of specific proteins or mRNA molecules. The use of multiple fluorochromes allows for correlation of different parameters with each other, as well as with cell size.

6.3 METABOLIC RESULTS OBTAINED IN CONTINUOUS SUSPENSION BIOREACTORS

6.3.1 Effects of Dilution and Perfusion Rates

6.3.1.1 Cell Concentration, Viability, and Growth Rate   Steady-state viable and total cell concentrations are shown as a function of dilution rate in Figure 6–3 for a hybridoma cell line (Miller et al. 1988a). The gradual decrease in total cell concentration with increasing dilution rate is similar to that observed by other investigators (Holmström 1968; Cohen and Eagle 1961; Low et al. 1987; Boraston et al. 1984; Tovey and Brouty-Boyé 1976) and contrasts with the constant cell mass characteristic of microbial systems. Part of this difference may be explained by the larger cell size generally observed at higher dilution rates (Cohen and Eagle 1961; Flickinger et al. 1987; Tovey and Brouty-Boyé 1976). This is illustrated by the more abrupt

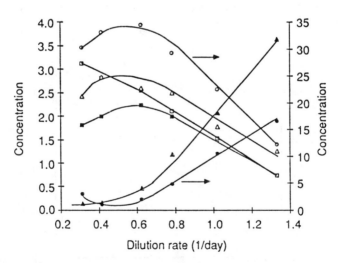

FIGURE 6–3 Effect of dilution rate on steady-state cell (millions of cells/mL) and metabolite (mM) concentrations in continuous hybridoma culture. Viable (solid squares) and total (open squares) cells; glucose (solid circles); lactate (open circles); glutamine (solid triangles); and ammonia (open triangles). The feed contained 22 mM glucose and 4.8 mM glutamine. The dissolved oxygen concentration was maintained at 50% of air saturation and the pH was controlled at 7.1 Reproduced with permission from Miller et al. (1988a).

decrease in total cell mass with increasing dilution rate reported by Tovey and Brouty-Boyé (1976). The viable cell concentration (see Figure 6–3) has a maximum at $D = 0.6$/day. The decrease in viable cell concentration at low dilution rates indicates a dramatic decrease in viability that can be attributed to increased maintenance requirements and low nutrient and/or high inhibitor concentrations. Similar viable cell curves have been observed for other cell lines (Low et al. 1987; Boraston et al. 1984; Mano et al. 1987). Several investigators (Griffiths and Pirt 1967; Moser and Vecchio 1967; Ray et al. 1989) have also observed maximum total cell concentrations at intermediate dilution rates.

The lower viability at low dilution rates (Figure 6–4) indicates that the specific growth rate of viable cells is larger than the apparent specific growth rate, which is equal to the dilution rate at steady state. The shape of the growth rate versus dilution rate curve suggests a limiting specific growth rate for viable cells, as proposed by Tovey and Brouty-Boyé (1976). As shown in Figure 6–4, the same limiting growth rate was observed for two hybridoma cell lines (Miller et al. 1988a; Boraston et al. 1984). Data for a third hybridoma cell line (Ray et al. 1989) fall on the same curve, even though the decrease in both total and viable cell concentrations at low dilution rates was much more extensive than that shown in Figure 6–3. Data for another hybridoma cell line (Low et al. 1987) and for L1210 cells (Tovey and Brouty-Boyé 1976 and 1979) indicate similar trends but with lower limiting values for the specific growth rate (i.e., less deviation from $\mu = D$).

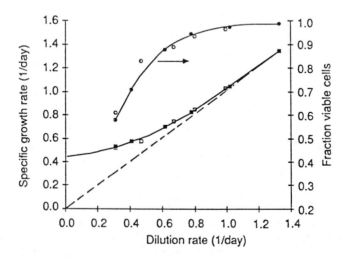

FIGURE 6–4 Specific growth rate (squares) and fraction of viable cells (circles) versus dilution rate in continuous hybridoma culture (solid symbols: Miller et al., 1988a; open symbols: data from Boraston et al., 1984). Reproduced with permission from Miller et al. (1988a).

6.3.1.2 Cell Metabolism   The essentially monotonic increase in glucose concentration with increasing dilution rate (see Figure 6–3) is typical for animal cell culture (Tovey and Brouty-Boyé 1976; Low et al. 1987; Mano et al. 1987). The increase in glucose concentration is more gradual than that predicted by the Monod model (Tovey and Brouty-Boyé 1976), which is consistent with the gradual decrease in cell concentration described above. In contrast to the monotonic nutrient curves, the lactate and ammonia concentrations reach a maximum value at a dilution rate near 0.6/day (see Figure 6–3). These maxima are similar to those observed by other workers (Low et al. 1987; Mano et al. 1987; Ray et al. 1989). The maximum in the lactate curve may be attributed to decreased yields of lactate from glucose at lower dilution rates (Low et al. 1987; Mano et al. 1987; Miller et al. 1988a). The decrease in the fraction of glucose metabolized to lactate at low residual glucose concentrations is consistent with the trends discussed in section 6.1.2.1. In contrast, Ray et al. (1989) reported a decrease in the yield of lactate from glucose at higher dilution rates for VX-7 hybridomas, even though the specific glucose consumption rate increased. It should be noted, however, that the residual glucose concentration was quite high at low dilution rates (7.4 mM at $D = 0.29$/day with a feed of 17.8 mM) for VX-7 cells; and that data presented for a second hybridoma cell line (VX-12) indicate a 20% greater specific glucose consumption rate and higher yield of lactate from glucose (1.8 versus 1.6) at a dilution rate of 0.62/day with the same culture conditions.

As may be expected from the above discussion, plots of specific nutrient consumption rate versus dilution rate generally exhibit curvature. This indicates changes in the use of various metabolic pathways at different conditions. The specific nutrient consumption rate generally increases monotonically with increasing D (Miller et al. 1988a; Low et al. 1987; Sinclair 1974; Flickinger et al. 1987; Tovey and Brouty-Boyé 1976), although the curves may be concave up, concave down, or a combination. In contrast to these results, minimum nutrient specific consumption rates have been observed at intermediate dilution rates for a few studies (glucose: Mano et al. 1987 and Ray et al. 1989; glutamine: Griffiths and Pirt 1967). Variations between nutrient consumption rate plots obtained by different investigators may reflect different nutrient feed concentrations. Some of the curvature observed may be due to changes in the cell size. Flickinger et al. (1987) were able to reduce the curvature and scatter in a plot of specific glutamine consumption rate versus apparent specific growth rate in a total cell recycle reactor by calculating the specific consumption rate in terms of cell surface area rather than in terms of cell number. The limited available information on specific oxygen consumption rate indicates a near-linear increase in oxygen consumption versus specific growth rate (Boraston et al. 1984) or dilution rate (Low et al. 1987). In spite of the variable nutrient consumption rates, it was observed that the specific activities of cytochrome oxidase and mitochondrial and cytoplasmic malate dehydrogenase were essentially independent of dilution rate (Sinclair 1980).

The steady-state cell concentration can be varied at a constant dilution rate by changing the concentration of the limiting nutrient (see section 6.3.2.1). Tovey et al. (1980) varied the dilution rate and glucose concentration in order to independently determine the effects of cell concentration and growth rate on intracellular concentrations of cyclic AMP (cAMP) and cyclic GMP (cGMP), which are thought to play important roles in the regulation of cell proliferation. They found that [cGMP] was independent of cell concentration, but increased 3.6-fold for a threefold increase in specific growth rate. In contrast, [cAMP] decreased with increasing cell concentration and increased only slightly with increasing growth rate. The two effects resulted in a decrease in [cAMP]/[c/GMP] at higher growth rates.

6.3.1.3 Product Formation   Most information on product formation in continuous culture has been obtained for monoclonal antibody production by hybridomas. Antibody production does not appear to be feedback inhibited, except for cell lines that produce very low levels (<1 mg/L) of antibody (Merten et al. 1985). This is consistent with results showing essentially constant specific antibody production rates for a 10-fold increase in antibody concentration at the same growth rate in continuous suspension culture (Fazekas de St. Groth 1983). There was also no correlation between antibody concentration and antibody production rate in a cell recycle reactor with antibody retention (Flickinger et al. 1987).

Hybridoma cells are formed by the fusion of an antibody-producing spleen cell and a myeloma cell. Hybridomas are often unstable with respect to antibody production and must be periodically recloned to select for highly producing cells. Loss of antibody productivity has been observed during extended periods in continuous culture (Miller 1987; Hu et al. 1987). This could be due to lower production by all cells or to a decrease in the fraction of cells producing antibody. Hu et al. (1987) observed an increase in the number of nonproducers but Birch et al. (1984) found no increase in the fraction of nonproducing cells after 60 to 90 generations in continuous culture with various limiting nutrients. The potential for instability provides an incentive for separating product formation from cell growth. This should be possible because antibody production is not directly associated with growth (see section 6.4.2.1). Significant antibody production often occurs after the viable cell concentration has started to decline in batch culture.

Miller et al. (1988a and 1987) observed that the specific (per viable cell) antibody production rate in continuous culture was higher at lower cell viabilities associated with lower medium flow rates (and hence, lower growth rates), extreme pH values, and nonoptimal (for cell growth) oxygen concentrations. This is consistent with the observation that the specific antibody production rate (calculated from data presented by van Wezel et al. 1985) was twice as high for cells grown in continuous culture with cell retention (at lower growth rate and viability) as it was for those grown without cell retention. Increased perfusion rates have been observed to increase the viable cell concentration and the specific antibody production rate (Velez et al. 1987), but no information was provided on the change in the fraction of viable cells. Birch et al. (1984) found that the apparent (per total cell) specific antibody production rate increased or decreased slightly with the cell growth rate depending on whether glucose, glutamine, or oxygen was the limiting nutrient. However, information provided elsewhere (Boraston et al. 1984) relating the effect of growth rate on viability for cells limited by oxygen indicates that the true (per viable cell) specific antibody production rate was much higher at lower growth rates, which were associated with lower cell viability. In contrast Ray et al. (1989) observed a maximum specific antibody production rate at a dilution rate of 0.46/day. Two other studies indicate that growth rate has no significant effect on the specific antibody production rate (Flickinger et al. 1987; Low et al. 1987) or the characteristics of the antibody molecules produced (Flickinger et al. 1987).

Increased specific antibody production rates at low cell viabilities suggest that hybridoma cells may increase antibody synthesis in response to stress. All cells are known to increase the production of some proteins (called heat shock proteins) in response to elevated temperatures and other stress factors (Schlesinger et al. 1982). The hypothesis that higher rates of antibody production are related to the heat shock phenomenon is supported by the observation that increased antibody production in vivo (in response to infection) is normally accompanied by inflammation and fever.

Tovey et al. (1973) investigated interferon induction by adding double-stranded RNA and DEAE-dextran to continuous cultures of mouse LS cells. Induction in glucose-limited cultures was followed by a transient decrease in viable cell concentration and a transient increase in lactate production. It was observed that interferon production in glucose-limited culture was not affected by changes in the dilution rate from 0.25–0.35/day. The interferon titer was 40% lower in medium that contained excess glucose. The authors suggest that interferon production may be subject to catabolite repression since the concentrations of lactic and keto acids from glucose catabolism were six times higher at the same dilution rate for cells grown in excess glucose.

6.3.2   Effects of Changes in the Culture Environment

6.3.2.1   Glucose and Glutamine Supply Rates

Steady-State Effects of Glucose and Glutamine Feed Concentrations.   Cells adjust their use of the metabolic pathways shown in Figure 6–1 in response to different culture conditions. For example, hybridoma cells grown at different glucose/glutamine feed ratios (Figure 6–5) metabolized glucose and glutamine in ratios ranging from 1.3 to 11 times as much glucose as glutamine (Miller et al. 1988d). The complementarity of the nutrients is evident from the nearly equal consumption and feed ratios over a wide range of feed concentrations, dilution rates (0.31–1.33/day, feed ratio = 4.5), dissolved oxygen concentrations (0.5% to 100% of air saturation, feed ratio = 2.9), and pH (7.1–7.7, feed ratio = 4.5). In most cases the nutrients were nearly exhausted, but the ratio did not change significantly at high dilution rates with ∼75% of both nutrients remaining. Deviations (open circles in Figure 6–5) observed at extreme pH values and for DO ≤ 0.5% saturation with air will be discussed in sections 6.3.2.2 and 6.3.2.3. As expected, the contributions from glycolysis and oxidative phosphorylation varied with the consumption ratio; but q_{ATP} was about the same (21.5 ± 2.5 × 10^{-9} mmol ATP per cell per day) for all the steady states (Miller 1987). The unique biosynthetic requirements for glucose and glutamine are indicated by deviations at very high and low feed ratios. The cells were limited by glucose at a feed ratio of 1.1 (consumption ratio = 1.3) and by glutamine at a feed ratio of 16.2 (consumption ratio = 11).

Lower yields of cells from glucose and glutamine have been observed at higher feed concentrations (Miller et al. 1989a and 1989b; Ray et al. 1989) and at higher dilution rates (with higher residual nutrient concentrations;

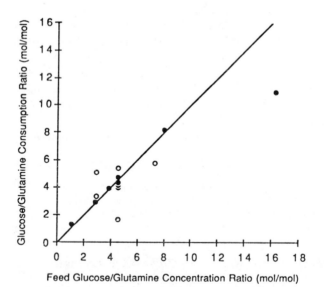

FIGURE 6–5 Effect of the glucose/glutamine feed concentration ratio on the steady-state nutrient consumption ratio for hybridomas grown in continuous culture at a variety of conditions. Points indicated by open circles were obtained at extreme pH values or dissolved oxygen concentrations below 0.5% saturation with air. Reproduced with permission from Miller (1987).

Miller et al. 1988a). In contrast, Tovey and Brouty-Boyé (1976) obtained a maximum yield of cells from glucose at a glucose feed concentration of 5.6 mM for continuous culture of L1210 cells grown at a dilution rate of 1.0/day. Lower apparent cell yields from glucose were obtained at feed concentrations of 2.8 mM and 11.1 mM glucose. Although glutamine consumption rates were not reported, the lower yield at 2.8 mM glucose suggests that glutamine consumption was limited by precursors derived from glucose or that production of some metabolic intermediates is much less efficient from glutamine than from glucose.

Effects of Glucose Additions.   Glucose addition to cells limited by glucose causes a large increase in the specific glucose consumption rate (Miller et al. 1989a; Krömer and Katinger 1982). The decrease in oxygen consumption (Figure 6–6) that followed a glucose step change (Miller et al. 1989a) was attributed to inhibition of oxidative phosphorylation by the increased rate of glycolytic ATP production. This is consistent with an increase in adenylate energy charge observed after a glucose pulse during Namalwa cell culture (Krömer and Katinger 1982). The plateau in specific glucose consumption rate during the first 24 hours after glucose addition (see Figure 6–6) suggests that glycolysis is also regulated by ATP in cultured cells, although at a much higher level than in normal quiescent cells (see section

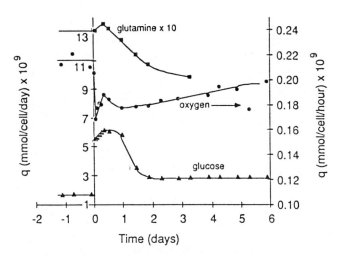

FIGURE 6–6 Changes in specific consumption rates for glucose (triangles), glutamine × 10 (squares), and oxygen (circles) in response to a glucose step change in continuous hybridoma culture. The glutamine feed concentration was constant at 4.8 mM. The reactor glucose concentration was increased by 8.3 mM and the feed glucose concentration was increased from 5.3–13.8 mM at time zero. The dilution rate was 0.54/day. The DO was maintained at 30% of air saturation and the pH was maintained at 7.2. Reproduced with permission from Miller et al. (1989a).

6.1.1). The lower steady-state specific oxygen consumption rate after the step change may be attributed to the higher steady-state glucose consumption rate. The magnitudes of changes in glucose consumption, oxygen consumption, and energy charge depend on the initial nutrient supply rates and the amount of glucose added (Miller et al. 1989a; Krömer and Katinger 1982). The approach to the new steady state is more rapid at higher dilution rates (Moser and Vecchio 1967).

Increased cell concentrations have been observed by a variety of investigators after step changes in glucose concentration (Moser and Vecchio 1967; Miller et al. 1989a) or glucose feed concentration (Pirt and Callow 1964; Tovey and Brouty-Boyé 1976). The cell concentration began to increase 12 hours after the glucose step change shown in Figure 6–6 (Miller et al. 1989a). The small change in glutamine consumption and the decrease in oxygen consumption immediately after glucose addition indicate less extensive glutamine oxidation, which suggests that additional glutamine is being used for production of biosynthetic intermediates. The subsequent decrease in q_{gln} was due to the increased cell concentration at a constant glutamine supply rate. The increases in cell concentration and estimated ATP production after the glucose addition were used to estimate a biosynthetic yield of 7.5×10^{10} cells/mol ATP. Maintenance accounted for about 60% of the ATP consumed at the specific growth rate of 0.66/day before and after the step change.

In contrast to the stimulatory effects described above, glucose addition inhibited P815Y mast cells growing in low-glucose medium at a dilution rate of 0.32/day (Graff et al. 1965). The decline in cell concentration following a 5.6 mM glucose step change was accompanied by sharp decreases in the fractions of cells synthesizing DNA and RNA. The feed pump was turned off for 16 days (eight days after the step change) to save the culture. The fraction of cells synthesizing RNA eventually returned to near 100%, but the fraction of cells synthesizing DNA (three-fourths of prestep value) and the cell concentration (two-thirds of the prestep value) were lower at the new steady state. The lower mast cell concentration after the glucose step increase is consistent with a gradual 25% increase in viable hybridoma cell concentration after a step decrease in the glucose feed concentration from 17.8 to 11.1 mM (Ray et al. 1989).

Effects of Glutamine Additions. The responses of hybridoma cell line AB2–143.2 to glutamine additions illustrate the effects of the culture environment on cell metabolism. Increased glutamine consumption rates, increased alanine and ammonia production rates, and a lower extent of glutamine oxidation were observed after all glutamine additions, but the effects on cell growth and consumption of glucose and oxygen were dependent on the initial nutrient supply rates (Miller et al. 1989b). The glutamate concentration changed only slightly after glutamine additions in spite of large changes in glutamine consumption, which suggests that the glutamine to glutamate step is rate-limiting for glutamine consumption (McKeehan 1986).

Under typical culture conditions with a feed of 3.0–4.8 mM glutamine and 14–22 mM glucose, the specific glutamine consumption rate increased by 50–80% after ~2.7 mM glutamine additions, but there was no immediate change in oxygen or glucose consumption (Miller et al. 1989b). Gradual increases in cell concentration began about two days after glutamine addition, and were preceded by small transient increases in the estimated ATP production rate. It appears that high rates of glycolysis prior to glutamine addition inhibited increased glutamine oxidation. Increased glutamine consumption at constant oxidation rates probably increased the supply of TCA cycle-derived biosynthetic intermediates, which induced a delayed increase in cell concentration.

In contrast, the cell concentration began to increase 12 hours after a 0.9 mM glutamine step change at initial feed concentrations of 0.9 mM glutamine and 14.3 mM glucose (feed concentration ratio = 16.2 in Figure 6–5; Miller et al. 1989b). As shown in Figure 6–7, the specific glutamine consumption rate increased fourfold immediately after glutamine addition. The rapid associated increases in g_{gluc} (specific glucose consumption rate) and q_{O2} suggest that glucose consumption was limited by intermediates derived

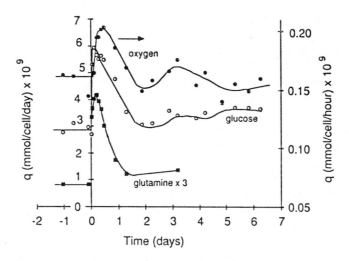

FIGURE 6–7 Changes in specific consumption rates for glucose (open circles), glutamine × 3 (squares), and oxygen (solid circles) in response to a 0.9 mM glutamine step change in continuous hybridoma culture. The glucose feed concentration was constant at 14.3 mM. The reactor glutamine concentration was increased by 0.9 mM and the feed glutamine concentration was increased from 0.9–1.8 mM at time zero. The dilution rate was 0.54/day. The DO was maintained at 50% of air saturation and the pH was maintained at 7.2. Reproduced with permission from Miller et al. (1989b).

from glutamine, and that oxygen consumption was limited by the glutamine supply rather than by glycolytic ATP production. The latter point is consistent with elevated consumption rates of branched-chain amino acids, which can be oxidized to TCA cycle intermediates. The lag between increases in q_{gln} and q_{O2} may be attributed to the buffering effect of the TCA cycle and associated pathways. Subsequent inhibition of oxygen consumption by the high rate of glycolysis was probably offset by increased ATP requirements for biosynthesis. The rapid increase in estimated ATP production due to simultaneous increases in glycolysis and oxidative phosphorylation after glutamine addition resulted in a large overshoot in the viable cell concentration. The biosynthetic yield from ATP and the fraction of ATP consumption attributed to maintenance were similar to those described above for the glucose step change.

6.3.2.2   pH   The responses of hybridoma cell growth and viability to pH step changes are shown in Figure 6–8 (Miller et al. 1988a). The steady-state values shown in Figure 6–8B indicate a broad optimum between pH 7.1 and 7.4 (7.7 after adaptation, as described below). Similar broad optima have been reported for other cell lines in continuous culture (Tajiri et al. 1987; Pirt and Callow 1964). One advantage of using continuous culture is illustrated by the adaptation to pH 7.7, shown in Figure 6–8A. The initial decrease in cell concentration was associated with increased specific glucose and glutamine consumption rates (Miller et al. 1988a). The subsequent recovery of cell concentration and nutrient consumption at pH 7.7 corresponded to an increase in $Y'_{lac/gluc}$ (apparent yield of lactate from glucose, mol/mol) from 1.5 to 1.8, which may have resulted in a lower intracellular pH. This suggestion is consistent with the observation that cultured cells can maintain their intracellular pH at a value different from that of the extracellular medium (Gillies et al. 1982). Moser and Vecchio (1967) have also reported a short-lived semistable steady state (with smaller yield of de novo cell synthesis) after a sudden change in pH.

The dramatic inhibition of cell growth at pH 6.8 (see Figure 6–8A) occurred in spite of increased residual glucose and glutamine concentrations (Miller et al. 1988a). Glutamine consumption increased at pH 6.8, but glucose consumption was inhibited. As a result, the glucose/glutamine consumption ratio decreased to 1.7 mol/mol at pH 6.8 compared to the feed ratio (and consumption ratio at pH 7.1–7.4) of 4.5 (Miller et al. 1988d). Decreased specific glucose consumption rates at lower pH values have also been reported for continuous culture of mouse L cells (Pirt and Callow 1964). The associated decrease in $Y'_{lac/gluc}$ at pH 6.8 (Miller et al. 1988a) suggests that glycolysis is directly inhibited by low pH. The concurrent increase in $Y'_{NH3/gln}$ observed at pH 6.8 may also be due to pH regulation or may be due to a higher rate of uncatalyzed hydrolysis at the higher glutamine concentration.

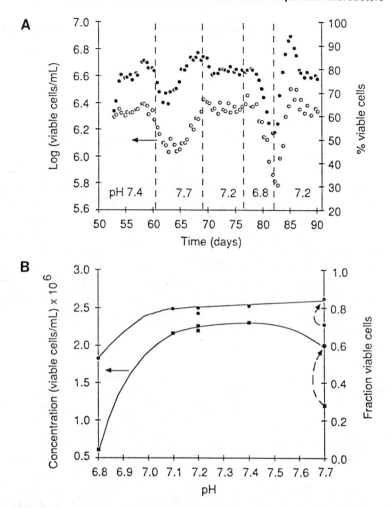

FIGURE 6–8 Effect of pH on hybridoma growth in continuous culture. The feed contained 22 mM glucose and 4.8 mM glutamine. The DO was controlled at 50% of air saturation and the dilution rate maintained at 0.52/day. (A) Response of viable cell concentration (open circles) and percent viability (solid circles) to pH step changes. (B) Steady-state viable cell concentration (squares) and fraction of viable cells (circles) versus pH. Values for pH 7.1 were interpolated from Figure 6–3 at a dilution rate of 0.52/day. Dashed arrows show the changes after adaptation at pH 7.7. Reproduced with permission from Miller et al. (1988a).

The antibody concentration was not significantly affected by the pH changes shown in Figure 6–8 (Miller et al. 1988a). This indicates that the specific antibody production rate is much higher at pH 6.8 and at pH 7.7 prior to recovery. This is consistent with increased antibody production in response to stress, as discussed in section 6.3.1.3. However, it should be noted that other cell types show optimal specific productivity at pH values within the range optimal for cell growth (Tajiri et al. 1987).

6.3.2.3 Dissolved Oxygen Concentration Hybridoma growth and metabolism in continuous culture are not strongly affected by dissolved oxygen concentrations ranging from 30% to 100% of saturation with air (Miller et al. 1987). Experiments with another cell line also showed little effect on the steady-state cell concentration between 20% and 80% DO (Tajiri et al. 1987). The ability of cells to adapt in continuous culture is evidenced by the smaller optimal range of 40–60% DO reported for the latter cell line in batch culture. The specific oxygen consumption rate for cell line AB2–143.2 was essentially the same for DO values between 10% and 100% of air saturation (Miller et al. 1987). Decreased oxygen consumption below 10% DO was partially offset by an apparent increase in the P/O (molecules of ATP produced per oxygen atom consumed) ratio from 2 to 3. The steady-state hybridoma concentration increased at lower DO values until it reached a maximum at 0.5% DO. Further decreases in the oxygen supply resulted in lower cell concentrations due to decreased glutamine oxidation. Cell viability increased as the DO was decreased. The viability continued to increase, after a transient decrease, even when the DO was decreased below 0.5%.

The effects of changes in the oxygen supply rate on the viable hybridoma cell concentration and fraction of viable cells are shown in Figure 6–9 (Miller et al. 1988b). The recovery in viability and partial recovery in cell concentration at 0.4% and 0.1% DO illustrate the adaptation of cells grown in continuous culture. The increased contribution from glycolysis at low DO values is illustrated by the increased glucose/glutamine consumption ratio shown in Figure 6–10 (Miller et al. 1988d). The increase in residual glutamine concentration to 2 mM at 0.1% DO accounts for the overshoot in cell concentration after the return to 10% DO (see Figure 6–9). The high residual concentration of glutamine at low DO and the small change in glutamate concentration support the suggestion made earlier, that glutaminase is the flux-limiting enzyme for glutamine entry into the TCA cycle. Glutamine consumption increased at lower DO values in spite of the decreased consumption of oxygen and branched-chain amino acids (Miller et al. 1988b). Increased glutamine consumption and reduced oxidation suggest that glutamine provided a larger fraction of biosynthetic intermediates at high $Y'_{lac/gluc}$ values.

6.3.2.4 By-product Concentration Lactate and ammonia are the major by-products produced by cultured cells and are potentially inhibitory to cell growth. A lactate pulse from 25 mM to 44 mM at pH 7.2 had no significant effect on hybridoma growth or nutrient consumption (Miller et al. 1988c). This is consistent with results obtained in batch culture (see section 6.1.2.4).

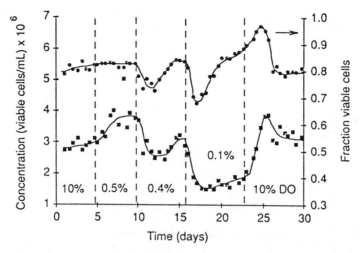

FIGURE 6–9 Responses of the viable cell concentration (squares) and the fraction of viable cells (circles) to changes in the reactor oxygen supply rate in continuous hybridoma culture. The cells were grown at a dilution rate of 0.54/day with automatic control of the DO at 10% of air saturation. After five days, the oxygen concentration in the reactor headspace was gradually reduced until the DO was 0.5%. After 10 days, the oxygen supply rate to the reactor was reduced by 35%; the DO fell to nearly zero before recovering to a steady-state value of 0.4%. A similar 50% decrease in the oxygen supply rate was implemented after 16 days. After 23 days, automatic oxygen control was resumed at 10% DO. The feed contained 13.9 mM glucose and 4.8 mM glutamine, and the pH was controlled at 7.2. Reproduced with permission from Miller et al. (1988b).

Inhibition of hybridoma growth was first observed at ammonia concentrations near 5 mM (Miller et al. 1988c). The viable cell concentration decreased by 13% and 24% after step changes from 2.9 to 5.4 mM and 3.2 to 9.6 mM ammonia, respectively. In both cases, however, the cells adapted to the new conditions and the viable cell concentration returned to prestep values even though the steady-state ammonia concentration was 8.2 mM after the larger step change. The recoveries were associated with a change in nitrogen metabolism. Alanine production increased and ammonia production decreased after ammonia addition, which suggests that the flux of glutamate to α-ketoglutarate via glutamate dehydrogenase (see Figure 6–1) decreased at the expense of that via the transaminase pathway. Alanine was the primary glutamine by-product at ammonia concentrations above about 5 mM. Glucose consumption and lactate production increased dramatically during growth inhibition by ammonia. Similar increases in q_{gln} were inhibited by ammonia, with more extensive inhibition at higher ammonia concentrations. Oxygen consumption also appears to be inhibited by ammonia at concentrations above 10 mM. For example, q_{O2} decreased by one order

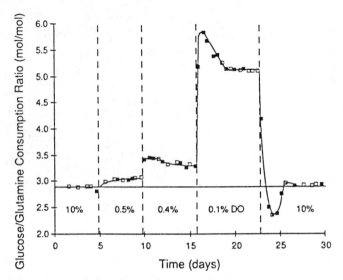

FIGURE 6–10 Changes in the glucose/glutamine consumption ratio in response to the changes in reactor oxygen supply described in Figure 6–9 (the glucose/glutamine feed concentration ratio is indicated by a horizontal solid line). The points indicated by open symbols were calculated using residual glutamine concentrations interpolated from Figure 4 in Miller et al. (1988b).

of magnitude due to combined inhibition by ammonia and a threefold increase in q_{gluc} after a pulse from 8.2 to 17 mM ammonia. As a result, 90% of estimated ATP production three days after the pulse was attributed to glycolysis. $Y'_{lac/gluc}$ increased to 2.6 (theoretical maximum is 2), which suggests that lactate production from glutamine may compensate for reduced mitochondrial oxidation of NADH derived from glutamine at low oxygen consumption rates.

6.3.2.5 Drug Concentration Continuous culture reactors with and without cell retention provide a controlled environment for evaluating the effects of cytotoxic drugs. Thayer et al. (1970) evaluated the effects of transient and continued exposure to amethopterin (methotrexate) on L1210 cells in suspension culture with cell retention. They observed a sharp decrease in the number of viable (ability to form colonies in soft agar) cells after a 1 μg/ml pulse of amethopterin, which was followed by rapid perfusion with amethopterin-free medium. Some increase in viable cell concentration was observed 24 hours after the pulse; but the log kill (initial/final cell concentration) was greater for a second pulse, indicating incomplete recovery. Recovery was inhibited by the rapid uptake of amethopterin and its subsequent slow release from cells into essentially amethopterin-free medium. The concentration of viable cells decreased by almost six orders of magnitude during the first 100 hours after a step change to 0.067 μg/ml amethopterin and then declined much more slowly. This suggests increased drug resistance; but when surviving clones were tested for sensitivity to amethopterin, they were no more resistant than the original culture. The authors also reported several biochemical effects of amethopterin. Methotrexate binds tightly to dihydrofolate reductase (dhfr). Dhfr activity dropped to zero after a 1 μg/ml pulse, and recovered slowly during perfusion with drug-free medium. Incorporation of deoxyuridine into DNA also dropped sharply after the pulse, although recovery to normal values was faster. In contrast, thymidylate synthetase activity increased after exposure to amethopterin.

Tovey and coworkers have extensively studied the effects of interferon on L1210 cells in chemostat culture. A decrease in cell concentration was first observed 12–24 hours after a step change to 6,000 units (U)/ml (Tovey et al. 1975), with shorter response times at higher dilution rates. The cell concentration decreased for nine days after the step change and then began to increase. Adaptation to interferon was also demonstrated during a subsequent step change to 6,400 U/ml at a dilution rate of 0.3/day (Tovey 1980). The cell concentration decreased by more than 80% after eight days before returning to the original concentration after 15 days. The effect of interferon on cell viability is negligible at high dilution rates, but becomes progressively more severe as the dilution rate is decreased (Tovey and Brouty-Boyé 1979). It was suggested that interferon reduces the growth rate, rather than directly killing cells, because the same relative initial decrease in apparent specific growth rate was observed after interferon addition at all dilution rates. Reduced viability at lower dilution rates may then be attributed to the lower cell viability generally observed at lower apparent specific growth rates in the absence of interferon (Tovey and Brouty-Boyé 1979; section 6.3.1.1).

Interferon also affects cell metabolism. A decrease in ^{3}H-thymidine incorporation into DNA was observed two to three hours after interferon addition (Tovey et al. 1975). However, this has been attributed to decreased thymidine uptake rather than a lower DNA synthesis rate because uptake of ^{3}H-deoxyadenosine was not affected until the decrease in cell multiplication after ~24 hours (Brouty-Boyé and Tovey 1978). The decrease in transport was selective since uptake of ^{3}H-uridine, ^{3}H-2-deoxy-D-glucose, ^{14}C-amino acids, and ^{3}H-deoxyadenosine remained at control values (Brouty-Boyé and Tovey 1978; Tovey et al. 1975). As shown in Figure 6–11, the first biochemical effect observed after interferon addition was a transient two- to fourfold increase in the concentration of cGMP after five to 10 minutes (Tovey et al. 1979; Tovey and Rochette-Egly 1980). This is consistent with similar activation of some immune system cells and neurons by interferon and agents that increase cGMP levels (Tovey et al. 1979). No increase was observed in the concentration of cAMP (see Figure 6–11) until the decrease in cell concentration after 24 hours, even though a rapid in-

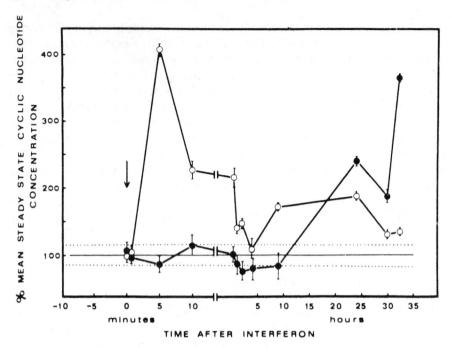

FIGURE 6–11 Effect of a step change in interferon concentration on the intracellular concentrations of cyclic AMP (solid circles) and cyclic GMP (open circles) in continuous culture of mouse L1210 cells at a dilution rate of 0.5/day. The horizontal solid line with broken lines above and below represents the steady-state mean values with standard deviations. The error bars represent the standard deviation of the replicates for a particular point. At time zero (indicated by the vertical arrow) interferon was injected into the reactor to give a concentration of 6,400 U/mL, and the feed interferon concentration was simultaneously increased from 0 to 6,400 U/mL. Reproduced with permission from Tovey and Rochette-Egly (1980).

crease in cAMP levels is observed two to three hours after interferon addition in batch culture (Tovey and Rochette-Egly 1980). Although prostaglandin E_1 and interferon have similar effects on cell multiplication and thymidine uptake, prostaglandin E_1 addition resulted in an immediate increase in cAMP levels and had only a small effect on cGMP (Tovey et al. 1979; Tovey and Rochette-Egly 1980).

6.4   MODELS OF CELL METABOLISM

6.4.1   Requirements of Models of Cellular Metabolism and Regulation

The complexity of growth and product formation by animal cells places constraints on the development of mathematical models. In many cases, data are available for growth conditions that are not well defined in terms of both the nature and concentration of the chemical species surrounding the cell. Models of the kinetics of cellular metabolism are thus limited not only by the lack of data, but also by the inability to identify which metabolic species are important. Thus, initial attempts to model cell growth involved simple unstructured models that predicted cell growth rates, nutrient uptake rates, and product formation. As is apparent from the foregoing discussions of glucose and glutamine metabolism, the complex interplay of carbon and energy metabolism in animal cells in response to changing environmental conditions means that such simple models will have limited applicability. These models do, however, provide order-of-magnitude estimates of uptake rates and can thus be employed in examining the relationship of intrinsic kinetics and mass transport in bioreactors. They are also useful in predicting metabolism for the limited range of metabolic parameters for which the data were obtained.

More complex, structured models are required to provide predictive capabilities when growth conditions are altered. Models capable of describing the regulation of glycolysis and glutaminolysis require details of the key regulatory elements in both pathways, ideally at the level of individual enzyme kinetics. With this information, metabolic fluxes can be calculated. Such detailed information is not in general available, but often useful estimates can be made. Comparable data to that available for *Escherichia coli* are required to develop single cell models, such as those of Shuler and Domach (1983). In some instances, insights can be gained by lumping analyses in which metabolites are combined into various metabolic pools that possess differing characteristic time scales. We shall review these various approaches to modeling hybridoma growth and metabolism.

6.4.2 Unstructured Constitutive Models

The simplest approach to modelling hybridoma growth is derived by analogy with the Monod equation for bacterial growth. The relationship between specific growth rate μ and limiting nutrient concentration has generally been modified to incorporate the inhibition attributed to lactate and ammonia. The limiting nutrient has been considered to be serum in the case of mouse-mouse hybridoma (CRL-1606) cells producing IgG (Glacken et al. 1988), glutamine in the case of an IgG-producing SP2/0-related cell line grown in DMEM medium with 25 mM glucose (Bree et al. 1988), or both glucose and glutamine (Miller et al. 1988a). The resulting expression for glutamine as limiting nutrient (Bree et al. 1988) is:

$$\mu = \mu_{max} \left(\frac{Gln}{K_{Gln} + Gln} \right) \left(\frac{K_A}{K_A + A} \right) \left(\frac{K_L}{K_L + L} \right)$$

where K_A and K_L are the inhibition constants for ammonia and lactate and K_{Gln} is the saturation constant for glutamine. Typical values for these constants have been reported as $K_{Gln} = 0.8$ mM glutamine, $K_A = 1.05$ mM ammonia, and $K_L = 8.0$ mM lactate.

Glacken et al. (1988) found that ammonia inhibition was noncompetitive. These authors also proposed that the inhibition constant (K_A) was inversely proportional to the ammonium ion concentration and proposed an unusual quadratic dependence for ammonium ion inhibition:

$$\mu = \mu_{max} \frac{S}{(S + K_S)\left(1 + \dfrac{A^2}{K'_A}\right)}$$

where S is the serum concentration (% FCS). K_S was found to be 0.93% FCS with $K'_A = 26$ mM² ammonium ion. Glacken et al. (1988) also examined the effect of glutamine, concluding that it behaved as a second growth-limiting substrate. When lactate inhibition was included, the complete model of hybridoma growth was of the form:

$$\mu = \mu_{max} \frac{S\,Gln}{(S + K_S)(Gln + K_{Gln})\left(1 + \dfrac{A^2}{K'_A}\right)\left(1 + \dfrac{L^2}{K'_L}\right)}$$

With this expression, values of the constants obtained from initial rate data were: $\mu_{max} = 0.061$/hour, $K_S = 1.6\%$ FCS, $K_{Gln} = 0.15$ mM, $K'_A = 45$ mM², and $K'_L = 12,000$ mM². The large value for K'_L indicates the rather small effect observed for lactate inhibition.

Bree et al. (1988) included a nutrient-dependent death term to account for the observed loss of viable cells:

$$r_{NV} = k_D \left(\frac{A}{K_{AD} + A}\right)\left(\frac{L}{K_{LD} + L}\right)\left(\frac{K_{GD}}{K_{GD} + Gln}\right)$$

where r_{NV} is the specific death rate of viable cells, which is assumed to be dependent on ammonia and lactate concentration and to be inhibited by glutamine. No independent data that could verify this rate expression for cell death are available. Constants were obtained by fitting total cell counts and fractional viability measurements to batch data.

A double-substrate-limited growth rate expression was proposed by Miller et al. (1988a) for a SP2/0 hybridoma, which incorporated ammonia and lactate inhibition. Growth was limited by glucose (G) and by glutamine:

$$\mu = \mu_{max} \left(\frac{G}{K_G + G}\right)\left(\frac{Gln}{K_{Gln} + Gln}\right)\left(\frac{K_A}{K_A + A}\right)\left(\frac{K_L}{K_L + L}\right)$$

Typical values of the constants, obtained from both batch and continuous cultures were: $\mu_{max} = 1.5$/day, $K_G = 0.15$ mM, $K_{Gln} = 0.15$ mM, $K_A = 20$ mM, and $K_L = 140$ mM.

Bree et al. (1988) considered the rate of glucose consumption to be zero order in glucose concentration, rather than the Monod form assumed above. The rate of glucose consumption included the following maintenance term:

$$r_G = -\frac{1}{Y_G} \mu n_v - m n_v$$

where m is of the order 1.25×10^{-8} mM glucose/day (per viable cell).

In the case of HL-60 and RPMI 8226 human cells grown in batch and continuous suspension culture, Taya et al. (1986) developed an expression that included inhibitory effects of ammonium ion, lactate, and glucose.

$$\mu = \mu_{max} \left(\frac{G}{(K_G + G)\left(1 + \frac{A}{k_A} + \frac{L}{k_L}\right) + \frac{G^2}{k_G}} \right)$$

6.4.2.1 The Specific Rate of Antibody Production Most experimental data suggest that the production of antibody by hybridoma cells is not associated with the growth rate of the cells. Glacken et al. (1988), employing initial rate studies, showed that the specific rate of MoAb production was growth associated up to a specific growth rate of 0.02/hour and independent of μ at growth rates greater than 0.02/hour, having a value of ~2.5 pg/cell/hour. These authors also observed a dependence of the specific production rate on the lactate concentration. In continuous cultures, Miller et al. (1988a) showed that the specific rate of antibody production increased sharply at low rates of growth (below 0.7/day) and decreased with increasing growth rate. A simple non-growth-associated model was proposed by Bree et al. (1988), which included a term describing glutamine inhibition of IgG production and a saturation term in cell number to account for the lag in antibody production observed in batch culture:

$$q_{IgG} n_v = k_p n_v \left(\frac{K_{GP}}{K_{GP} + Gln}\right)\left(\frac{n_T}{n_T + K_{NTi}}\right)$$

Constitutive expressions such as that proposed above need to be independently verified, using data from both batch and continuous cultures. No comparable glutamine inhibition effect on antibody production has been reported by others.

The first-order dependence of antibody production on viable cell concentration can be illustrated by examining the behavior of the integrated form of the batch rate expression:

$$\frac{d(MoAb)}{dt} = \beta n_v$$

$$MoAb (t) = \beta \int_0^t n_v \, dt$$

where β is assumed to be constant. Renard et al. (1988) and Ozturk et al. (1988) have shown that their batch data on murine hybridoma lines follow the integrated equation very well. The value of the constant β was found to be independent of serum level in the culture medium for both studies, although Renard et al. (1988) did observe a variation of 25% in β for different basal media. Luan et al. (1987) and Miller et al. (1988a) observed a linear relationship between the final antibody concentration and the integral of the viable cell concentration curve for batch cultures with a wide range of initial feed concentrations and supplemental nutrient additions. All these results suggest that antibody production is nongrowth associated. However, in batch culture the period over which the specific growth rate varies is small and it is difficult to observe the range of specific growth rates possible with continuous culture.

6.4.2.2 Yield Coefficients   In relating the uptake rates of the various nutrients, yield coefficients can be defined for cell number, lactate, ammonia, and antibody. Determination of the intrinsic yield coefficients requires a knowledge of the carbon flows through the glycolytic and glutaminolytic pathways. Since lactate can be produced from glutamine metabolism, apparent yield coefficients that relate lactate production solely to glucose consumption can thus only provide indications of relative fluxes through the operative pathways. Typical apparent yield coefficients and representative values are:

$Y'_{n/gln}$ (number of viable cells/mole glutamine) 6.3×10^8 cells (per mM glutamine).

$Y'_{n/glu}$ The glucose yield may vary with the glucose concentration and growth conditions.

$Y'_{lac/glu}$ (mole lactate produced/mole glucose consumed) 1.5–2.0 mol/mol. May be greater than 2.0 if glutamine is converted to lactate. May be lower than 1.5 if glucose concentration is very low and is primarily consumed by the pentose phosphate pathway.

$Y'_{amm/gln}$ (mole ammonia produced/mole glutamine consumed) 0.3–0.7 mol/mol. Deamidation of glutamine to glutamate gives a value of 1.0. Deamination to α-ketoglutarate can provide a second mole of ammonia. The lower measured values suggest incorporation of amino groups via transamidation and transamination reactions.

$Y'_{O2/gln}$ (mole O_2 consumed/mole glutamine consumed) Oxygen consumption results from glutamine entering the TCA cycle and producing reducing equivalents (e.g., NADH).

$Y'_{O2/glu}$ (mole O_2 consumed/mole glucose consumed) Oxygen consumption results from the entry of electrons into the mitochondria via the malate-aspartate shuttle and entry of pyruvate.

The specific nutrient uptake rates (q_i, moles (per 10^6 cells) per hour) can be related from the following mass balances, where Y now represents the intrinsic yield rather than the apparent yield defined above:

$$q_{glu} = \frac{\mu}{Y_{n/glu}} + \frac{q_{lac}}{Y_{lac/glu}} + m_{glu}$$

$$q_{O2} = \frac{q_{gln}}{Y_{O2/gln}} + \frac{q_{glu}}{Y_{O2/glu}} + m_{O2}$$

$$q_{gln} = \frac{\mu}{Y_{n/gln}} + m_{gln}$$

6.4.3 Simple Structured Models

In order to explain the observed rates of glucose and glutamine uptake, oxygen consumption, and amino acid metabolism profiles, details of the metabolism must be included in more sophisticated models. The regulation of glucose and glutamine metabolism cannot be adequately described by simple models which do not consider energy (ATP) metabolism or any intracellular detail. Several simple structured models have been proposed. Suzuki et al. (1988) proposed that hybridoma cells arrested in the G1 phase of the cell cycle produce antibody at a rate considerably higher than cells in other phases of the cycle. For myeloma and lymphoid cells, data suggest that the maximum antibody synthesis rate occurs in late G1 or early S phase (Garatun-Tjeldstø et al. 1976). As the specific growth rate decreases, the time the cell spends in the G1 phase increases, while the times the cell requires to traverse the S, G2, and M phases are approximately constant. Cells may be arrested at a point late in the G1 phase, known as the restriction point, unless signaled to continue. Suzuki et al. (1988) argue that the fraction of viable cells arrested at the restriction point (where maximum immunoglobulin synthesis occurs) increases as the specific growth rate decreases. Several models for the fraction of arrested cells were examined, and the assumptions that gave the best agreement with data from the literature were: "the fraction of arrested cells is close to zero at high and intermediate growth rates, and increases with decreasing growth rate. Arrested cells produce antibody at a greater rate than cycling cells." Based on this model, methods that arrest cells might be expected to yield high antibody production rates. Thus, cells exposed to drugs (e.g., thymidine), oxygen depletion, isoleucine deprivation, etc., might be expected to yield higher rates than nonarrested cells. This hypothesis remains to be tested, however.

The maximum rate of antibody synthesis has been examined by Savinell et al. (1989). Based on the rates of mRNA synthesis, the velocity of RNA polymerase II and the spacing of the nucleotides, these authors provide an upper limit for antibody synthesis of 2,300–8,000 antibody (Ab) molecules/cell/sec. Most reported values for a variety of cell lines lie in the range 17–8,000 Ab molecules/cell/sec.

6.4.4 More Complex Structured Models

Several approaches can be taken to introduce more structure into models of cellular behavior. Of particular interest is the regulation of carbon flow from the glycolytic and glutaminolytic pathways through the TCA cycle and to the formation of various amino acids (e.g., alanine). One approach to this can be derived from "metabolic control theory," developed by Kacser and Burns (1973), and developed further by Kell and Westerhoff (1986). The objective of this approach is to identify the rate-controlling pathways or steps, based on the extent to which the various enzyme-catalyzed steps influence the overall flux in a complex metabolic network. These authors have developed various theorems that govern the behavior of the flux-control coefficients and the elasticity coefficients in branched and cyclic pathways. The difficulty in applying this to mammalian cell culture is the need to experimentally determine the foregoing coefficients. This requires some understanding of the in vivo kinetics of the individual enzymes in a pathway.

A second approach has been recently proposed by Liao and Lightfoot (1988a). It depends on the time-scale characteristics of the metabolic system being examined. By identifying characteristic reaction paths, useful information can be obtained by phase plots of intermediate pairs or combinations of these. This information can be used to identify rate-limiting reactions under in vivo environments. A generalized approach, following along the same lines, is the *lumping* technique. Liao and Lightfoot (1988b) have applied the technique to general biochemical reaction networks where time scale separation exists. Groups of metabolic intermediates can be combined into "pools" (the weighted sum of individual components) and a simplified description of the process results, without loss of the key metabolites. While the rigorous lumping technique proposed by Liao and Lightfoot has not been employed to model hybridoma metabolism, a structured model with lumped metabolic pools has been proposed by Kompala and co-workers (Batt and Kompala 1987; Kompala et al. 1987). Simulations of steady-state and dynamic behavior using this model showed good agreement with the trends observed experimentally by Miller et al. (1988a and 1989a) for continuous hybridoma culture. The present availability of dynamic kinetic information and the increasing information that will become available from dynamic intracellular measurements (e.g., using NMR techniques) should make lumping an attractive approach. The alternative, developing complete kinetic descriptions of all enzymes in the metabolic pathway, is likely to be a formidable task. Although this has been attempted for *E. coli*, the large number of constants involved in such models and the coupling of the resulting mass balance equations for even this simple organism present difficulties.

REFERENCES

Adamson, S.R., Behie, L.A., Gaucher, G.M., and Lesser, B.H. (1987) in *Commercial Production of Monoclonal Antibodies* (Seaver, S.S., ed.), pp. 17–34, Marcel Dekker, New York.

Akatov, V.S., Lezhnev, E.I., Vexler, A.M., and Kublik, L.N. (1985) *Exp. Cell Res.* 160, 412–418.

Ardawi, M.S.M., and Newsholme, E.A. (1983) *Biochem. J.* 212, 835–842.

Ardawi, M.S.M., and Newsholme, E.A. (1984) in *Glutamine Metabolism in Mammalian Tissues* (Häussinger, D., and Sies, H., eds.), pp. 235–246, Springer-Verlag, Berlin and Heidelberg.

Avison, M.J., Hetherington, H.P., and Shulman, R.G. (1986) *Annu. Rev. Biophys. Biophys. Chem.* 15, 377–402.

Balaban, R.S., and Heineman, F.W. (1989) *Mol. Cell. Biochem.* 89, 191–197.

Barton, M.E. (1971) *Biotechnol. Bioeng.* 13, 471–492.

Batt, B.C., and Kompala, D.S. (1987) ACS 194th National Meeting, New Orleans, LA, September 3, 1987, MBTD Paper no. 145.

Birch, J.R., and Edwards, D.J. (1980) *Develop. Biol. Stand.* 46, 59–63.

Birch, J.R., Thompson P.W., Lambert, K., and Boraston, R. (1984) ACS National Meeting, Philadelphia, PA, August 27, 1984.

Boraston, R., Thompson, P.W., Garland, S., and Birch, J.R. (1984) *Develop. Biol. Stand.* 55, 103–111.

Brand, K., von Hintzenstern, J., Langer, K., and Fekl, W. (1987) *J. Cell. Physiol.* 132, 559–564.

Bree, M.A., Dhurjati, P., Goeghegan, R.F., and Robnett, B. (1988) *Biotechnol. Bioeng.* 32, 1067–1072.

Brosemer, R.W., and Rutter, W.J. (1961) *Exp. Cell Res.* 25, 101–113.

Brouty-Boyé, D., and Tovey, M.G. (1978) *Intervirology* 9, 243–252.

Butler, M., and Spier, R.E. (1984) *J. Biotechnol.* 1, 187–196.

Cohen, E.P., and Eagle, H. (1961) *J. Exp. Med.* 113, 467–474.

Cooper, P.D., Burt, A.M., and Wilson, J.N. (1958) *Nature* 182, 1508–1509.

Cooper, P.D., Wilson, J.N., and Burt, A.M. (1959) *J. Gen. Microbiol.* 21, 702–720.

Darmaun, D., Matthews, D.E., Desjeux, J-F., and Bier, D.M. (1988). *J. Cell. Physiol.* 134, 143–148.

Dhainaut, F., Gerbert-Gaillard, B., and Maume, B.F. (1987) *J. Biotechnol.* 5, 131–138.

Dodd, N.J.F. (1984) in *Magnetic Resonance in Medicine and Biology* (Foster, M.A., ed.), pp. 66–91, Pergammon Press, Oxford.

Donnelly, M., and Scheffler, I.E. (1976) *J. Cell. Physiol.* 89, 39–51.

Eagle, H. (1973) *J. Cell. Physiol.* 82, 1–8.

Eigenbrodt, E., Fister, P., and Reinacher, M. (1985) in *Regulation of Carbohydrate Metabolism* Vol. II. (Beitner, R., ed.), pp. 141–179, CRC Press, Boca Raton, FL.

Fazekas de St. Groth, S. (1983) *J. Immunol. Methods* 57, 121–136.

Feder, J., and Barker, G.E. (1987) AIChE Annual Meeting, November 18, 1987, Paper no. 161E.

Feder, J., and Tolbert, W.R. (1985) *Am. Biotechnol. Lab.* Jan/Feb, 24–36.

Fee, J.A. (1981) in *Oxygen and Life* Royal Society of Chemistry Special Publication no. 39, pp. 77–97, Royal Society of Chemistry, London.

Fernandez, E.J., and Clark, D.S. (1987) *Enz. Microbiol. Technol.* 9, 259–271.

Fleischaker, R.J., and Sinskey, A.J. (1981) *Eur. J. Appl. Microbiol. Biotechnol.* 12, 193–197.

Flickinger, M.C., Goebel, N.K., McNeil, D., et al. (1987) ACS 194th National Meeting, September 3, 1987, MBTD Paper no. 151.

Foster, M.A. (1984) *Magnetic Resonance in Medicine and Biology* pp. 28–65, Pergammon Press, Oxford.

Frame, K.K., and Hu, W.-S. (1985) *Biotechnol. Lett.* 7, 147–152.

Garatun-Tjeldstø, O., Pryme, I.F., Weltman, J.K., and Dowben, R.M. (1976) *J. Cell. Biol.* 68, 232–239.

Gillies, R.J., Ogino, T., Shulman, R.G., and Ward, D.C. (1982) *J. Cell Biol.* 95, 24–28.

Glacken, M.W., Adema, E., and Sinskey, A.J. (1988) *Biotechnol. Bioeng.* 32, 491–506.

Glacken, M.W., Fleischaker, R.J., and Sinskey, A.J. (1983) *Trends Biotechnol.* 1, 102–108.

Glacken, M.W., Fleischaker, R.J., and Sinskey, A.J. (1986) *Biotechnol. Bioeng.* 28, 1376–1389.

Graff, S., Moser, H., Kastner, O., Graff, A.M., and Tannenbaum, M. (1965) *J. Natl. Cancer Inst.* 34, 511–519.

Griffiths, J.B., and Pirt, S.J. (1967) *Proc. R. Soc. B* 168, 421–438.

Gstraunthaler, G., Harris, H.W., Jr., and Handler, J.S. (1987) *Am. J. Physiol.* 252, C239–C243.

Haveman, J., and Hahn, G.M. (1981) *J. Cell. Physiol.* 107, 237–241.

Holmström, B. (1968) *Biotechnol. Bioeng.* 10, 373–384.

Hu, W-S., and Dodge, T.C. (1985) *Biotechnol. Prog.* 1, 209–215.

Hu, W.-S., Frame, K.K., and Sen, S. (1987) AIChE Annual Meeting, November 18, 1987, Paper no. 160E.

Iio, M., Moriyama, A., and Murakami, H. (1985) in *Growth and Differentiation of Cells in Defined Environment* (Murakami, H., Yamane, I., Barnes, D.W., et al., eds.), pp. 437–442, Kodansha, Tokyo, and Springer-Verlag, Berlin.

Kacser, H., and Burns, J.A. (1973) (Davies, D.D., ed.), *Symp. Soc. Exp. Biol.* 27, 65–104.

Katinger, H.W.D., and Scheirer, W. (1982) *Acta Biotechnol.* 2, 3–41.

Katz, L.A., Koretsky, A.P., and Balaban, R.S. (1987) *FEBS Letters* 221, 270–276.

Kekonen, E.M., Jauhonen, V.P., and Hassinen, I.E. (1987) *J. Cell Physiol.* 133, 119–126.

Kell, D.B., and Westerhoff, H.V. (1986) *FEMS Microbiol. Rev.* 39, 305–320.

Kilburn, D.G., Lilly, M.D., Self, D.A., and Webb, F.C. (1969) *J. Cell Sci.* 4, 25–37.

Kilburn, D.G., and van Wezel, A.L. (1970) *J. Gen. Virol.* 9, 1–7.

Kimura, T., Iijima, S., and Kobayashi, T. (1987) *J. Ferment. Technol.* 65, 341–344.

Knighton, D.R., Hunt, T.K., Scheuenstuhl, H., et al. (1983) *Science* 221, 1283–1285.

Kompala, D.S., Bentley, W.E., and Batt, B.C. (1987) AIChE Annual Meeting, New York, NY, November 16, 1987, Paper no. 158P.

Kovacevic, Z., and Morris, H.P. (1972) *Cancer Res.* 32, 326–333.

Krömer, E., and Katinger, H.W.D. (1982) *Develop. Biol. Stand.* 50, 349–354.

Lanks, K.W. (1986a) *J. Cell. Physiol.* 126, 319–321.

Lanks, K.W. (1986b) *Exp. Cell Res.* 165, 1–10.

Lanks, K.W., and Li, P.-W. (1988) *J. Cell. Physiol.* 135, 151–155.

Lavietes, B.B., Regan, D.H., and Demopoulos, H.B. (1974) *Proc. Natl. Acad. Sci. USA* 71, 3993–3997.

Lazo, P.A. (1981) *Eur. J. Biochem.* 117, 19–25.

Lee, A.S. (1987) *Trends Biol. Sci.* 12, 20–23.

Leist, C., Meyer, H.-P., and Fiechter, A. (1986) *J. Biotechnol.* 4, 235–246.

Liao, J.C., and Lightfoot, E.N. (1988a) *Biotechnol. Bioeng.* 31, 847–854.

Liao, J.C., and Lightfoot, E.N. (1988b) *Biotechnol. Bioeng.* 869–879.

Low, K., and Harbour, C. (1985) *Develop. Biol. Stand.* 60, 73–79.

Low, K.S., Harbour, C., and Barford, J.P. (1987) *Biotechnol. Tech.* 1, 239–244.

Luan, Y.T., Mutharasan, R., and Magee, W.E. (1987) *Biotechnol. Lett.* 9, 691–696.

Mano, T., Taya, M., Taniguchi, M., and Kobayashi, T. (1987) *J. Ferment. Technol.* 65, 425–429.

McKeehan, W.L. (1986) in *Carbohydrate Metabolism in Cultured Cells* (Morgan, M.J., ed.), pp. 111–150, Plenum Press, New York.

McLimans, W.F., Blumenson, L.E., and Tunnah, K.V. (1968) *Biotechnol. Bioeng.* 10, 741–763.

Merten, O-W., Reiter, S., Himmler, G., Scheirer, W., and Katinger, H. (1985) *Develop. Biol. Stand.* 60, 219–227.

Miller, W.M. (1987) *A Kinetic Analysis of Hybridoma Growth and Metabolism.* PhD thesis, University of California, Berkeley.

Miller, W.M., Blanch, H.W., and Wilke, C.R. (1986) ACS 192nd National Meeting, Anaheim, CA, September 11, 1986.

Miller, W.M., Blanch, H.W., and Wilke, C.R. (1988a) *Biotechnol. Bioeng.* 32, 947–965.

Miller, W.M., Wilke, C.R., and Blanch, H.W. (1987) *J. Cell. Physiol.* 132, 524–530.

Miller, W.M., Wilke, C.R., and Blanch, H.W. (1988b) *Bioprocess Eng.* 3, 103–111.

Miller, W.M., Wilke, C.R., and Blanch, H.W. (1988c) *Bioprocess. Eng.* 3, 113–122.

Miller, W.M., Wilke, C.R., and Blanch, H.W. (1988d) ACS 196th National Meeting, Los Angeles, CA, September 29, 1988, MBTD Paper no. 128.

Miller, W.M., Wilke, C.R., and Blanch, H.W. (1989a) *Biotechnol. Bioeng.* 33, 477–486.

Miller, W.M., Wilke, C.R., and Blanch, H.W. (1989b) *Biotechnol. Bioeng.* 33, 487–499.

Mizrahi, A., Vosseller, G.V., Yagi, Y., and Moore, G.E. (1972) *Proc. Soc. Exp. Biol. Med.* 139, 118–122.

Moreadith, R.W., and Lehninger, A.L. (1984) *J. Biol. Chem.* 259, 6215–6221.

Morgan, M.J., and Faik, P. (1986) in *Carbohydrate Metabolism in Cultured Cells* (Morgan, M.J., ed.), pp. 29–75, Plenum Press, New York.

Moser, H., and Vecchio, G. (1967) *Experientia* 15, 120–123.

Nakashima, R.A., Paggi, M.G., and Pedersen, P.L. (1984) *Cancer Res.* 44, 5702–5706.

Newsholme, P., Gordon, S., and Newsholme, E.A. (1987) *Biochem. J.* 242, 631–636.

Olander, C.P. (1972) *Am. J. Physiol.* 222, 45–48.

Ozturk, S.S., Lee, G.M., Huard, T.K., and Palsson, B. (1988) ACS 196th National Meeting, Los Angeles, CA, September 28, 1988, MBTD Paper no. 123.

Packer, L., and Fuher, K. (1977) *Nature* 267, 423–425.

Pardridge, W.M., Duducgian-Vartavarian, L., Casanello-Ertl, D., Jones, M.R., and Kopple, J.D. (1981) *Am. J. Physiol.* 240, E203–E208.

Parks, D.R., Lanier, L.L., and Herzenberg, L.A. (1986) in *Handbook of Experimental Immunology* Vol. 1, Fourth Edition (Weir, D.M., Herzenberg, L.A., Blackwell, C., and Herzenberg, Lenore A., eds.), pp. 29.1–29.21, Blackwell Scientific, Oxford.

Parshad, R., Sanford, K.K., Jones, G.M., Price, F.M., and Taylor, W.G. (1977) *Exp. Cell Res.* 104, 199–205.

Paul, J. (1965) in *Cells and Tissues in Culture* (Wilmer, E.N., ed.), pp. 239–276, Academic Press, New York.

Pederson, P.L. (1978) *Prog. Exp. Tumor Res.* 22, 190–274.

Peraino, C., Bacchetti, S., and Eisler, W.J. (1970) *Science* 169, 204–205.

Pirt, S.J., and Callow, D.S. (1964) *Exp. Cell Res.* 33, 413–421.

Radlett, P.J., Telling, R.C., Stone, C.J., and Whiteside, J.P. (1971) *Appl. Microbiol.* 22, 534–537.

Raivio, K.O., and Seegmiller, J.E. (1973) *Biochim. Biophys. Acta* 299, 283–292.

Randerson, D.H. (1985) *J. Biotechnol.* 2, 241–255.

Ray, N.G., Karkare, S.B., and Runstadler, Jr., P.W. (1989) *Biotechnol. Bioeng.* 33, 724–730.

Reitzer, L.J., Wice, B.M., and Kennell, D. (1979) *J. Biol. Chem.* 254, 2669–2676.

Renard, J.M., Spagnoli, R., Mazier, C., Salles, M.F., and Mandine, E. (1988) *Biotechnol. Lett.* 10, 91–96.

Renner, E.D., Plagemann, P.G.W., and Bernlohr, R.W. (1972) *J. Biol. Chem.* 247, 5765–5776.

Reuveny, S., Velez, D., Macmillan, J.D., and Miller, L. (1986a) *J. Immunol. Methods* 86, 53–59.

Reuveny, S., Velez, D., Miller, L., and Macmillan, J.D. (1986b) *J. Immunol. Methods* 86, 61–69.

Rheinwald, J.G., and Green, H. (1974) *Cell* 2, 287–293.

Richter, C., Park, J-W., and Ames, B.N. (1988) *Proc. Natl. Acad. Sci. USA* 85, 6465–6467.

Roth, E., Ollenschlager, G., Hamilton, G., et al. (1988) *In Vitro* 24, 696–698.

Ryan, W.L., and Cardin, C. (1966) *Proc. Soc. Exp. Biol. Med.* 123, 27–30.

Savinell, J.M., Lee, G.M., and Palsson, B. (1989) *Bioprocess Eng.* 4, 231–234.

Schlesinger, M.J., Ashburner, M., and Tisseries, A., eds. (1982) *Heat Shock, from Bacteria to Man.* Cold Spring Harbor Laboratory, Cold Spring Harbor, New York.

Self, D.A., Kilburn, D.G., and Lilly, M.D. (1968) *Biotechnol. Bioeng.* 10, 815–828.

Shirai, Y., Hashimoto, K., Yamaji, H., and Kawahara, H. (1988). *Appl. Microbiol. Biotechnol.* 29, 113–118.

Shuler, M.L., and Domach, M.M. (1983) in *Foundations of Biochemical Engineering* (Blanch, H.W., Papoutsakis, E.T., and Stephanopolous, G., eds.), *ACS Symposium Series* 207:93–134.

Sinclair, R. (1974) *In Vitro* 10, 295–305.

Sinclair, R. (1980) *In Vitro* 16, 1076–1084.

Spier, R.E. (1982) *J. Chem. Technol. Biotechnol.* 32, 304–312.

Stoner, G.D., and Merchant, D.J. (1972) *In Vitro* 7, 330–343.

Suleiman, S.A., and Stevens, J.B. (1987) *In Vitro Cell. Develop. Biol.* 23, 332–338.

Sutherland, R.M. (1988) *Science* 240, 177–184.

Suzuki, E., Sayles, G.D., and Ollis, D.F. (1988) AIChE Annual Meeting, Washington, DC, November 29, 1988, Paper no. 137L.

Tajiri, D., Webster, J., and Terando, J. (1987) ACS 194th National Meeting, New Orleans, LA, September 1, 1987, MBTD Paper no. 81.

Taya, M., Mano, T., and Kobayashi, T. (1986) J. Ferment. Technol. 64, 347–350.

Taylor, W.G. (1983) in *Uses and Standardization of Vertebrate Cell Cultures* (*In Vitro Monogr. no. 5*) (Patterson, M.K., ed.), pp. 58–70, Tissue Culture Association, Gaithersburg, MD.

Thayer, P.S., Himmelfarb, P., and Roberts, D. (1970) *Cancer Res.* 30, 1709–1714.

Thomas, J.N. (1986) in *Mammalian Cell Technology* (Thilly, W.G., ed.), pp. 109–130. Butterworths, Boston.

Thorpe, J.S., Murdin, A.D., Sanders, P.G., and Spier, R.E. (1987) ACS 194th National Meeting, New Orleans, LA, September 3, 1987, MBTD Paper no. 147.

Tolbert, W.R., Feder, J., and Kimes, R.C. (1981) *In Vitro* 17, 885–890.

Tovey, M.G., (1980) *Adv. Cancer Res.* 33, 1–37.

Tovey, M.G. (1985) in *Animal Cell Biotechnology* Vol. 1 (Spier, R.E., and Griffiths, J.B., eds.), pp. 195–210. Academic Press, Orlando, FL.

Tovey, M., and Brouty-Boyé, D. (1976) *Exp. Cell Res.* 101, 346–354.

Tovey, M., and Brouty-Boyé, D. (1979) *Exp. Cell Res.* 118, 383–388.

Tovey, M., Brouty-Boyé, D., and Gresser, I. (1975) *Proc. Natl. Acad. Sci. USA* 72, 2265–2269.

Tovey, M.G., Mathison, G.E., and Pirt, S.J. (1973) *J. Gen. Virol.* 20, 29–35.

Tovey, M.G., and Rochette-Egly, C. (1980) *Ann. N.Y. Acad. Sci.* 350, 266–278.

Tovey, M.G., Rochette-Egly, C., and Castagna, M. (1979) *Proc. Natl. Acad. Sci. USA* 76, 3890–3893.

Tovey, M.G., Rochette-Egly, C., and Castagna, M. (1980) *J. Cell. Physiol.* 105, 363–367.

van Wezel, A.L., van der Velden-de Groot, C.A.M., de Haan, H.H., van den Heuvel, N., and Schasfoort, R. (1985) *Develop. Biol. Stand.* 60, 229–236.

Velez, D., Reuveny, S., Miller, L., and Macmillan, J.D. (1987) *J. Immunol. Methods* 102, 275–278.

Visek, W.J., Kolodny, G.M., and Gross, P.R. (1972) *J. Cell. Physiol.* 80, 373–382.

Wice, B.M., Reitzer, L.J., and Kennel, D. (1981) *J. Biol. Chem.* 256, 7812–7819.

Williamson, J.R., and Cooper, R.H. (1980) *FEBS Lett.* 117, K73–K85.

Zielke, H.R., Ozand, P.T., Tildon, J.T., Sevdalian, D.A., and Cornblath, M. (1978) *J. Cell. Physiol.* 95, 41–48.

Zielke, H.R., Sumbilla, C.M., Sevdalian, D.A., Hawkins, R.L., and Ozand, P.T. (1980) *J. Cell. Physiol.* 104, 433–441.

Zielke, H.R., Sumbilla, C.M., and Ozand, P.T. (1981) *J. Cell. Physiol.* 107, 251–254.

Zielke, H.R., Zielke, C.L., and Ozand, P.T. (1984) *Fed. Proc. Fed. Am. Soc. Exp. Biol.* 43, 121–128.

Zimber, A., and Topping, D.C. (1970) *Fed. Proc. Fed. Am. Soc. Exp. Biol.* 29, 428 (Abstract).

PART III

Anchorage-Dependent Cell Supports

Fixed Immobilized Beds for the Cultivation of Animal Cells

Bryan Griffiths
Denis Looby

Immobilization of cells infers that cells are prevented from freely mixing with the aqueous phase of the culture system. This is a natural phenomenon in vivo where cells are architecturally organized in tissues and organs and perfused by lymph, blood, etc. In in vitro systems, immobilization is primarily used to (a) increase the stability and, thus, culture time of the cell, and (b) increase the process intensity of the culture.

These steps are needed to overcome the low catalytic activity of animal cells in culture and to lower recovery costs by increasing the unit product concentration. There may be other advantages such as reduced requirement for nutritional factors, more efficient utilization of media, and constant rather than feed-and-starve environmental conditions. Many immobilized systems are dynamic; i.e., the cell and its support material are suspended in the growth medium by stirring or fluidization techniques (e.g., microcarriers). In this chapter, only the nondynamic, or fixed (packed) bed immobilization systems will be described; i.e., the cell and its support is stationary and the medium is dynamic. Anchorage-dependent cells attached to a substrate are, of course, already immobilized, and thus amenable to perfusion methods since cell wash out will not occur. However, the problem for this type of cell is the ability to increase both process intensity and volumetric scaleup. Suspension cells are more difficult to immobilize, but many systems have been devised based on entrapment (e.g., textured ceramic surfaces, porous beads, and sponges) and immurement (e.g., hollow fibers, gel encapsulation, and membranes). Once suspension cells are immobilized it has generally proven easier to scaleup process intensity than it has with attached cells.

The main focus of this chapter is the description of the fixed bed immobilized culture, its development, advantages and disadvantages, and its role in current-day animal cell technology.

7.1 GENERAL PRINCIPLES

Fixed (packed) bed reactors have been in existence since the turn of the century but have not been widely used for immobilizing microorganisms except in wastewater treatment (trickle filters) and vinegar production (Acetifiers) (Atkinson and Mavituna 1983).

7.1.1 Advantages and Disadvantages of Fixed Bed Reactors

The advantages and disadvantages of fixed bed reactors for the immobilization of microorganisms are given in Table 7–1. The main reasons why they have not been more widely used in bacterial fermentations are poor oxygen transfer, excessive biomass build-up, and difficulties in recovering biomass from the bed. These limitations are not necessarily a problem with

TABLE 7–1 Advantages and Disadvantages of Fixed Bed Reactors

Advantages
  Higher unit cell density and productivity.
  Protection from shear.
  Media can be changed easily and product separated from cells.
  Continuous removal of inhibitory metabolites may increase cell productivity.
  Maintenance of long-term productivity.
  Cells are in direct contact with the media.
  Large scale-up potential.
Disadvantages
  More suited to adherent microorganisms than suspension.
  Accumulation of biomass can lead to blockage and channeling in the bed.
  Recovery of biomass difficult.
  Fixed beds are unsuitable for highly aerobic fermentations due to low mixing and
    mass transfer characteristics.

From Atkinson and Mavituna 1983; Denac and Dunn 1988; van Brunt 1986; Payne et al. 1987; Heinrich and Rehm 1981.

animal cells because they have a much lower oxygen requirement than bacteria; blockages due to biomass build-up can be easily overcome by optimizing the bed channel size (Looby and Griffiths 1987). Also, since most products from animal cells are secreted into the culture media, biomass recovery is not always necessary.

7.1.2 Reactor Configurations

Fixed bed reactors have the virtue of being simplistic and are usually columns packed with support particles. Microorganisms are immobilized either on the surface or throughout the support, with culture media flowing past. Fixed bed reactors are operated in several modes (Figure 7–1). In its simplest form, the medium passes straight through the bed from the reservoir to a harvest vessel (see Figure 7–1,X). A more efficient medium utilization is obtained when the medium is recycled through either a reservoir, which acts as an environmental chamber (see Figure 7–1,Y), or through a membrane oxygenator (see Figure 7–1,W). This is typical for animal cells that have a low biosynthetic capacity. A more complex configuration is to have continuous or semicontinuous harvesting and medium feed coupled with recycling (see Figure 7–1,Z). This allows full utilization of the media with maintenance of optimum conditions for long-term cell productivity.

7.1.3 Support Particles

Carriers can be divided into two main groups, organic and inorganic (Table 7–2). Inorganic carriers can be further divided into native (unmodified) or derivitized, i.e., with organic groups attached to their surfaces by coupling (Kolot 1981a). There are two main methods of immobilizing organisms, i.e., entrapment and carrier binding. Entrapment occurs where organisms are trapped in the interstices of porous particles, within gels, or behind membranes. Carrier binding occurs where there is a direct binding of the organisms to the carrier by physical adsorption or by ionic and/or convalent bonds (Vega et al. 1988).

There are a wide variety of potential supports for fixed bed reactors (see Table 7–2). The most suitable carriers would have as many of the characteristics as possible outlined in Table 7–3. They should be mechanically strong to survive long-term culture and to prevent bed compression in a tall reactor; this can be a problem with soft supports such as gels and foams. Carriers should have a high surface area per unit volume with a high biomass loading capacity. In this case, porous materials offer a distinct advantage due to their very high surface area-to-volume ratio. It is important, however, to have the optimal pore size for particular microorganisms, because the surface area of the support is inversely proportional to pore radius (Kolot 1981b). For microorganisms that reproduce by fission, the optimal pore size is considered to be in the range of one to five times the major dimensions

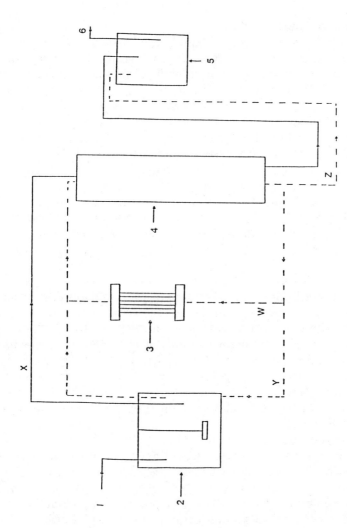

FIGURE 7-1 Fixed bed reactor systems (alternative configurations), (1) media in, (2) reservoir vessel, (3) membrane oxygenator, (4) fixed bed reactor, (5) harvest vessel, and (6) product out. Normal plug flow perfusion (X) from reservoir, through bed, to harvest vessel. Modifications to increase performance include; recirculation of media from reservoir to bed (Y) with bleed-off (continuous or intermittent) to harvest vessel (Z); recirculation of media through membrane oxygenator (W).

TABLE 7–2   Carriers Suitable for Immobilizing Organisms

Organic
   Protein: Gelatin
   Polysaccharides: Cellulose, agar, agarose, Carrageenan Sephadex
   Synthetic polymers: Polystyrene, polyurethane, acrylamide
Inorganic
   Native: Glass, silica, ceramics, alumina, steel
   Derivatized: Surface treated for cell attachment

From Kolot 1981a; Spier 1985.

TABLE 7–3   Characteristics of an "Ideal" Carrier for Fixed Bed Reactors

Packing arrangement to give maximum homogeneity and distribution of fluids.
High biomass loading capacity.
Maximum surface area per unit volume.
Optimum diffusion distance from flowing media to center of particle.
Mechanical stability.
Calculable surface area.
Specific shapes and sizes.
Uniform size distribution.
Nontoxic.
Reusable and in situ cleanable.
Autoclavable/steam sterilizable.
Inexpensive.

From Kolot 1981a; Spier 1985; Nunez and Lema 1987.

of the cell (Messing and Opperman 1979). The optimum particle size for porous carriers has to take into account the diffusion distance from the nutrient source to the center of the particle, while at the same time, have a packing arrangement that allows homogeneous distribution of culture media.

7.1.4 Fluid Dynamics and Scaleup

Fixed bed reactors are essentially plug flow, which means that mixing is axial with little significant radial- or back-mixing. This in turn means that concentration gradients of nutrients, waste metabolites, and product will build up along the length of the column (Figure 7–2), and puts a severe limitation on the height that a bed can be scaled up. In animal cell culture, scaleup to 30 l (bed volume) has been achieved (Whiteside et al. 1979). However, in theory it has been reported that scaleup to 4.3 m^3 (bed height of 1.76 m) could be achieved, based on a cell density of 7.5 × 10^6 cells/ml (bed volume) in beds of 3 mm diameter glass spheres (Spier 1985).

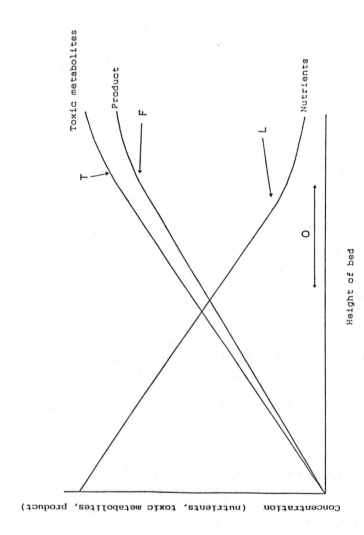

FIGURE 7-2 Concentration gradients in a fixed bed system. Theoretical considerations: (L) limiting nutrient concentration, (T) limiting toxic metabolite concentration, (F) loss of specific product expression due to negative feedback, (O) optimum bed height within the region.

7.1.5   Suitability of Fixed Bed Reactors for Animal Cells

Fixed bed reactors are ideally suited to animal cell products, many of which are nongrowth associated and secreted into the culture medium. The advantages over nonimmobilized (free cell) systems are that the culture period can be extended by continuous media perfusion and more controlled environmental conditions for maintenance of cells in a stationary phase. In batch cultures, cells show a rapid decline after reaching stationary phase, which is the period when nongrowth-associated products are expressed. Thus immobilization will greatly extend the production phase. Also, the decreased growth rates, associated with steady-state immobilized cell growth, often leads to increased product formation.

7.2   HISTORICAL DEVELOPMENTS IN ANIMAL CELL IMMOBILIZATION

The need to increase the surface area of the substrate to obtain higher unit cell densities has been recognized for some time. The innovations of Earle and co-workers include the use of cellophane (Evans and Earle 1947; Earle et al. 1950), glass rings (Earle et al. 1951a), and glass helices (Earle et al. 1951b and 1953) in attempts to develop mass cell culture techniques. The use of glass helices was further developed by McCoy et al. (1962), by using a continuous perfusion of medium. They used an ion-exchange column (1.2 $\times$ 75 cm) with a top reservoir (100 ml capacity) that, when fully packed had a void volume of 40 ml. Twelve of these units were serviced from a single reservoir (22 l). Perfusion through the packed bed was initially very low (70 ml/24 hours) but was increased to maintain the pH above 7.1 and the glucose level above 1.7 mg/ml, to 4 l/24 hours (4 volumes/hour). Their work, although small scale, demonstrated the feasibility of using packed beds of glass helices for mass cell cultivation, and the advantages of continuous medium perfusion to keep the culture physiologically stable.

Many matrices have subsequently been used for the packed bed immobilization of cells (see Spier 1985, and Table 7–2). The most commonly used matrix has been the glass sphere, an idea that probably originated with Gey (1933). The next report on the use of glass spheres came from Robineaux et al. (1970), but the object of their work was not mass cell growth but a closed circuit perfusion apparatus to study the effect of diphenylhydantoin (DPH) on cell growth and respiration. The vessel had a void volume of 125 ml, and 5 mm diameter glass beads were used. Flow rates were initially 1 l/hour (i.e., eight volumes per hour) and increased to over 2 l/hour (i.e., 16 volumes per hour). These rapid perfusion rates should have allowed high densities of L cells to be obtained, but this is unknown since the object of the experiment was to measure tritiated thymidine (^{3}H-Tdr) uptake in cells treated with DPH.

The use of glass spheres for mass cell cultivation was initiated by Rudiger and co-workers (Wohler et al. 1972; Rudiger 1975). Their aim was to develop a large, single unit system for the growth of human diploid cells (HDC), which was an early attempt to move from multiple to unit process systems for anchorage-dependent cells. Their initial system was based on 800 ml of 3.5–5 mm glass spheres, which gave a surface area of 1 m², i.e., sufficient for 10^9 HDC, in a funnel. The apparatus was termed the Perlacell culture vessel and was commercially available in sizes from 1–5 l (void volume 40%). The perfusion rate from 24 hours and longer was 0.75 volumes/day, and medium was added from the top. The culture period was eight to 10 days and the cells were harvested with trypsin/EDTA with a yield of 2×10^9 HDC per 3 l (1,300 ml) vessel from an inoculum of 5×10^7. An interesting development in technique was to only partially trypsinize the cells and refill the apparatus with fresh medium. By these means, 10^9 cells were harvested every three days for several weeks. This was an early and novel example of the fed-batch technique. The use of a funnel device, which they state as being important for distribution of fresh medium and for preventing pH gradients, of course, results in an uneven flow rate through the packed bed. Claims that this is a large-scale unit process may now seem to be overrated, but at that time the only realistic alternative was roller bottles. Microcarriers (van Wezel 1967) had not yet been optimized and novel devices based on rolled plastic films (House et al. 1972) and multiple plates (Schleicher and Weiss 1968) were still very experimental and gave variable results.

Fixed beds of immobilized cells had been used for obtaining tissue or even organ-like growth of cells, which allowed differentiation (Leighton 1951) for studying cells in a physiologically stable environment by medium perfusion and as a mass cell culture technique. With this background of knowledge and experience, the time had come to determine whether these techniques would be suitable for scale up to industrial-type processes.

7.3   FIXED BEDS IN A PRODUCTION PROCESS

7.3.1   Glass Sphere Reactors

A variety of fixed bed concepts have been developed and evaluated (Table 7-4), but the most used method is based on glass spheres. The popularity of glass spheres is based on the fact that they are a cheap, reutilizable substrate to which cells preferentially attach. In addition, they have a regular and stable packing geometry and conform to the geometry of the vessel and leave a regular channel size running through the bed.

Also, the bioreactor requires a low cost outlay. Comparisons of glass bead reactors to other scaleup systems have been made (Griffiths et al. 1987a and 1987b).

TABLE 7–4 Matrices Used in Fixed Bed Reactors for Animal Cells

Cellophane	Polystyrene spirals
Glass helices	Plastic films
Glass rings	Stainless steel springs
Glass spheres	Ceramics
Glass tubing	Polyurethane foam
Glass rods	Fibers

Systems described thus far have been aimed at producing cells in large numbers. The assessment of glass spheres for the manufacture of a cell product was initiated by Spier and co-workers (Spier and Whiteside 1976). Initial studies established that height-to-diameter ratios between 1 and 22, sphere diameter between 2 mm and 7 mm, and the shape of the vessel bottom had little detectable effect on cell growth (Spier and Whiteside 1976). The circulation rate of medium was critical, however, and a linear flow rate of 2 cm/min was found to be optimal.

From this preliminary work using 30 cc of 3 mm diameter glass spheres, the system was scaled up to 100 l and used for the production of FMDV from BHK monolayer cells (Whiteside et al. 1979 and 1980; Whiteside and Spier 1981).

The 100 l propagator had 50 kg of 3 mm diameter spheres that gave a surface area in excess of 20 m^2. Growth was monitored by glucose utilization, and virus production by lactate dehydrogenase (LDH) concentration. The system worked well but had a lower (approximately 50%) productivity than 10 and 1 l reactors for cell growth and, thus, virus production. The reason for this was attributed to a reduction in aeration rates during scaleup. Increased aeration rates did increase yields to some degree but were still not comparable with that obtained at 10 l, for instance (Whiteside et al. 1980). An examination of the packed beds showed a significant cell density gradient, which was overcome by introducing the inoculum at the top and bottom, or by using shallow beds (Whiteside and Spier 1985).

Work on developing glass sphere systems for human diploid fibroblasts (HDF) with the possible aim of producing interferon was going on concurrently with the above studies for FMDV. Burbidge (1980) described a 10 l system based on 3 mm diameter spheres (a range was tried and 2–3 mm found optimal; 1 mm size produced interstitial retention problems) for the production of a range of HDF. The cells were harvested by trypsin and mechanical agitation and comparable yields to flask-type cultures were obtained. Metabolic studies, including glucose and glutamine utilization, and lactate production were performed in this system (Robinson et al. 1980) with the conclusion that cells behaved identically in glass sphere reactors to stationary flasks. This system was subsequently scaled up to 200 l (unpublished results).

The only other reports on using fixed beds of glass spheres were some preliminary studies by Griffiths et al. (1982) for Herpes simplex virus (HSV), and Brown et al. (1985, 1988) for tPA. The work with HSV was not extended due to problems of harvesting the cells containing intracellular virus. Both in situ lysis and harvesting the infected HDC proved difficult because the 3 mm sphere bed acted as a depth filter and became blocked with only a low percentage recovery of cells and/or virus. Mechanical agitation, including the forced sparging of air through the bed, was not considered a sensible option in view of the large scaleup that was necessary if the system was found suitable.

Brown et al. (1985) developed the mass culturing technique (MCT), which would enable a wide variety of cell types to be grown. The growth chamber consisted either of hollow fibers (for suspension cells) or packed glass beads of 1–3 mm diameter for anchorage-dependent cells. Medium is recirculated through this chamber via an oxygenator. Using glass beads with an aspect ratio of no greater than 2:1 was considered appropriate and high linear medium flow rates were needed to prevent oxygen gradients. Cell densities of 1.3×10^{10} viable cells have been recovered from the 12 l bioreactor, and the system has been kept continuously running for over 1 year (Brown et al. 1988) for tPA production from the Bowes melanoma cell line. It has also been found possible to keep the cells in protein-free media once the culture is established and steady-state conditions, as measured by O_2, CO_2, pH, glucose, lactic acid, ammonia, lactate dehydrogenase, and product levels, is established.

Long-term cultures obviously depend upon the growth rate of cells exceeding the death rate since it is unrealistic to expect cells to remain viable for periods of months. In many systems, this equilibrium between growth and death rate is not constant and the culture needs periodic recovery periods in which enriched media is used.

In all these studies, cylindrical and not funnel-shaped beds were used, as previously deemed necessary (Rudiger 1975). This change was made possible by getting even medium distribution over a wide surface area. To do this an initial layer of large glass spheres (5–7 mm) was used (Whiteside and Spier 1985; Griffiths et al. 1982) or chromatography column dispersion plates were introduced at the top and bottom of the beds (Brown et al. 1988). The other factor was to keep aspect ratios below 2:1.

In conclusion, glass sphere reactors were suitable for industrial processes over 100 l volume. The reason they have not been more widely used is (a) other anchorage-dependent systems, especially microcarrier, improved at a very rapid rate; (b) the need to grow hybridoma (suspension) cells has been a major influence in the 1980s and glass beads are not suitable for this type of cell; and (c) the solid glass sphere is a relatively low cell density system and priority is increasingly on developing high density systems.

7.3.2 Alternative Matrix Materials

Variations on the glass sphere have been tried, for example, borosilicate glass tubing, 6 mm long with a 4 mm diameter (2.4 mm inner bore) (Harms and Wendenburg 1978), and stainless steel springs (Merk 1982). The former system was run at the 2 l volume scale for HDF cells and used a gas permeator of silicone rubber capillary tubes for aeration of the circulating medium. The medium was pumped at 3 l/hour (i.e. 1.5 volume/hour) and a harvest of 1–2 $\times$ 10^9 cells (from 20 l medium) obtained. As described for other systems, it was difficult to harvest the cells from the bed but this fact was used to reinitiate the culture with fresh medium. Again, this was a feasibility study demonstrating its practicality for mass cell culture with the advantage of a more homogeneous environment than stationary flasks, etc.

Stainless steel spirals (0.6 cm long) were chosen from among 200 matrices that were evaluated because they had a good surface area-to-volume ratio, good media run-out properties, and were easy to clean (Merk 1982). The system was scaled up to 250 l for the production of human fibroblast interferon.

7.3.3 Multiple Glass Tube Reactors

Another approach to having good flow-through characteristics and thus reducing gradients, anoxic pockets, etc., is to use long lengths of glass tubing. This method was initially reported by Santero (1972), developed by Chemap as the Gyrogen (Girard et al. 1980), and as the Multi-Tube system (Corbeil et al. 1979) for vaccine production. The concept is to have a cartridge of glass tubing fixed into a template at either end, or with spacers, and placed into a horizontal cylindrical vessel through which environmentally controlled medium is perfused. Small scale units can be placed in roller bottle machines, larger ones are mounted in a special rotating device. Unit sizes range from 15,000 cm^2 (Bellco Multi-Tube) to 34 m^2 (Gyrogen). As a large-scale unit process, the system was never accepted, presumably because of its complexity, high price, and especially its lack of versatility. However, the smaller systems are in limited use as a commercial process for vaccine production (Dugre et al. 1987).

7.3.4 Plate Reactors

Another immobilization technique to provide increased surface areas within a unit volume is the plate reactor. These consist of stacks of parallel plates in either a horizontal or vertical mode and have been used in bioreactors of 200–300 l (Molin and Hedin 1969; Schleicher and Weiss 1968). However, they do have serious limitations because vertical plates are difficult to inoculate evenly and confluent sheets of cells have a tendency to slide off. Horizontal plates trap media and need tilting devices, and of course only the top side of each plate can be used. A large-scale system that overcomes many of these disadvantages is the plate heat exchanger (Burbidge 1980; Burbidge and Dacey 1984; Griffiths et al. 1987a). Plate reactors have been briefly reviewed (Griffiths 1988).

7.3.5 High Cell Density Reactors

During the past few years, with the realization that animal cell products were gaining increasingly important commercial value, many novel bioreactor systems have been developed. The aim of these systems has been twofold, first, to have a 50 to 100 times higher cell density than conventional stirred-tank reactors, and second, with the use of perfusion, to greatly extend the period of culture into weeks or months with a physiologically stable environment. To do this, various techniques of immurement and entrapment have been used (Griffiths 1988). The fixed bed systems include the Opticell, Hollow fiber reactor, and the Bellco bioreactor. These are all commercially available, so well-reported systems will only be briefly described here.

7.3.5.1 Ceramic Matrices Ceramics are cheap, inert materials suited to use as both a cell substrate and as an immobilization matrix. A small-scale (50 ml) ceramic matrix bed reactor was reported by Marcipar et al. (1983) for the entrapment of hybridomas. The unit consisted of 20 g of a ceramic matrix known as "Biogrog A" in which 7×10^7 cells were immobilized. The medium flow rate was 1.3 cm/min and four to five times increase in MoAb was achieved compared to flask culture on a unit medium volume basis.

The use of a ceramic matrix has been commercially developed in the Opticell bioreactor (Lydersen et al. 1985; Berg and Bodeker 1988). Initially, a ceramic cartridge consisting of square 1 mm channels running longitudinally was produced to provide a high surface area for anchorage-dependent cells (40 cm²/cm). Units were available with up to 12 m² of surface area. Subsequently, the ceramic surface was changed from a smooth to a rough, porous finish in which suspension cells became entrapped. The equipment is marketed as a complete "turn-key" apparatus with full computer control of environmental parameters. Its continuous running performance over many weeks, with an average daily production of 100–300 mg MoAb makes it a useful production system. Although scaleup is by multiple cartridges, both the size and the number that can be simultaneously handled is being constantly improved (e.g., 210 m² units are now available).

7.3.5.2 Hollow Fiber Systems The potential of using hollow fiber filtration devices for high-density cell growth was first recognized by Knazek et

al. (1972). Thin ultrafiltration fibers bundled together in a cylindrical cartridge provide a large surface area for perfusion of nutrients into the extra
capillary space in which cells can be maintained at densities of over 10^8 ml.
The system has problems with pressure drop and concentration gradients
along the length of the cartridge (reviewed by Tharakan and Chau 1986a),
with the result that scaleup beyond a few hundred milliliters is difficult.
Various modifications have been made in an attempt to overcome this
limitation, including cross flow of nutrients, flat bed systems to reduce the
nutrient path length (Feder and Tolbert 1983), and the Acusyst system (Tyo
et al. 1988). The latter uses cyclical pressure pulses to circulate the medium
alternatively into and out of the capillary fibers and is currently the most
successful technique for hollow fiber reactors. This technology is mainly
aimed at immobilizing suspension (hybridoma) cells in a fixed volume compartment with medium perfusion through the filter material, which holds
back the cells. Ultrafiltration grade fibers are the most popular because the
product (antibody) is also held with the cells and accumulates to high concentrations. However, filtration grade fibers cause less problems with blockage and gradients (Brown et al. 1985) because they allow a higher flux rate.
The use of hollow fibers for anchorage-dependent cells is still very limited,
largely due to the initially used cellulose acetate fibers being unsuitable for
cell attachment. However, the use of polypropylene fibers (Ku et al. 1981)
or coating with poly-D-lysine (Tharakan and Chau 1986b) makes this
method feasible for such cells.

An example of modifications of this type of technology being used for
the commercial production of cell products are the In Vitron static maintenance reactor (Tolbert et al. 1988) and the membroferm (Scheirer 1988),
a sandwich of flat membranes with different molecular weight cut-offs.

7.3.5.3 Airlift Bioreactor A 3 l airlift bioreactor (without draft tube) is
packed with up to 12 cartridges that contain a matrix of stainless steel coils.
Each cartridge has a surface area of 3,400 cm^2 and a perfusion rate of 0.1
volume/hour is used. It has been used to grow a recombinant CHO cell
expressing 1L-2 receptor protein continuously for over 12 months, producing 34 mg of protein per day (Familletti and Fredericks 1988).

7.4 OPTIMIZATION OF GLASS SPHERE REACTORS

In the above description of the historical and production developments, it
is repeatedly stated that the unit established the feasibility, or the practicality, of using fixed beds, particularly glass sphere, as a mass cell culture
technique with production potential. Why then is this not a more widely
used technique today and why has there been no follow-up publications by
the groups cited? Obviously there are a variety of reasons but it seems clear

to us that (1) the system has not been critically studied and optimized, and (2) there were definite process problems that had to be overcome if the system was to work at the industrial scale. Problems of reduced efficiency with scaleup and blockage of beds and harvesting have been reported. In addition, the questions of whether it is best to use upward or downward flow perfusion, whether 2–3 mm spheres are indeed optimal, if borosilicate glass is the best substrate, or if the surfaces should be derivitized are largely conjectural. On the assumption that these fixed beds were by no means homogeneous, a careful study of the various parameters in fixed bed culture were undertaken (Looby and Griffiths 1987).

7.4.1 Homogeneity of Fixed Glass Sphere Beds
Using a medium recycle system, as described in Figure 7–1, with an upward medium flow through a 3 mm glass bead column, cell counts per unit area were lower than that obtained in either roller bottles or microcarrier culture (Table 7–5). When aliquots of beads were removed from various parts of the bed for cell counting, a significant cell density gradient was apparent (Figure 7–3). The density was greatest at the bottom of the bed and decreased up the bed. There was also some horizontal variation. The reasons for this were believed to be due to one or more of the following effects: (1) inoculation from the base caused a filtering effect up the column; (2) cells settled down the column before attachment was complete; and (3) the medium perfused upward.

7.4.2 Modifications that Improved Homogeneity
The direction of medium flow had no appreciable effect. However, when cells were inoculated via a central perforated tube running the total vertical length of the bed, a far more uniform cell density was achieved (Figure 7–4A), and the total cell yield was increased by 50%. Fibronectin-coated beads were also used in an attempt to achieve a more rapid attachment of cells and thus reduce inoculation gradients. This modification gave similar results

TABLE 7-5 Comparison of Cell Yields in Different Systems

System	Cell Yield, $\times$ 10^5 cm^2
Glass bead,[1] 1.5 kg	0.91
Glass bead,[1] 5.0 kg	0.84
Roller bottle	2.00
Microcarrier[2]	1.51

[1]Three millimeter borosilicate glass spheres.
[2]Cytodex 3 microcarriers.
Data from Looby and Griffiths 1987.

0.66	0.55	0.54
0.74	0.64	0.73
0.96	0.88	1.10
1.05	1.30	0.98

3mm
5 Kg

Cells/cm^2 x 10^5

Average Cells/cm^2 0.84 x 10^5

Cells/Kg   4.4 x 10^8

Cells/ml* 1.7 x 10^6

FIGURE 7-3 The distribution of Vero cells in a 5 kg reactor (bed height 20 cm) packed with 3 mm spheres. The cell seed was introduced into the bottom of the bead bed. (I) inoculation direction, (*) bead bed void volume. Values within the figure matrix are localized cell counts in various regions of the bed. Average values for the whole bed (per area per volume) as summarized.

(see Figure 7-4B) to the central inoculation technique, indicating that non-homogeneous growth was due to inadequate inoculation procedures. The result, using 7 mm beads (see Figure 7-4C), was unexpected since a very high cell density per unit area was achieved (twofold that of 3 mm beads). This almost made up for the 60% lower surface area per unit volume of the larger beads. This experiment was carried out to determine whether a filtration effect was occurring with the smaller channel size produced by 3 mm beads.

7.4.3 Optimum Bead Size

Following the result with 7 mm spheres, a comparative study of beads with diameter sizes from 2-8 mm was made (Table 7-6). The cell density per unit area increased in beads up to 5 mm diameter and remained constant up to 8 mm. Five millimeter diameter beads were thus considered optimal since they allowed a significantly higher total cell yield (by bed or void medium volume) than any other size used.

This result, and the magnitude of the difference in cell yield, was unexpected in view of earlier work that had consistently found sizes in the range of 2-3 mm to be the best (Burbidge 1980; Brown et al. 1985), with little difference outside this range (Spier and Whiteside 1976). The attraction of using the smallest diameter possible to maximize surface area is obvious.

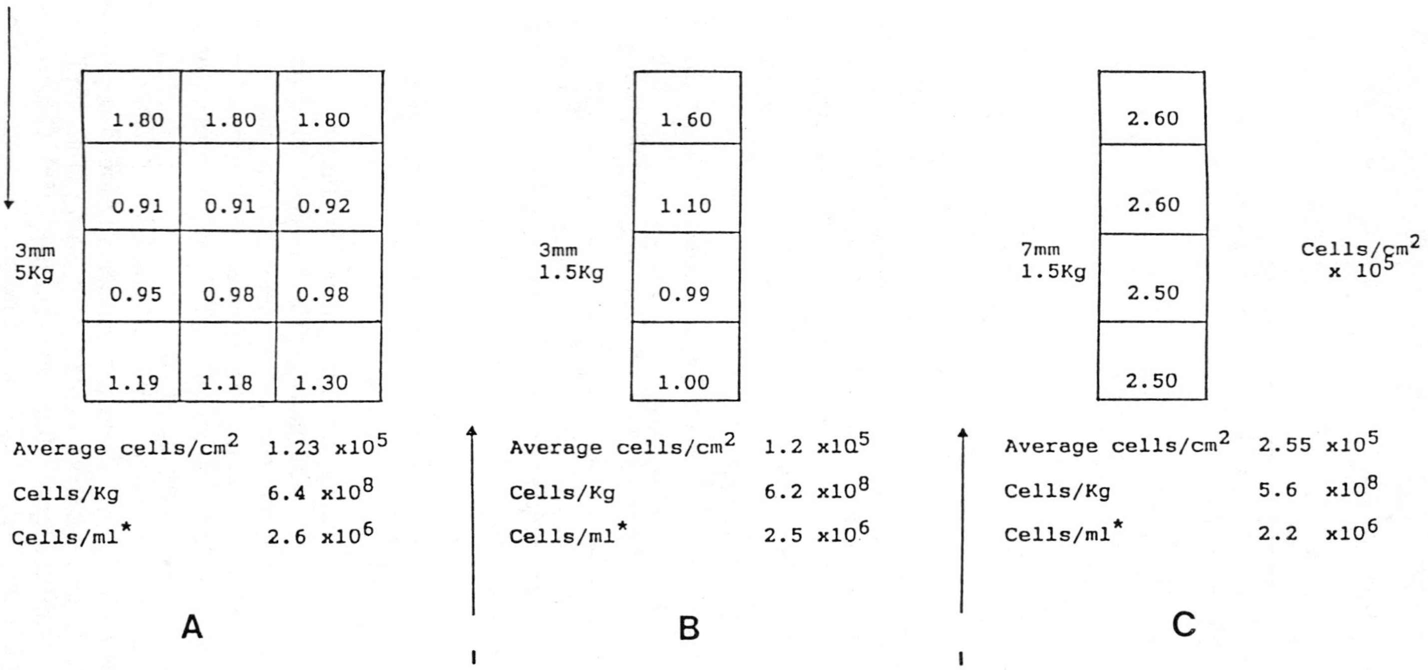

1.80	1.80	1.80
0.91	0.91	0.92
0.95	0.98	0.98
1.19	1.18	1.30

3mm
5Kg

	Average cells/cm^2	1.23 ×10^5
	Cells/Kg	6.4 ×10^8
	Cells/ml*	2.6 ×10^6

A

1.60
1.10
0.99
1.00

3mm
1.5Kg

Average cells/cm^2	1.2 ×10^5
Cells/Kg	6.2 ×10^8
Cells/ml*	2.5 ×10^6

B

2.60
2.60
2.50
2.50

7mm
1.5Kg

Cells/cm^2
× 10^5

Average cells/cm^2	2.55 ×10^5
Cells/Kg	5.6 ×10^8
Cells/ml*	2.2 ×10^6

C

FIGURE 7–4 The distribution of Vero cells in glass sphere (3 or 7 mm diameter) at 1.5 kg (0.96 l) or 5 kg (3.2 l) scale reactors (bed height 20 cm). (A) Three millimeter spheres with inoculations from the top through a centrally placed perforated tube; (B) 3 mm beads precoated with fibronectin; and (C) 7 mm beads. Values are as explained in legend to Figure 7–3. (I) inoculation direction, (*) bead bed void volume.

TABLE 7-6 Growth of Cells on Spheres of Different Diameter

Bead Diameter, mm	Approximate Channel Size, mm^2	Surface Area,[1] m^2/L	Count, $\times 10^5/cm^2$	Count,[2] $\times 10^6/ml$	Count,[3] $\times 10^6/ml$	Count, $\times 10^8/kg$
2	0.16	1.22	0.86	2.6	1.05	6.7
3	0.32	0.81	0.78	1.6	0.63	4.0
4	0.68	0.64	1.25	2.0	0.80	5.0
5	1.0	0.50	2.5	3.1	1.25	8.0
6	1.4	0.39	2.2	2.1	0.86	5.6
7	1.8	0.33	2.4	1.98	0.79	5.0
8	2.2	0.29	2.3	1.7	0.67	4.3

[1]Estimated available surface area due to contact between spheres (i.e., approximately 70% of total surface area).
[2]Cells per milliliter bead bed void volume.
[3]Cells per milliliter bead bed volume.
Data from Looby and Griffiths 1987.

One can only conjecture that differences were not detected previously due to (1) less efficient inoculation procedures, and/or (2) the lower aspect ratio used (<1:2) compared to these studies, which did not allow the differences to be so noticeable. The efficiency of 5 mm beads is demonstrated in Figure 7–5, which shows only small cell density gradients even though this work was carried out in an eight-times larger bed volume than was previously used.

A comparison of cell densities per kilogram shows a twofold increase from those originally obtained (see Figure 7–3). The reason for the efficiency of 5 mm beads is not fully understood. It is probably due to the resultant increase in channel size. This size was estimated to be 1 mm² for 5 mm

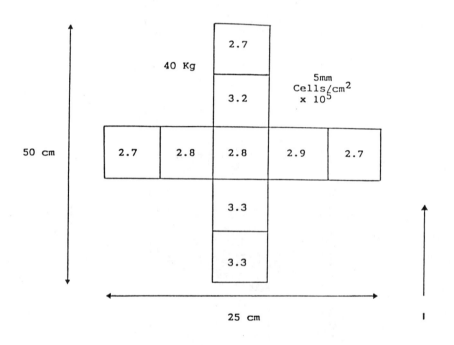

CELL YIELD

Average cells/cm^2	2.9×10^5
Cells/Kg	9.3×10^8
Cells/ml*	3.7×10^6

FIGURE 7–5 Scaleup with 5 mm spheres showing good homogeneity in a bed of 25 cm diameter, 50 cm height (40 kg/25 l capacity). Cells were inoculated into the bottom of the bead bed. Values as explained in the legend to Figure 7–3. (I) inoculation direction, (*) bead bed void volume.

beads, a threefold increase over channels made by 3 mm beads (Table 7–6). The more open structure of a 5 mm bed would predictably allow a more homogeneous distribution of the growth medium at lower, less shearing, perfusion rates. It would also reduce channelling effects, especially those caused by infiltration of cell mass between the beads. It is of interest to note that the ceramic matrix of the Opticell Bioreactor has 1 mm² channels (Lydersen et al. 1985).

In conclusion, an optimized fixed bed of solid glass beads is capable of giving higher cell densities than both roller bottles and microcarriers on a surface area basis. However, even though a fixed bed of solid glass is more efficient than a roller bottle on a yield per total volume basis, it cannot match the microcarrier system. It is thus an efficient, but low-process intensity, bioreactor. The main advantage over microcarriers is that it is capable of continuous operation over many months.

7.5 POROUS PACKING MATERIALS FOR HIGH CELL DENSITY CULTURE

Immobilization of anchorage-dependent cells is very straightforward, as indicated by the methods previously described. The task is far more difficult for suspension cells that must be trapped within a suitable matrix, which then allows free perfusion of medium for efficient nutrition of the cells, and mass transfer of oxygen, without washing out the cells. This was initially achieved using the ceramic Opticell system (Berg and Bodeker 1988) but this method does not fulfill the need for unit scaleup. A substrate that will entrap suspension cells will by its very nature have a large surface area and will therefore enable a scale up of process intensity (unit cell density) of anchorage-dependent cells. Thus, the aim of finding a suitable matrix is advantageous for both suspension and attached cells, and may even be a first step toward developing a universal cell system!

Considerable progress has been made in establishing fluidized beds with porous matrix particles (Hayman et al. 1987; Nilsson et al. 1986), but finding a similar substrate for fixed beds is proving more difficult.

7.5.1 Sponge Matrices
Murdin et al. (1987a and 1987b) evaluated a range of bed materials including stainless steel wool, nylon wool, sintered glass discs, and polyester foam. The latter material, cut into 3–5 mm cubes was the most efficient for trapping cells. Good production of antibody was obtained from a 150 ml packed bed, equivalent to that in a 175 cm² flask and 1500 ml airlift fermenter. Scaleup potential is not known, but the limiting factor is going to be adequate medium perfusion rather than physical factors such as foam compression.

A similar material, polyurethane (PU) sponge has also been evaluated. PU sponges have a reticulated macroporous structure with pores ranging in diameter from 100–1000 μm. The pores are only partially interconnected and thus form plentiful pockets in which cells are entrapped. In addition, PU is inert and inexpensive. Details of a 300 ml bioreactor based on PU foam matrices have been published (Lazar et al. 1987 and 1988). The system is a 300 ml culture vessel in which 5 mm cubes of PU sponge are packed. Cells are entrapped in the foam matrix and immobilized while medium is continuously circulated through the bed by an airlift pump in the 700 ml cell-free chamber. Hybridoma cells have been successfully cultured in this system, 95% becoming entrapped within 2 hours, and with a perfusion rate of 1.51/day, antibody levels were in the range of 150–200 μg/ml giving a productivity of 225–300 mg per 1 reactor per day. Critical factors were the airflow rate and medium dilution rate.

This culture technology is suitable for both anchorage-dependent and independent cells, is able to support a high cell density as judged by glucose/oxygen utilization rates and antibody secretion, and is able to run for extended periods of time (over 30 days). Although of undoubted success in currently used small scale systems, its scaleup potential has yet to be proven.

7.5.2 Porous Glass Sphere Reactor

The description of the solid glass bead reactor in sections 7.4.1 to 7.4.3 indicated that despite higher cell growth per unit area than a roller bottle, the glass bead reactor was a low process intensity system compared to microcarriers (Looby and Griffiths 1988). Spheres have a low surface area-to-volume ratio and the fact that 5 mm beads are the most suitable size further decreases the total surface area compared to using beads with a 1–3 mm diameter. In addition, the system is only appropriate for anchorage-dependent cells. For these reasons the fixed bed glass bead reactor has not achieved widespread use in commercial processes despite its many positive features.

One method of increasing process intensity is to substitute the solid beads with a porous, or open matrix bead. This concept is proving very successful for porous microcarriers (Hayman et al. 1987; Nilsson et al. 1986) but gelatin/collagen based beads are not suited to a fixed bed configuration. However, after investigating many candidates, the Siran sintered glass carriers (Schott Glaswerke, FR Germany) have been considered to have considerable potential.

The sintered glass was strong enough to allow packing into a fixed bed, it was available in a range of diameters up to the 5 mm target size, and had a very large surface area per unit volume (74 m²/L) caused by the open porosity construction. A range of pore sizes is available in the range 60–300 μm, giving a pore volume up to 60%. A comparison of the physical characteristics of 5 mm diameter solid and porous spheres is given in Table

7-7 to illustrate the huge increase in surface area and therefore potential process intensity.

The textured surface of the Siran spheres caused by the pores is shown in Figure 7-6A, and the interconnecting open pores of the spheres is shown in Figure 7-7A. Anchorage-dependent cells such as the MRC-5, Vero, and

TABLE 7-7   Characteristics of Solid and Porous Glass Spheres Suitable for Fixed Bed Reactors

Characteristics	Solid	Porous
Average diameter (mm)	5.0	5.0
Total surface area (m²/l)	0.7 (0.5)[1]	74.0
Extrasphere void volume (ml/l)	400.0	400.0
Intrasphere void volume (ml/l)	—	360.0
Total void volume (ml/l)	400.0	760.0
Pore diameter (μm)	—	60–300
Bed channel size[2] (mm², approximately)	1.0	1.0

[1]Actual surface area available due to contact between spheres is given in the parenthesis.
[2]Cross-sectional area of the channel running through the bed caused by the packing of the spheres.
Data from Looby and Griffiths 1988.

A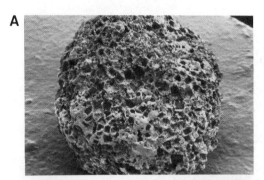

B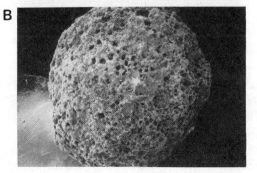

FIGURE 7-6 Scanning electron micrographs of porous glass spheres. (A) Whole sphere ($\times$10). (B) Cells on outside of sphere ($\times$12).

A   B

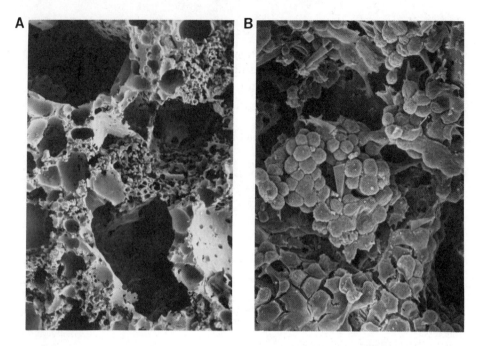

FIGURE 7–7 Scanning electron micrograph of porous glass spheres. (A) Internal structure of sphere ($\times$135). (B) Cells immobilized in the porous structure of the spheres ($\times$470).

GPK attach on both the outside of the sphere (see Figure 7–6B) and into the porous interior (see Figure 7–7B). Initial studies showed that cells penetrated less than 30% of the sphere radius. However, if cells penetrated 1 mm into the sphere then this would represent 80% (60 m^2) of the total surface area available; i.e., total penetration within the sphere is relatively unimportant and demonstrates the advantage of spheres in maintaining a short diffusion path for nutrients.

Better penetration occurs when cells are added to desiccated, rather than moist, spheres (H.W.D. Katinger, personal communication).

Another very big advantage of the porous bead over the solid bead is that suspension cells become entrapped in the system and are thus equally suitable for both anchorage-dependent and suspension cells in a bioreactor of identical design and configuration to that used for solid beads (see Figure 7–1).

The results shown in Table 7–8 are for anchorage-dependent cells (GPK, Vero, and CHO) and show a 10-fold higher cell density than that obtained with solid spheres and also show that higher cell yields can be achieved when cells are inoculated into dry instead of damp bead beds. The OKT3 hybridoma cell line was also grown in this system with less than 5% washout,

TABLE 7–8 Growth of Attached Cells on Glass Spheres

	Cell Yield,[1] $\times$ 10[6]/ml	
	Porous	Solid
Cell line	Fixed bed (5.0 mm)	Fixed bed (5.0 mm)
GPK[2]	13.0	1.3
Vero[2]	14.0	1.5
CHO[3]	20.0	

[1]Cell yields are expressed per unit reactor (fixed bed) volume.
[2]Cells inoculated into damp bead beds.
[3]Cells inoculated into dry bead beds.
Data from Looby and Griffiths 1988 and Looby et al. 1990.

TABLE 7–9 Growth and Monoclonal Antibody (MCAB) Production of Suspension Cells (Hybridomas) in Different Culture Systems

Culture System	Cell Yields,[1] $\times$ 10[6]/ml	MCAB, μg/ml[1]
OKT3 Cells		
Fixed bed (porous sphere)[2]	10.0	225.0
Airlift	1.0	16.0
Stirred	1.3	18.0
CIE3 Cells		
Fixed bed (porous sphere)[3]	40.0	385[4]

[1]Cell and MCAB yields are expressed per unit reactor volume.
[2]Cells inoculated into damp bead beds.
[3]Cells inoculated into dry bead beds.
[4]A total of 2.2 g of MCAB was produced from an 800 ml bed operated in repeated feed and harvest mode for 15 days (140 l medium used).
Data from Looby and Griffiths 1988, Racher et al. 1990, and Looby et al. 1990.

achieving an eight- to 10-fold increase in cell density over conventional stirred and airlift reactors (Table 7–9). More recent work with more optimized conditions has demonstrated further improvements in cell yield up to 4.0×10^7/ml for mouse hybridoma (CIE3) Table 7–9. All this work has far been limited to the 1 l scale (Looby and Griffiths 1988 and 1989; Racher et al. 1990) but it should allow scaleup in size analogous to that obtainable with solid spheres. Thus, a 100 l bed volume, has the potential to support 4×10^{12} cells.

There has often been conjecture about whether it would be possible to have a universal culture system. However, due to differences between cells (suspension and attached), growth characteristics and product generation kinetics (growth- or nongrowth-associated secretion) was never thought to be a realistic goal. This opinion may have to be revised in view of the enormous potential of porous particles in both fluidized and fixed beds.

7.6 CONCLUSION

Fixed beds of immobilized cells have a long history of use both for mass cell culture and the study of cells in a three-dimensional configuration. The fact that it has not been widely used for commercial production processes is due to a combination of factors. These include development of the more efficient microcarrier system, intrinsic problems with gradients and low unit productivity, and limited use with suspension cells. These disadvantages have outweighed the advantages of process simplicity, a cheap reusable bioreactor system and easy scale-up. However, with the availability of porous glass spheres, this method now has high potential value. The use of porous spheres increases unit productivity by at least 10- to 20-fold, makes the system suitable for both suspension and anchorage-dependent cells, and retains the volumetric scaleup potential to at least 200 l. It can be considered a universal system with potential for both volumetric and process intensity scaleup.

REFERENCES

Atkinson, B., and Mavituna, F. (1983) *Biochemical Engineering and Biotechnology Handbook*, pp. 579–669. MacMillan Publishers Ltd., England.

Berg, G.J., and Bodeker, B.G.D. (1988) in *Animal Cell Biotechnology* Vol. 3 (Spier, R.E., and Griffiths, J.B., eds.), pp. 321–335, Academic Press Ltd., London.

Brown, P.C., Costello, M.A.C., Oakley, R., and Lewis, J.L. (1985) in *Large-Scale Mammalian Cell Culture* (Feder, J., and Tolbert, W.R., eds.) pp. 59–78, Academic Press, Inc., Orlando, FL.

Brown, P.C., Figueroa, C., Costello, M.A.C., Oakley, R., and Maciukas, S.M. (1988) in *Animal Cell Biotechnology* Vol. 3 (Spier, R.E., and Griffiths, J.B., eds.), pp. 251–262, Academic Press Ltd., London.

Burbidge, C. (1980) *Develop. Biol. Stand.* 46, 169–172.

Burbidge, C., and Dacey, I.K. (1984) *Develop. Biol. Stand.* 55, 255–259.

Corbeil, M., Trudel, M., and Payment, P. (1979) *J. Clin. Microbiol.* 10, 91–95.

Denac, M., and Dunn, I.J. (1988) *Biotechnol. Bioeng.* 32, 159–173.

Dugre, R., Corbeil, M., and Boulay, M. (1987) in *Modern Approaches to Animal Cell Technology* (Spier, R.E., and Griffiths, J.B., eds.), pp. 579–586, Butterworths, Guildford, England.

Earle, W.R., Bryant, J.C., and Schilling, E.J. (1953/4) *Ann. N.Y. Acad. Sci.* 58, 1000–1011.

Earle, W.R., Evans, V.J., and Schilling, E.L. (1950) *J. Natl. Cancer Inst.* 10, 943–967.

Earle, W.R., Evans, V.J., Sanford, K.K., Shannon, J.E., Jr., and Waltz, H.K. (1951a) *J. Natl. Cancer Inst.* 12, 563–567.

Earle, W.R., Schilling, E.L., and Shannon, J.E. (1951b) *J. Natl. Cancer Inst.* 12, 179–193.

Evans, V.J., and Earle, W.R. (1947) *J. Natl. Cancer Inst.* 8, 103–119.

Familletti, P.C., and Fredericks, J.E. (1988) *Bio/Technology* 6, 41–44.

Feder, J., and Tolbert, W.R. (1983) *Sci. Am.* 248, 24–31.

Gey, G.O. (1933) *Am. J. Cancer* 17, 752–758.

Girard, H.C., Sutcu, M., Erdem, H., and Gurhan, I. (1980) *Biotechnol. Bioengr.* 22, 477–493.

Griffiths, J.B. (1988) in *Animal Cell Biotechnology* Vol. 3 (Spier, R.E., and Griffiths, J.B., eds.), pp. 179–220, Academic Press Ltd., London.

Griffiths, J.B., Cameron, D.R., and Looby, D. (1987a) in *Plant and Animal Cells: Process Possibilities* (Webb, C., and Mavituna, F., eds.), pp. 149–161, Ellis Horwood, Chichester, England.

Griffiths, J.B., Cameron, D.R., and Looby, D. (1987b) *Develop. Biol. Stand.* 66, 331–338.

Griffiths, J.B., Thornton, B., and McEntee, I. (1982) *Develop. Biol. Stand.* 50, 103–110.

Harms, E., and Wendenburg, J. (1978) *Cytobiologie* 18, 67–75.

Hayman, E.G., Ray, N.G., and Runstadler, P.W., Jr. (1987) in *Bioreactors and Biotransformations* (Moody, G.W., and Baker, P.B, eds.), pp. 132–140. Elsevier, London and New York.

Heinrich, M., and Rehm, H.J. (1981) *Eur. J. Appl. Microbiol. Biotechnol.* 11, 139–145.

House, W., Sheaner, M., and Maroudas, N.G. (1972) *Expt. Cell Res.* 71, 293–296.

Knazek, R.A., Guillino, P.M., Kohler, P.O., and Dedrick, R.L. (1972) *Science* 178, 65–67.

Kolot, F. (1981a) *Process Biochem.* Aug./Sept., 2–9.

Kolot, F. (1981b) *Process Biochem.* Oct./Nov., 30–46.

Ku, K., Kuo, M.J., Delenti, J., Wildi, B.S., and Feder, J. (1981) *Biotechnol. Bioeng.* 23, 79–95.

Lazar, A., Silberstein, L., Mizrahi, A., and Reuveny, S. (1988) *Cytotechnology* 1, 333–338.

Lazar, A., Reuveny, S., Mizrahi, A., et al. (1987) in *Modern Approaches to Animal Cell Technology* (Spier, R.E., and Griffiths, J.B., eds.), pp. 437–448, Butterworth & Co. Ltd., Guildford, England.

Leighton, J. (1951) *J. Natl. Cancer Inst.* 12, 545–561.

Looby, D., and Griffiths, J.B. (1987) in *Modern Approaches to Animal Cell Technology* (Spier, R.E. and Griffiths, J.B., eds.), pp. 342–352, Butterworths, Guildford, England.

Looby, D., and Griffiths, J.B. (1988) *Cytotechnology* 1, 339–346.

Looby, D., and Griffiths, J.B. (1989) in *Advances in Animal Cell Biology and Technology for Bioprocesses* (Spier, R.E., and Griffiths, J.B., eds.), pp. 342–352, Butterworths, Guildford, England.

Looby, D., Racher, A.J., Griffiths, J.B., and Dowsett, A.B. (1990) in *Physiology of Immobilized Cells* (De Bont, J.A.M., Visser, J., Mattiason, B., and Tramper, J., eds.), pp. 255–264, Elsevier, Amsterdam.

Lydersen, B., Putnam, J., Bognar, E., et al. (1985) in *Large Scale Mammalian Cell Culture* (Feder, J. and Tolbert, W., eds.), pp. 39–58, Academic Press Inc., Orlando, Fl.

Marcipar, A., Henno, P., Lentwojt, E., Roseto, A., and Broun, G. (1983) *Ann. N.Y. Acad. Sci.* 413, 416–420.

McCoy, T.A., Whittle, W., and Conway, E. (1962) *Proc. Soc. Exp. Biol. Med.* 109, 235–237.

Merk, W.A.M. (1982) *Develop. Biol. Stand.* 50, 137–140.

Messing, R.A., and Opperman, R.A. (1979) *Biotechnol. Bioeng.* 21, 49–58.

Molin, O., and Heden, C.G. (1969) *Prog. Immunobiol. Stand.* 3, 106–110.

Murdin, A.D., Thorpe, J.S., and Spier, R.E. (1987a) in *Modern Approaches to Animal Cell Technology* (Spier, R.E. and Griffiths, J.B., eds.), pp. 420–436, Butterworths, Guildford, England.

Murdin, A.D., Thorpe, J.S., Kirkby, N., Groves, D.J., and Spier, R.E. (1987b) in *Bioreactors and Biotransformations* (Moody, G.W. and Baker, P.B., eds.), pp. 99–110, Elsevier, London.

Nilsson, K., Buzsaky, F., and Mosbach, K. (1986) *Bio/Technology* 4, 989–990.

Nunez, M.J., and Lema, J.M. (1987) *Enzyme Microbiol. Technol.* 9, 642–651.

Payne, G.F., Shuler, M.L., and Brodelius, P. (1987) in *Large Scale Cell Culture Technology* (Lydersen, B.K., ed.), pp. 193–229, Hanser Publishers, Munich, Germany.

Racher, A.J., Looby, D., and Griffiths, J.B. (1990) *J. Biotech.* 15, 129–146.

Robineaux, R., Lorans, G., and Beare d'Augeres, C. (1970) *Rev. Eur. Etudes Clin. Biol.* 15, 1066–1071.

Robinson J.H., Butlin, P.M., and Imrie, R.C. (1980) *Develop. Biol. Stand.* 46, 173–181.

Rudiger, H.W. (1975) in *Methods in Cell Biology* Vol. 9 (Prescott, D.M., ed.), pp. 12–23, Academic Press Inc., New York.

Santero, G.G. (1972) *Biotechnol. Bioeng.* 14, 753–775.

Scheirer, W. (1988) in *Animal Cell Biotechnology* Vol. 3 (Spier, R.E., and Griffiths, J.B., eds.), pp. 263–281, Academic Press Ltd., London.

Schleicher, J.B., and Weiss, R.E. (1968) *Biotechnol. Bioeng.* 10, 617–624.

Spier, R.E. (1985) in *Animal Cell Biotechnology* Vol. 1 (Spier, R.E., and Griffiths, J.B., eds.), pp. 243–263, Academic Press Inc., London.

Spier, R.E., and Whiteside, J.P. (1976) *Biotechnol. Bioeng.* 18, 649–657.

Tharakan, J.P., and Chau, P.C. (1986a) *Biotechnol. Bioeng.* 28, 329–342.

Tharakan, J.P., and Chau, P.C. (1986b) *Biotechnol. Bioeng.* 28, 1064–1071.

Tolbert, W.R., Srigley, W.R., and Prior, C.P. (1988) in *Animal Cell Biotechnology* Vol. 3 (Spier, R.E., and Griffiths, J.B., eds.), pp. 373–393, Academic Press Ltd., London.

Tyo, M.A., Bulbulian, B.J., Zaspel, B., and Murphy, T.J. (1988) in *Animal Cell Biotechnology* Vol. 3 (Spier, R.E., and Griffiths, J.B. eds.), pp. 357–371, Academic Press Ltd., London.

van Brunt, J. (1986) *Bio/Technology* 4, 505–510.

van Wezel, A.L. (1967) *Nature* 216, 64–65.

Vega, J.L., Clausen, E.C. and Gaddy, J.L. (1988) *Enzyme Microbiol. Technol.* 10, 390–402.

Whiteside, J.P., and Spier, R.E. (1981) *Biotechnol. Bioeng.* 23, 551–565.

Whiteside, J.P., and Spier, R.E. (1985) *Develop. Biol. Stand.* 60, 305–311.

Whiteside, J.P., Whiting, B.R., and Spier, R.E. (1980) *Develop. Biol. Stand.* 46, 187–189.

Whiteside, J.P., Whiting, B.R., and Spier, R.E. (1979) *Develop. Biol. Stand.* 42, 113–119.

Wohler, W., Rudiger, H.W., and Passarge, E. (1972) *Exp. Cell Res.* 74, 571–573.

Microcarriers for Animal Cell Biotechnology: An Unfulfilled Potential

R.E. Spier
N. Maroudas

8.1 TIMES PAST

Toon van Wezel (1935–1986) was an engineer. He was charged with making polio vaccines for the Dutch health authorities. The polio virus was grown in Green Monkey kidney cells, which in turn were grown in roller bottles. Scale up of this multiple process was by increasing the number of bottles. As an engineer, van Wezel would have been appalled by such a requirement. His and the thoughts of his colleague van Hemert turned to consider the alternatives. They would have known about the column systems of Earle and McCoy, (Spier 1980), and would have inferred that the conditions at the top of the column would be different from those at the bottom. To obviate such an inhomogeneity and to provide an equivalent environment for all the cells in the bioreactor, it would be necessary to mix the system of surfaces to attain homogeneity. This was achieved initially with the Diethylaminoethyl (DEAE) Sephadex A50 particle, which had been originally designed as a column packing (Van Wezel 1967). An unnamed virus (probably polio) was grown on the cells on the microcarriers.

The attractiveness of the method was obvious. However, when one of the authors of this chapter (R.E.S.) attempted to coat the microcarriers with nitrocellulose in 1970 and then grow an avian embryo culture, the results were disappointing. Nevertheless, on returning to the United Kingdom in 1973, his first thoughts were to achieve success with the method. So like many others before and after, he made the trip to Bilthoven and was well received by van Wezel, who explained the practical aspects of the technique, which eventually led the author to publish a paper detailing how the beads could be coated with serum and made to support the growth of BHK21 C13 cells for the production of FMDV (Spier and Whiteside 1976); a process implemented by I.F.F.A. Merieux (Meignier et al. 1980), who also used it for polio virus production (Montagnon et al. 1984). This was but a beginning. It was clear that the technique was sensitive and the conditions of the culture had to be carefully defined and implemented for success. The question that was always on van Wezel's lips was, "Have you got a better microcarrier he could try?"

One early response to van Wezel's question was to lead to polystyrene microcarriers and to important developments in the surface treatment of such materials, which would make them compatible for cell attachment and growth. The other author (N.M.) wrote (Maroudas 1973a) that, "In preliminary experiments, polystyrene beads were . . . treated with the spark discharge. They readily supported growth." This was the first published report of the use of an electrical corona discharge for the treatment of plastic for the purposes of cell culture. Since that time several brands of polystyrene microcarrier beads have become available. It is of interest that the polystyrene surface is negatively charged, which is in contrast to the positive charge on the sephadex beads used by van Wezel, (see section 8.3.3 for a further discussion).

Later work at the Massachusetts Institute of Technology by Levine et al. (1979), who developed a dextran microcarrier with a lower charge density, paved the way for more reliable operations with microcarrier cultures. From this time there have been many further developments of different microcarriers, and the number of cell types that have been grown on their surfaces has become legion (for reviews, see van Wezel 1985; details of bioreactor techniques, see Butler 1988 for details of the different types of microcarriers, see Nilsson 1988 for an exposition of cells that have been grown on microcarriers and the effects of surface charge density on different cell systems).

The events related in these reviews have amply justified van Wezel's engineering vision, and large-scale microcarrier cultures are now commonplace.

But cultures based on growing cells on the surface of smooth spheres have their limitations. A swollen bead diameter of 200 μm gives a volume of 4×10^{-6} ml/bead, or assuming a 40% void space in randomly packed spheres, a maximum of some 150,000 carriers per milliliter. From a bead surface area of 1.2×10^{-3} cm^2 and a cell density of 1×10^5 cells/cm^2, about

2×10^7 cells can be accomodated in a volume of 1 ml. This is about 100 times lower than the number of cells, (diameter, 10–15 μm), which can be made to fill the same space (2×10^9). It is because of this low (1%) utilization of available space that, in our view, microcarriers have yet to fulfill their true potential for high density cell culture. Even the figure of 1% utilization is based on a hypothetical bed of packed spheres (60% solids). But practical microcarrier cultures are normally run at a much lower volume ratio of beads to medium, about 3%. The higher solid contents lead to difficulties in mixing, viscosity, and aggregation, while the alternative of running the system as a packed bed results in liquid channeling and bed blocking. (It is unlikely that bead-bead collisions, which occur with greater frequency at higher bead concentrations, result in cell damage since bead-bead contact frequently results in the beads being held together by the cells.) The conventional use of low density of beads means that the bioreactor volume is only used to 0.1% of its potential, which indicates that there is untapped potential for the development of improved microcarrier techniques. A second restriction is that cells that have become adapted to the surface-independent mode of growth do not grow on the surfaces of the conventional sephadex microcarriers. This calls for the next development in microcarrier techniques; three-dimensional microcarriers.

We use the term "three-dimensional microcarriers" to indicate carriers that not only contain cells on the outer surfaces of the particles but cells are also present within the body of the particle. This can be done by either of two ways: first by encapsulating the cells inside a microporous gel (this only works for cells in which growth is not anchorage dependent, such as hybridoma cells); and second, a carrier is needed for cells in which growth or product synthesis depends on cell attachment and spreading over an extended surface. For the latter the microcarrier is made "macroporous" by folding or convoluting the sphere to produce a carrier particle with increased internal surface and internal cavities of about 50–150 μm in diameter. The classical macroporous microcarrier is the gelatin emulsion bead described by Nilsson (1988).

8.2 THREE-DIMENSIONAL MICROCARRIERS

Virus vaccines are administered at levels of 1–10 μg/dose. By contrast, a therapeutic material, such as MoAb or enzyme, could be used at levels of milligrams or grams per dose. The need to establish high intensity, high product-volume generating cultures of animal cells is thus made evident. Such cultures become practicable when highly productive animal cells can be held at high concentrations for extended times. Three-dimensional microcarriers offer this potential but concurrently present a new set of problems to the animal cell culturist.

Cells need to be made productive at concentrations similar to those that exist in tissues in the body. How, then, does the biology of the cells change in comparison to the conventional type of cell culture operation that holds cells at 1/100–1/1000 of the concentration found in the body?

Although the dominant philosophy behind the development of microcarriers has been to provide all the cells of the culture with as identical a microenvironment as practicable, the implementation of high cell concentration cultures will, of necessity, preclude the establishment of such a condition. What are the implications of the deliberate creation of such inhomogeneities?

In a three dimensional carrier, more cells come in contact with more surface than they do with the conventional spherical bead. It is also clear that the way the cell performs is crucially dependent on how it interacts with the contiguous surfaces (Emerman and Bissel 1988; Edelman 1985; Maroudas 1972 and 1973b). Of what materials should such carriers be made? How should their surfaces be treated and with what materials should they be coated?

The effects of anchorage, i.e., of surface attachment and spreading, is worth deeper investigation than has been applied so far. Nilsson (1988) states that "even suspension-type cells, for instance, hybridomas, tend to give increased product yields when attached." One of the authors (N.G.M.) found that 99.9% of the primary cells sampled from a freshly dissociated whole mouse embryo were anchorage dependent for growth (Maroudas and Schmitt 1973). Penman and Ben Ze'ev found that some protein synthesis was stimulated by attachment alone whereas DNA synthesis required both attachment and spreading (Ben Ze'ev et al. 1980). These examples suggest that anchorage dependence is a powerful mechanism that should be further studied for its probable role in the stimulation of cell growth and the production of differentiated products.

Carriers can be held in either static packed beds through which nutrient fluids are pumped or they can be fluidized as particulate suspensions. What is the optimum configuration that will achieve lasting cultures with the fewest technical problems?

It is often necessary to recover the cells from a cell culture to provide a seed culture for the next level of scaleup. How could this be achieved for a three-dimensional system?

How are such cultures to be controlled and optimized? What parameters should be monitored and how should they be related to achieve performance? How are process models to be used to improve process performance?

How can such cultures be presented to the licensing authorities in a way in which they can be convinced that the culture is under control, that known reactions are proceeding according to plan, and that the product produced is homogeneous and does not contain materials likely to damage health?

How can bioreactor systems be made that are compatible with the three-dimensional carriers and that combine the characteristics of simplicity, robustness, flexibility, and availability? (The latter requirement includes the freedom from the need to license patented materials or procedures and to be independent of any particular supplier for a component of the system.)

8.3 THE PROBLEMS EXPOUNDED

It should be recognized at the outset that the definition of the problems that exist in the area of microcarrier culture serves to provide a framework on which work, which will lead to advances in knowledge and capability, can be focused. In some of the areas there is both a measure of understanding and existing practice, yet it is also clear that the dominant technology of the stirred tank, be it an impeller or gas bubbles, yet holds sway. Twenty years of successful operation cannot be passed off at a stroke. Like the implementation of any new technique, it will be the most modern applications that will be in the forefront in the efforts to introduce a new system. It will only be from success in these systems that traditional operations will come under scrutiny. But such an achievement can only come about if progress is made on the resolution and analysis of the situation that pertains when animal cells at high concentrations are held in culture for extended times. This recreation of the in vivo situation in vitro is not only the objective of the animal cell culturist of old. It is also the primary goal of researchers who wish to run productive systems using cellular material that has a long generation time, and which has a biosynthetic potential that has not yet come up to the rates attributed to prokaryotic systems.

8.3.1 Changes in Cell Biology

When cells are held at high local cell concentrations, many aspects of their biology changes. It is well known that cells do not grow well unless they are seeded into the growth system above a minimum cell concentration (Rein and Rubin 1968; Rubin and Rein, 1967). The implications of such an effect for the design of the nutrient medium was explored by Eagle and his colleagues (Eagle and Piez 1962; Eagle 1965). These workers discovered that as the seeding cell concentration increased, it was possible to lower the concentration of certain amino acids in the medium formulation. The implication of these studies was that cells had to be bathed in a minimum concentration of a metabolite in order to thrive. Such a concentration was dependent on both the exogenous source and the rate at which the metabolite could be synthesized by the cell and secreted into the surrounding medium to maintain that minimum concentration. More recently, many discoveries of materials have been made that influence the properties of both cells local to the secreting cell (paracrinal action) and on the secreting cell itself (autocrinal action). Such materials are often the products of cellular versions of oncogenes, and they exert significant effects on the rate at which cells grow and differentiate (reviewed by Spier 1988).

In the case of three-dimensional microcarriers, it is important to recognize that the high local cell concentration can be vitiated by a low concentration of microcarriers in the culture. This effect could also be achieved in perfusion cultures in which the rate of change of medium was the factor that controlled the concentration of metabolites in the medium. Additionally, one would have to consider the microenvironment that would exist within each particle. The concentration of materials that readily diffuse would be subject to modification, yet other, higher molecular weight materials, would be less subject to such effects. Also, one can expect effects due to cell-cell contacts, although the significance of such interactions has not yet been determined. There exists in the microcosm of the three-dimensional microcarrier a situation that lies between what occurs in the tissues of the body and what prevails in cell cultures. New approaches are needed to investigate this, and novel techniques will enable us to use such information to advantage.

8.3.2   Gradients and Their Implications

Within the confines of a three-dimensional mass of cells, each molecular species will move according to its concentration gradient and to the energy the cell expends in controlling that gradient. Clearly, cells that do not acquire nutrients due to their position within the cell mass will not thrive, nor will they secrete materials of biotechnological value. The question that arises from such a consideration is the delineation of the maximum distance a cell can be from a nutrient source and still perform its necessary functions. A corollary to this question is if a cell is deprived of nutrient and dies, what is the effect of its death on the neighboring cells?

Oxygen is a material whose concentration in cell culture fluids is determined by its solubility coefficient and its partial pressure in the gas phase in contact with the liquid phase. Animal cells in culture seem to require exogenous supplies of oxygen, although the amount they need varies with the extent of their metabolic activity. Oxygen requirement is also cell-type dependent (Spier and Griffiths 1983). While the cells remove oxygen from the liquid phase, the rate at which oxygen can be supplied to the respiring cell mass can become the limiting factor determining the size (diameter) of the cell mass. There are a number of ways in which size can be predicted.

By a histological examination of the tissues of the body, it is possible to determine the relation between the distance between adjacent capillaries and the number of cell diameters that are interposed. By examining lymph tissues, for example, it is possible to discern that the number of cells between adjacent blood vessels will likely be in the region of 10 to 20 (Leeson et al. 1985).

By calculating the area of all the capillaries in the body it is possible to determine the height of the cell layer that would result if all the cells of the body were spread uniformly over the area, which is defined by the surface area of the capillaries. The results of one such calculation yielded a value of 21 cell diameters (256 μm) (McCullough and Spier 1990a).

By constructing a model of a cell mass based on layers of cells situated on the top of a volume of medium, it is possible to calculate the thickness of a cell sheet that can be oxygenated. From such a model it can be shown that 1 ml of oxygenated medium (7 ppm oxygen) can support 16,000 layers of cells for 1 sec, provided that the rate of oxygen diffusing through the cells is not taken into account. When the rate of diffusion is taken into account in conjunction with the assimilation of oxygen, then a value of about 30 layers of cells can be calculated by using a simple numerical method (McCullough and Spier 1990b).

It was shown by an analytical method that a result identical to that derived by the numerical approximation could be achieved (Murdin et al. 1987).

The identification of actively metabolizing cells held in an agarose gel was found to determine the distance that active cells could be maintained below the gel/air interface. The results from this empirical test correlated well with those obtained from analytical methods (Wilson and Spier 1988).

From the above methods and calculations it may be concluded that cells can remain viable when they are not more than about 30 cell diameters away from a source of dissolved oxygen whose concentration is defined as being in equilibrium with air at normal temperature and pressure (NTP) as the gas phase. It can be inferred from this that a three-dimensional particle of less than 0.6 mm diameter would provide an environment that would contain adequate amounts of oxygen for all the cells in the particle. A corollary to this is that if the particle had a diameter larger than 0.6 mm, there would likely be a necrotic zone in the center of the particle whose effects would need to be defined. Clearly, a particle that only allowed cells access to the space within 0.3 mm of its surface would also provide a practicable environment for cell culture.

Other nutrients would be expected to have their own concentration gradients, though this can be expected since a nutrient concentration in a culture fluid that has a 5 mM concentration and a molecular weight of 300 would be present at 1,500 ppm (about 200 times more concentrated than oxygen). Again, the cells at the outside of the particle would be exposed to both the cell growth activators and inhibitors in the culture fluid, while the cells embedded in the particle would be more likely to experience only the growth factors made by the cells around them (including the ones they make themselves). In this sense, a gradient effect has turned into a qualitative difference whose consequences need to be determined.

Gradients due to the consumption of polypeptide growth factors are probably even more crucial than those due to consumption of oxygen, because the concentrations of these hormones are so low relative to their rates of consumption. Following the work of Rubin and Rein (1967), a theory of growth limitation through "short-range diffusion gradients" of consumable polypeptide growth factors was enunciated (Maroudas 1974). (These factors were later to be known as the insulin-like growth hormones.) The equations of Fick and Nernst are in the correct form for steady-state diffusion with consumption and replace the equation of Rubin and Rein, which does not allow for consumption. Further calculations based on the latest available data from the literature (Maroudas and Fuchs 1987) for growth factor size, concentration, and rate of receptor internalization suggest a remarkably steep gradient of hormone concentration over a distance of 60 μm. Experimental measurements of activated nuclei in mice revealed that gradients of nuclear activity existed in the neighborhood of capillaries; moreover, these gradients were even shorter than had been calculated; about 20 μm as opposed to 60 μm. These ultrashort range gradients of only one or two cell diameters refer to growth-stimulated cells with visibly activated nuclei (Maroudas and Wray 1985). They are an order of magnitude steeper than those of the 256 μm gradients described above for oxygen. However, in addition to the consumption of growth factors, it is also necessary to take into account the local cellular production of the self-same materials as a result of the higher local cell concentrations. It is possible that such effects balance out. Evidently, additional experimental data are needed to resolve these questions.

It could be that processes leading to a more differentiated cell are switched on when particular gradients are established. In this case secondary metabolism (Spier and Bushell 1990) may ensue with enhanced commitment. It follows that the control of such gradients could be of significance in attempts to obtain maximized productivity from cells held in three-dimensional microcarriers.

8.3.3 Cell-Surface Interactions

It is something of a paradox that the two most widely used microcarrier beads have surface charges of opposite sign (DEAE sephadex A50, being positively charged and polystyrene being negatively charged). Gelatin beads (Nilsson 1988) are either slightly negative or slightly positive, depending upon the isoelectric point of the material used and the method of preparation. Thus, at first sight it appears that surface charge does not correlate with cell adhesion. However, a few simple and reliable rules have been established empirically. First, actual cell culture dishes (tissue culture grade) usually have surface energy of 56 ergs/cm^2 compared to 72 ergs/cm^2 for water (Maroudas 1973a and 1975a). Now, most plastics in their native state have surface energies of 31 to 43 ergs/cm^2, and these surfaces usually are not acceptable for cell adhesion. Yet if such surfaces are treated by the corona discharge process, their surface energy rises to 56 ergs/cm^2 or higher and cell adhesion rises. This indicates that cell adhesion depends primarily on surface energy and the high surface energy of glass and ceramics explains their performance in cell attachment studies.

The unusually high surface energy of tissue culture polystyrene has prompted one of the authors (N.G.M.) to speculate that the important chemical groups introduced by the tissue culture treatment must be carboxylic acid groups, which explains the negative charge. Likewise, the introduction of sulphonic acid groups into the polystyrene surface will also markedly increase cell adhesion and spreading (Maroudas 1975b and 1977a; Spier 1980). Additionally, the hypothesis that the corona discharge leads to the development of carboxylic groups in the polystyrene was confirmed by meticulous work at the research laboratory of the Corning Tissue Culture Company (Ramsey et al. 1984). The existence of surface sulphonic acid groups and the absence of surface hydroxyl groups has also been confirmed (Maroudas 1976; Addadi et al. 1987).

The attachment of a negatively charged cell to a negatively charged surface runs counter to electrostatic theory. However, if the interaction is due to Lewis acids then the short range electron donor-acceptor interactions, as in hydrogen bonds, could lead to cell attachment. In the case of the uncharged dextran material, the Flory polymer exclusion principle (Maroudas 1975a, 1975b, 1975c, and 1979) (where hydrophilic polymers have a low mutual interaction because of an entropic phenomenon called "polymer exclusion") requires that the polymers be modified by a charged group before they become acceptable for cell attachment. By contrast with the charged surfaces, the neutral gelatin beads seem to require attachment factors such as fibronectin (Nilsson 1988), although collagen beads do not require this material if the collagen is in an undenatured form (denatured collagen does require fibronectin to achieve cell adhesion) (Maroudas 1977b). Although cells in vivo are surrounded by an extracellular matrix of polymers, it is not possible to reproduce this environment on a large scale due to the cost. Thus, reproducible factor-free systems are likely to become the dominant technique.

The further investigation of the relationship between the nature of the microcarrier surface and attaching cells can be effected in a system that measures the surface forces, which limit cell attachment and cause cell detachment (Crouch et al. 1985).

8.3.4 Bioreactor Configuration

Three-dimensional carriers can be arrayed in either of two basic configurations. One would be based on a packed static bed while the other would be fluidized. The advantages and disadvantages are set out in the lists below.

8.3.4.1 Advantages of a Packed Bed

1. Hydrodynamics is the minimum to ensure the provision of adequate supplies of nutrients.
2. High efficiency consumption of nutrients.
3. Freedom from back mixing.
4. Attrition of particles minimized.
5. Seeding of particles with cells is simple and not dependent on the homogeneous distribution of the particles.
6. The separation of carriers and fluids is simple.
7. Apparatus is generally simple and robust.
8. The carrier particles may be sterilized in situ.
9. It is possible to use a wider range of materials for the particles since they do not have to be fluidized.
10. The system is fail safe if there are small changes in the fluid flow rates.
11. The relatively low flow rates allow cells to use the outer extremities of the particles.
12. Flexibility in particle diameter can be used to keep channels between particles open.
13. Since all the particles are held in part of the bioreactor, it is possible to obtain higher local cell concentrations in the cell-containing section of the bioreactor system.
14. It is less necessary to add chemicals to minimize the effects of shear.
15. Foam control can be effected off line.

8.3.4.2 Advantages of a Fluidized System

1. Gradients do not occur across (or down) the reactor.
2. The system is homogeneous and therefore easier to monitor and control.
3. Higher mass transfer rates can be expected between the bulk fluids and the particles as a result of greater hydrodynamic activity.
4. Scaleup can be achieved without increasing concentration gradients.
5. It is not possible for cells to block the channels between the particles, which could create further inhomogeneities.
6. A sample of the bulk of the reactor represents the reactor's contents.
7. As the scale increases, the advantages are made more apparent.

On balance it would appear that at smaller scales of operation it could be more advantageous to use a packed bed system, while for larger scales it becomes worthwhile to solve some of the technical problems inherent in the fluidized method.

8.3.5 Cell Recovery

In cases when the microcarrier is composed of material that can be enzymically hydrolized without damaging the cells, the options are open and clear. Both dextran and collagen can be hydrolized, thereby freeing the contained cells in the process (Butler 1988). Materials such as porous glass, polystyrene, sintered metals, or ceramics do not lend themselves to such procedures. It may be possible to use a combination of enzymic and vibrational techniques similar to those used by Spier and co-workers to liberate cells from bioreactors based on packed beds of glass spheres (Spier et al. 1977). Other techniques based on exhaustively washing the beds with calcium sequestering agents (ethylene glycol-bis(β-aminoethyl ether)) (EGTA) could also be used alone or in conjunction with other reagents. In the event that insufficient cells could be realized by these or other techniques, it would still be practicable to use an alternative system, such as a cytogenerator, specifically for the final stage bioreactor.

8.3.6 Control Strategies

Control begins with the quality assurance assays on all the raw materials that are to be introduced into the bioprocess. This would include the cell inoculum, the medium, any complex mixture of biochemicals, and, in the case of microcarriers, the quality of the particular batch of material provided. Three assurances are required. The material must be free from contaminating microorganisms; the material must be suitable for the task allotted, and the composition of the material has to be what is alleged to be.

The next level of control occurs at the bioreactor. First, physicochemical controls are now the norm, and the parameters of temperature, pH, and dissolved oxygen are routinely monitored and controlled in most bioreactor systems. Off-line measurements of biomass, product, and nutrients and/or metabolites such as glucose, lactate, glutamine, ammonia, and the composition of the gas phase for oxygen and carbon dioxide are also much in evidence. The relationship between these parameters has been achieved through a number of model systems. Some, based on lumped parameters, which regard the cell as a black box and consider the raw materials and products only, are derived from Michaelis-Menton kinetics (Miller et al. 1988), while others are based on a structured segregated approach and consider the cell, its position in the cell cycle, and describe the condition of the culture as the sum of the conditions of all the cells of that culture (Faraday et al. 1989). Although such models can represent bioreactor runs that have taken place (most of the parameters used in the model have been derived from the self-same runs), they are not generally adequate for predicting future culture runs. For this reason they have not been used extensively to optimize the parameters that they contain.

There is little doubt that new techniques in measurement such as online high pressure liquid chromatography (HPLC), Fourier analysis of infrared absorption spectra, and fluorescent spectrum analysis will yield valuable data on which to base further and more adequate models from which a sounder base for predicting and optimizing strategies will emerge. In addition to these measurements, it will also be necessary to determine the controlling parameters involved in controlling the physiological profile of cells that are held in concentrated masses within the pores of a three-dimensional microcarriers. One can expect that there will be discovered key chemicals whose concentration determines the quality of the metabolism under such conditions. From such discoveries more detailed models will be able to serve the investigator to greater effect than at present.

Artificial intelligence will play a role in the control strategies. Given models whose predictive capability can be relied upon, the application of intelligent machine-operated strategies will provide commercially efficient processes of consistent performance. It will also be possible to present the licensing authorities with the kind of information about the process that will provide them the degree of comfort they seek before they will accept a particular process operation. The situation wherein artificial intelligence can be brought to run a process means de facto that the process is under control and that the quality of the product is thereby most closely defined.

The establishment of increasingly stringent control procedures betokens the transfer of a subject area from that of a black art to that of an engineering discipline; we are clearly not there yet, but continued progress will achieve the level of understanding and capability that will make this objective a reality.

8.3.7   Licensing Products from Systems Using Three-Dimensional Microcarriers

The product manufacturer has to convince the licensing authorities that the product is safe, efficacious, and consistent. The product is deemed to be safe when it has been shown to have less than 100 pg DNA/dose and to be innocuous in a variety of safety tests with animals and cell cultures. Efficacy is determined in relation to materials that are already available. Within the definition is a component that relates the properties of the putative product with its potential benefit to society. It is in the area of consistency of the production process that we have to consider the particular case of the three-dimensional microcarrier. Rules that delineate consistency have not been laid down with finality. In culture systems that involve bottles, observations on the quality of the cell sheet in the bottle is often taken as an index of consistency of performance. With systems in which the cells are not seen directly, it has often proven practicable to operate a parallel culture that can be examined in more detail or even harvested while the production culture is left inviolate. Alternatively, as suggested above, consistency can be inferred if the process to be so defined by a process model and operated on according to that model, thus making a clear signal to the agency that there existed a body of knowledge about the process that was describable in a mathematical (or other) model. Whatever method is chosen or accepted, a debate is necessary to establish the parameters of performance that determines the consistency of operation. Such parameters could be bulk culture observations, which could depend on analyses of particles replete with cells that have been retrieved from the bulk culture or on observations made in situ by, perhaps, fiber optic probes located within the body of a porous particle.

In addition to the above considerations, it will also be necessary to show that the material of the particle does not become incorporated into the product. All materials are soluble to some extent in aqueous salt-containing solvents. Matrices composed of plastic materials can leak out either unreacted monomers or the materials used in plasticizers. Organic polymers, such as collagen, gelatine, or polysaccharides, may contain impurities that could be allerogenic, immunoactive, or oncogenic. Trace elements from metal or ceramic composites should be considered when using more solid support materials and product definitions. This would include an elemental analysis, which would be regarded as necessary for all cases requiring a license application. In general, the tighter the case that can be made in defining the process and the materials in the product, the closer one moves to the issuance of a license to produce.

8.3.8 Bioreactor Design

Bioreactors for both packed bed and fluidized particle options already exist in a variety (over 57) of manifestations (Griffiths 1988). There is a somewhat more complex equipment requirement in the fluidized bed case, which gives rise the possibility of novel flow distribution equipment that, in turn, leads to the possibility of patentable equipment. Such is also the case for the composition of microcarriers. This means that the cell culturist has to address the prospects and the problems that arise as a result of working in an area where there is patentable property, where there is precedence in proprietary procedures, materials, and equipment, and where the prospect of the ownership of intellectual property rights determines the investigations that will and will not proceed.

When a microcarrier becomes a marketable device it can be subject to a patent, which limits the sale of that proprietary material to the patent holder or to the subtending licensee. There is the prospect, however, that materials similar to those on the exclusive market could be made in-house, in which case it would be unlikely that the exact composition, as defined in any patent, would be infringed. A further point is that if a manufacturer became dependent on a particular supplier, then a degree of uncertainty is engendered; uncertainty of the price, the availability, the continuity of supply in the event that commercial events terminate the life of the supplying company, and the prospect of competition for the supply of the basic materials with competing production companies. On the one hand, companies producing the proprietary product must design a material that excels in the application and that would require the purchaser to make an uneconomically large investment in a production system to arrive at a position which they can buy off-the-shelf. On the other hand, the purchaser may consider cheaper in-house alternatives that can be subjected to local control. Indeed, this could be a selling point for the product due to assertions regarding the advantages that could be alleged to accrue in the in-house material, vis a vis, that which would have to have been bought in. The local control could also be advantageous in that it may be necessary to effect small changes to the three-dimensional microcarrier structural materials in order to achieve particular process objectives. The control of the surface properties of the carrier could be crucial to the performance of a particular cell line. Thus, it would be a marked advantage to be able to tailor the carrier for such a unique purpose.

Whatever the nature of the carrier, the equipment in which it is to function should be as simple and robust as is required to achieve the particular process type chosen. The areas of greatest flexibility are in the degree to which the equipment is instrumented and the use to which the data generated by the instruments are applied. The instrumentation and computer adjunct (combining both in hardware and software) can cost considerably more than the vesselry and ancillary piping. It is important to determine the definition of what is expected from the equipment before it is specified since it is possible that any one set of equipment can achieve much more sophisticated activities than was originally foreseen. These latter features could cause considerable waste of time and effort, so it is important to spend time developing a clear specification of the equipment that is to hang off the basic bioreactor container before the latter is purchased.

8.3.9 Review of Problems

Each of the problem areas presents a challenge for the animal cell culturist. The solution to the problem presents opportunities to acquire intellectual property or at least an advantage in the race to bring a product to the market place. None of these problems is insoluble, and when the solutions have been tested and tried the systems that result will present a severe challenge to other microbial systems that presently hold sway by virtue of their purported efficacy.

8.4 CURRENT MANIFESTATIONS OF THREE-DIMENSIONAL MICROCARRIERS

When it becomes difficult to keep up with the literature it is a sign that the field is advancing at breakneck rate. Such is the case with three-dimensional microcarriers. It began with calcium alginate-based (Rosevear and Lambe 1983; Nilsson and Mosbach 1980) systems that were a spin-off from plant cell technology and ethanol production from yeast systems but which at 37 °C and in phosphate buffers were too labile for protracted use. The development of the alternative agarose bead (Nilsson and Mosbach 1987a) showed some promise in the small scale (Nilsson and Mosbach 1987b), but the difficulties inherent in the scaleup and the cumbersome nature of the preparatory procedures made the technique less than suitable (Nilsson et al. 1987). While these two gel based systems were in contention, the development of the hollow sphere system Rupp (1985) showed the potential for growing cells in composite masses. Indeed, it was also shown that the production of clumps of cells could occur without the need for an encasing membrane (Reuveny et al. 1987). Although the latter system may be applicable to a unique cell line handled in a particular way, it could still be possible to determine how to achieve such natural clumps and how to control their size and metabolism for other cell lines as required.

More recent developments involve the emergence of the agarose-agar composite beads and the development of collagen/gelatin bead systems. The latter may be made wholly of polymer or alternatively could be an adjunct between the biopolymer and a weighting particle (Runstadler and Cernek 1988).

8.4.1 The Calcium Alginate System

Calcium alginate is a useful starting point for immobilization. The gel forms under mild conditions of pH and temperature and the materials of the gel do not interfere with the physiological activities of the cell. However, the beads formed are not stable and large molecules such as antibodies do not diffuse out from the gel matrix. The calcium is easily removed from the gel by complexing agents such as phosphates and EDTA and the beads are also susceptible to weakening at 37 °C. They are useful for the entrapment phase and can be coated (see section 8.4.4).

Recently, calcium alginate has been used in a different mode. The cells were suspended in calcium chloride with a viscosity enhancer and the mixture was dropped into alginate. In this way the alginate formed a skin around the viscous drop. The negative charge on the alginate was then neutralized by poly-lysine so that some control of the porosity and permeability of the membrane could be achieved. Such capsules are presently under investigation (Wang 1989).

8.4.2 Agarose

Although the agarose beads did not fall apart at the temperatures normally used, they were difficult to form without incorporating air into them. A further problem was that it was necessary to hold the cells and gel at a temperature in excess of 40 °C during the bead formation stage, so that in large-scale operations the cells could be damaged during the holding period. Also, particular care had to be taken with regard to the source of the oil used in the immobilization coupled with the awkward procedures needed to remove the oil and wash the cell-containing beads once they had formed. These difficulties were amplified when attempts were made to scale up the process. Consequently, this approach remains a laboratory operation.

8.4.3   Agarose-Alginate

Attempts to overcome the limitations of the alginate and agarose systems were made. In this system a composite gel of agarose and alginate was made and the alginate portion of the composite was removed, leaving a porous agarose bead. Carageenan gum can substitute for the agarose in this system. Cell viability may be maintained and the porous particle that results is permeable to blue dextran, molecular weight 2,000,000 Da (Miles and Reading 1988).

8.4.4   Hollow Capsules

The hollow capsule technology has as its chief attraction the prospect that, by choosing the materials of the capsule appropriately, it would be possible to control the ingress and egress of materials from the inner space on the basis of size, charge, or some other definable property. (This would create a mimic of the cell to house the cells!) One possibility that has been explored extensively is that of the poly-lysine-poly arginine variety. This material retains large molecular weight materials and is highly positively charged; it is not generally regarded as a satisfactory route to a flexible biotechnological system.

A second system (see section 8.4.1 above) involves the creation of a membrane of alginate (negatively charged) and poly-lysine (positively charged). Other combinations are those based on chitosan and polyphosphate, chitosan-alginate (Rha and Kim 1989), or chitosan and tripolyphosphate. Fibrin, collagen, or gelatin composites can also be used and for anchorage-dependent cells and fibronectin or laminin can be incorporated into the membrane (Wang 1989).

8.4.5   Collagen and/or Gelatin

Recreation of the in vivo environment in vitro is an apt approach for the cultivation of animal cells ex vivo. Therefore, the use of collagen and/or its degraded derivative, gelatin, is a natural progression from the status quo in vivo. There are at least 10 identifiable collagens, and gelatin is defined as a product on the basis of its viscosity and gel strength when set. Such materials are not found as pure materials in nature, especially in the region of the cells. Additional proteins, glycoproteins, and lipoglycoproteins coexist in the extracellular matrix and a number of such substances have been identified (e.g., fibronectin, laminin, chondroitin sulphate, heparan, chondronectin, epibolin, fetuin, integrin, and heparan sulphate).

Collagen is soluble at the extremes of pH and when made into a particle is isopycnic with water. In a bioreactor such a material does separate from its surrounding medium and so the perfusion of fresh medium through the reactor requires special sieving devices (clogging), filters (blinding), separators (variable performance), or centrifuges (complexity and cost). A simple solution to such a problem is to embed the particle of collagen within a material that increases the density of the whole particle. There are two manifestations of such an approach.

8.4.5.1 Weighted Collagen I There are two aspects to the use of porous collagen particles as the carriers of animal cells in culture. One involves the cross-linking of the collagen to preserve its integrity. This can be achieved by agents such as formaldehyde, glutaraldehyde, and other bifunctional agents. The weighting particles can be made of any material that is not toxic to animal cells and that can be prepared in the form of a powder with particle dimensions of less than 50 μm and a density of four to 10. Such materials as titanium, titanium oxide, stainless steel, or high-density ceramics would serve in this capacity.

When cells held at high local cell concentrations (Figure 8–1) in such particles are compared to those perfused in a chemostat there is evidence to suggest that the physiology of the cells has changed so that the specific cell productivity increases with increase in the medium perfusion rate. The amount of ammonia produced per mole of glutamine consumed is lower in the case where the cells are held at high local concentrations compared to when they are held in a chemostat (Ray et al. 1989).

8.4.5.2 Weighted Collagen II Gelatin can also be made into particles (Figure 8–2), and the particles can be made to carry weights and are, hence, fluidized (Runstadler and Cernek 1988; Nilsson 1989). Such porous particles are available and their implementation should reflect the differences between particles that have the capacity to house many more cells than does the conventional microcarrier culture. It is a general principle that the number of cells that can be grown and supported in a bioreactor is roughly dependent on the volume of medium that is presented to them. Thus, the major advantage of the porous particles is made apparent when the perfusion rate of the medium is adjusted to reflect the increased number of cells present, since oxygen could be a limiting nutrient. A further advantage of such beads is that the inocula can be small (four cells per microcarrier has been quoted) (Nilsson 1989), although it has been stated that such particles are only likely to be suitable for less than 50% of the cell lines extant (Nilsson 1989).

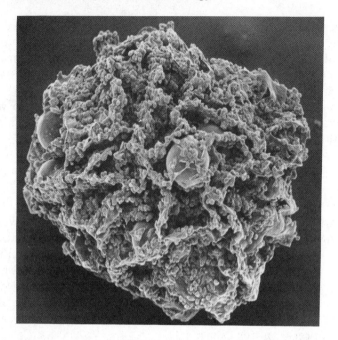

FIGURE 8–1 Cells growing in weighted cross-linked collagen particles. Reproduced with permission from J.B. Griffiths and D. Looby, Center for Applied Microbiology Research, Porton, Salisbury, Wiltshire, UK.

8.4.6 Others
Animal cells can grow and flourish in many environments. The materials that have been used to house such cells reflect this situation. A clear resolution of the opportunities must be achieved and a system that is

- cheap,
- flexible,
- has high productivity,
- is reliable,
- robust,
- readily available,
- controllable in *all* its aspects, and
- free from regulatory agency constraints,

is needed. Some efforts to achieve this have been presented earlier, although others are in the offing.

8.4.6.1 Twisted Ribbons
The convoluted polystyrene in the form of twisted ribbons can be used in either the packed bed or in the fluidized

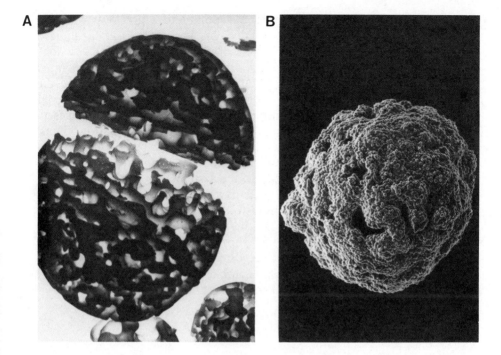

FIGURE 8-2 Cultisphere particles of gelatin (A) without cells, showing the nature of the porous structure, and (B) colonized by cells. Reproduced with permission from K. Nilsson, Biolytica.

state. Such materials have been shown to be capable of supporting cells at a concentration of 1×10^7/ml (Kadouri et al. 1988). This type of system sits between the static and dynamic systems considered above (see section 8.3.4) and reflects the wealth of opportunities that exist for critical system evaluation investigations.

8.4.6.2 Polyurethane Foam An open lattice of polyurethane (PU) foam can also be used to grow and exploit animal cells in culture (Lazar et al. 1987). Such a support material can also be used in the static bed configuration and in a fluidized state. In its availability, inexpensiveness, and flexibility it begins to fulfill some of the salient criteria delineating an acceptable carrier for cells.

8.5 CONCLUSIONS

Three-dimensional microcarriers can provide a framework within which animal cells can both grow and maintain themselves in a sufficiently viable state to be able to manufacture product materials in quantities that are economically attractive. The essential question as to which system is preferred will require much effort to ascertain. This work is necessary to achieve comparisons that will attain the respect of the animal cell biotechnology community. It requires that each of the systems to be compared is operated at its optimal condition. This in itself is a major problem, because the optimum for any one system has to be defined. Such can only be achieved by much painstaking experimentation to the point where any changes to the system will not improve the efficiency of product generation. Thus, keeping the cell inocula and the medium (at least as far as the basal composition is concerned) constant, it is possible with an optimized process to determine the relative productivities of the systems.

There is little doubt that such systems as described in this chapter will have a major role to play in the industrial processes that are coming into being. (It is difficult to supplant an existing process since regulatory approval is expensive and the increase in efficiency may not justify the additional expenditure.) The extent to which three-dimensional microcarriers will penetrate the conventional stirred tank/airlift/two-dimensional microcarrier/packed bed operations will be an exciting area to keep under observation.

REFERENCES

Addadi, L., Moradian, J., Shay, E., Maroudas, N.G., and Weiner, S. (1987) *Proc. Natl. Acad. Sci. USA* 84, 2732–2736.

Ben Ze'ev, A., Farmer, S.R., Penman, S. (1980) *Cell* 21, 365–372.

Butler, M. (1988) in *Animal Cell Biotechnology* Vol. 3 (Spier, R.E., and Griffiths, J.B., eds.), pp. 284–305, Academic Press, London.

Crouch, C.F., Fowler, H.W., and Spier, R.E. (1985) *J. Chem. Technol. Biotechnol.* 35B, 273–281.

Eagle, H. (1965) *Science* 148, 42–51.

Eagle, H., and Piez, K. (1962) *J. Exp. Med.* 116, 29–43.

Edelman, G.M. (1985) *Annu. Rev. Biochem.* 54, 135–169.

Emerman, J.T., and Bissell, M.A. (1988) in *Advances in Cell Culture* (Maramorosch, K., and Sato, G.H., eds.), pp. 137–160, Academic Press, London.

Faraday, D.B.F., Hayter, P., Kirkby, N.F., and Spier, R.E. (1989) *A Mathematical Model of the Cell Cycle of Hybridoma Cultures* Presented to the Engineering Foundation, Cell Culture Engineering II, Santa Barbara, CA, December 1989.

Griffiths, J.B (1988) in *Animal Cell Biotechnology* Vol. 3 (Spier, R.E., and Griffiths, J.B., eds.), pp. 179–221, Academic Press, London.

Kadouri, A., Scher, D., Maroudas, N.G. (1988) *Cytotechnology* 1, 301–307.

Lazar, A., Reuveney, S., Mizrahi, A., et al. (1987) in *Modern Approaches to Animal Cell Technology* (Spier, R.E., and Griffiths, J.B., eds.), pp. 437–448, Butterworths, Guildford, England.

Leeson, C.R., Leeson, T.S., and Paparo, A.A. (1985) *An Atlas of Histology*, pp. 139–145, Saunders Company, London.

Levine, D.W., Wang, D.I.C., and Thilly, W.G. (1979) *Biotechnol. Bioeng.* 21, 821–845.

Maroudas, N.G. (1972) *Exp. Cell Res.* 74, 337–342.

Maroudas, N.G. (1973a) in *New Methods in Cell Biology and Biophysics* (Payne, R., and Smith, R., eds.) Wiley, London.

Maroudas, N.G. (1973b) *Exp. Cell Res.* 81, 104–110.

Maroudas, N.G. (1974) *Cell* 3, 217–219.

Maroudas, N.G. (1975a) *Nature* 244, 353–355.

Maroudas, N.G. (1975b) *J. Theor. Biol.* 49, 417–424.

Maroudas, N.G. (1975c) *Nature* 254, 695–696.

Maroudas, N.G. (1977a) *J. Cell Physiol.* 90, 511–520.

Maroudas, N.G. (1977b) *Nature* 267, 183.

Maroudas, N.G. (1979) in *Cell Shape and Surface Architecture* (Revel, J.P., and Fox, C.F., eds.) pp. 511–520, Liss, New York.

Maroudas, N.G., and Fuchs, A. (1987) in *Ocular Circulation and Neovascularisation* (Ben Ezra, D., and Ryan, S.J., eds.)

Maroudas, N.G., and Schmitt, C.M. (1973) *Cell Differen.* 2, 243–245.

Maroudas, N.G., and Wray, S. (1985) *Connective Tissue Res.* 13, 217–225.

McCullough, K.C., and Spier, R.E. (1990a) *Monoclonal Antibodies in Biology and Biotechnology: Theoretical and Practical Aspects,* p. 281, Cambridge University Press, Cambridge, England.

McCullough, K.C., and Spier, R.E. (1990b) *Monoclonal Antibodies in Biology and Biotechnology: Theoretical and Practical Aspects*, p. 283, Cambridge University Press, Cambridge, England.

Meignier, B., Mougeot, H., and Favre, H. (1980) *Develop. Biol. Stand.* 46, 249–256.

Miles, B.J., and Reading, A.H. (1988) U.K. Patent Application GB 2201966A.

Miller, W.M., Blanch, H.W., and Wilke, C.R. (1988) *Biotechnol. Bioeng.* 32, 947–965.

Montagnon, B., Vincent-Falquet, J.C., and Fanget, B. (1984) *Develop. Biol. Stand.* 55, 37–42.

Murdin, A.D., Wilson, R., Kirkby, N.F., and Spier, R.E. (1987) in *Modern Approaches to Animal Cell Technology* (Spier, R.E., and Griffiths, J.B., eds.), pp. 353–364, Butterworths, Guildford, England.

Nilsson, K. (1988) *Biotechnol. Genet. Eng. News* 6, 403–439.

Nilsson, K. (1989) Presented to the Second Meeting of the Japanese Association for Animal Cell Technology, Tsukuba, Ibaraki, Japan. November 1989.

Nilsson, K., Birnbaum, S., Buzaky, F., and Mosbach, K. (1987) in *Modern Approaches to Animal Cell Technology* (Spier, R.E., and Griffiths, J.B., eds.), pp. 492–503, Butterworths, Guildford, England.

Nilsson, K., and Mosbach, K. (1980) *FEBS Lett.* 118, 145–150.

Nilsson, K., and Mosbach, K. (1987a) *Develop. Biol. Stand.* 66, 183–188.

Nilsson, K., and Mosbach, K. (1987b) *Develop. Biol. Stand.* 66, 189–193.

Ramsey, W.S., Hertl, W., Nowlan, E.D., and Binkowski, N.J. (1984) *In Vitro* 20, 802–808.

Ray, N.G., Tung, A.S., Runstadler, P., and Vournakis, J.N. (1989) Presented at the Cell Culture II Engineering Meeting of the American Engineering Foundation, Santa Barbara, CA, December 1989.

Rein, A., and Rubin, H. (1968) *Exp. Cell. Res.* 49, 666–678.

Reuveny, S., Lazar, A., Mizrahi, A., et al. (1987) in *Modern Approaches to Animal Cell Technology* (Spier, R.E., and Griffiths, J.B., eds.), pp. 724–737, Butterworths, Guildford, England.

Rha, O.-K., and Kim, S.K. (1989) Presented at the Cell Culture II Engineering Meeting of the American Engineering Foundation, Santa Barbara, CA, December 1989.

Rosevear, A., and Lambe, C.A. (1983) in *Topics in Enzyme and Fermentation Biotechnology* Vol. 7 (Wiseman, A., ed.), pp. 13–37, Ellis Horwood Ltd., Chichester, England.

Rubin, H., and Rein, A. (1967) in *Growth Regulating Substances for Animal Cells in Culture* (Defendi, V., and Stoker, M., eds.), pp. 51–66, Wistar Institute Press.

Runstadler, P.W., and Cernek, S.R. (1988) in *Animal Cell Biotechnology* Vol. 3 (Spier, R.E., and Griffiths, J.B., eds.), pp. 306–321, Academic Press, London.

Rupp, R. (1985) in *Large-Scale Mammalian Cell Culture* (Feder, J., and Tolbert, W.R., eds.), pp. 19–38, Academic Press, London.

Spier, R.E. (1980) in *Advances in Biochemical Engineering* Vol. 14 (Fiechter, A., ed.), pp. 119–162, Springer Verlag, Berlin.

Spier, R.E. (1988) in *Animal Cell Biotechnology* Vol. 3 (Spier, R.E., and Griffiths, J.B., eds.) pp. 30–55. Academic Press, London.

Spier, R.E., and Bushell, M.E. (1990) in *BioMedia*, Biolytica, Lund, Sweden (In press).

Spier, R.E., and Griffiths, J.B. (1983) *Develop. Biol. Stand.* 55, 81–92.

Spier, R.E., and Whiteside, J.P. (1976) *Biotech. Bioeng.* 18, 659–667.

Spier, R.E., Whiteside, J.P.W., and Bolt, K. (1977) *Biotechnol. Bioeng.* 19, 1735–1738.

Van Wezel, A.L. (1967) *Nature* 216, 64–65.

Van Wezel, A.L. (1985) in *Animal Cell Biotechnology* Vol. 1 (Spier, R.E., and Griffiths, J.B., eds.), pp. 266–283, Academic Press, London.

Wang, H. (1989) Presented at the Cell Culture Engineering II Meeting of the American Engineering Foundation, Santa Barbara, CA, December 1989.

Wilson, R., and Spier, R.E. (1988) *Enzyme Microbiol. Technol.* 10, 161–164.

Hydrodynamic Effects on Animal Cells in Microcarrier Bioreactors

Matthew S. Croughan
Daniel I.C. Wang

For the culture of animal cells that will adhere to surfaces, growth on the surface of microcarriers appears promising for industrial operations. First developed by van Wezel (1967), the microcarrier technique can provide a homogeneous culture environment with high cell densities and simple medium/cell separation (Nahapetian 1986).

Many researchers have employed the microcarrier technique and have noted its advantages. Nonetheless, animal cells on microcarriers are especially susceptible to damage from fluid-mechanical forces. This susceptibility results from the lack of a protective cell wall, the relatively large size of animal cells, and the lack of individual cell mobility. Anchored cells cannot freely rotate or translate; they therefore cannot reduce the net forces and torques experienced upon exposure to fluid-mechanical forces.

In microcarrier cultures, agitation is required for cell-liquid mass transport, gas-liquid mass transport (oxygenation), and liquid-phase mixing. For

The authors wish to acknowledge the financial support from the National Science Foundation under the Engineering Research Center (ERC) Initiative to the Biotechnology Process Engineering Center under the cooperative agreement CDR-88–03014.

maximum cell growth, adequate mass transfer must be achieved with little or no detrimental effects from hydrodynamic forces. Although this condition can be readily attained in low-density laboratory cultures, it becomes more difficult to attain as cell densities or culture volumes are increased. Successful scale up to high-density, large-volume cultures will require a thorough understanding of the hydrodynamic and mass transport phenomena.

Accordingly, hydrodynamic phenomena in microcarrier cultures has recently arisen as an area of intense study. Research in this area spans the fields of cell biology and turbulent fluid mechanics. This chapter will review our current understanding of this multidisciplinary topic.

9.1 METHODS OF INVESTIGATION

9.1.1 Flow Fields and Cell Deformation

When shear flow occurs over a cell anchored to a surface, the microscopic flow field is affected by the protrusion of the cell. The cell experiences a net torque created by the flow around its circumference. This torque is countered by the adhesive force between the substrate and the cell surface.

Hyman (1972a and 1972b) attempted to solve for the microscopic flow field around a cell protruding into a linear shear field. The relevant Reynold's number is given by

$$Re = Yh^2/\nu \qquad (9.1)$$

where Y is the undisturbed shear rate, h is the cell height (hemispherical shape assumed), and ν is the kinematic fluid viscosity. Because currently available flow visualization techniques can not resolve to better than 20–30 μm, any theoretical solution to the microscopic flow field cannot currently be experimentally verified with actual cells.

Shear effects on anchored animal cells are often investigated through experiments with laminar shear fields in specially constructed flow devices (Crouch et al. 1985; Stathopoulos and Hellums 1985; Sprague et al. 1987). In such experiments, the shear effects have been generally correlated with the undisturbed wall shear stress or shear rate. This approach is taken not only because very little is known about the microscopic fluid mechanics, but also because the microscopic flow field is coupled to the cell shape and cell deformation mechanics. The cell shape will be determined, at least in part, by the local flow field and hydrodynamic forces. To determine the cell shape and the local flow field, one must simultaneously solve for both the fluid motion and the cell deformation mechanics. The deformation mechanics of nucleated cells are only currently being elucidated (Sato et al. 1987a, 1987b; Cheng 1987).

In nearly all microcarrier cultures, the flow field is turbulent. The cells are exposed not only to shear forces, but also to normal forces. The effects of normal forces on animal cells have, to our knowledge, never been quantitatively investigated. The undiscovered role of normal forces may account for the typically poor agreement between shear effects in laminar flow fields and global hydrodynamic effects in stirred tanks, such as reported by Rosenberg et al. (1987). This poor agreement may also arise due to the limited understanding of turbulence. For microcarrier cultures in turbulent bioreactors, one can only roughly estimate the magnitude and direction of the hydrodynamic forces on the cells.

To investigate the mechanisms of hydrodynamic damage in turbulent bioreactors, information should be obtained both through direct experiments with turbulent bioreactors and through translation of results from experiments with well-defined hydrodynamic forces. There is currently a limited ability to define or mimic the turbulent flow in a stirred bioreactor. As such, direct experiments in the turbulent bioreactors are indispensable with regard to the development of scale-up criteria and a mechanistic understanding of hydrodynamic damage. Future advances in fluid mechanics will hopefully allow one to readily translate results between laboratory flow devices and large-scale turbulent bioreactors.

9.1.2 Assessment of Hydrodynamic Effects

Hydrodynamic effects on growth are frequently assessed in terms of cell number and not cell mass or volume. This is undoubtedly due, at least in part, to the tradition of cell number measurements in animal cell culture. It is probably also due to the difficulty in accurately measuring the mass or volume of cells attached to microcarriers.

The term "growth" generally refers to increases in total cell number; the term "net growth" refers to increases in viable cell number. The observed growth rate generally refers to the difference between the specific total, or actual, growth rate and the specific death rate.

Growth can be assessed not only through cell number measurements, but also by monitoring the incorporation of radioactive DNA precursors, such as thymidine (Dewey et al. 1981; McQueen et al. 1987; Chittur et al. 1988). However, when there is excessive cell lysis, such as what occurs under excessive agitation, the radioactive precursor technique can have complications (Aherne et al. 1977). Specifically, if a cell is labelled as it progresses through the S phase, but then lyses before the end of the labelling period, that particular labelling event is lost. In many agitation studies, death and lysis compete with growth, and the phenomena described above can lead to complications in the total growth rate calculation. Furthermore, it is often not practical to label cells while they are actually in the culture vessel and under the influence of agitation. The cost of the label for an entire culture is prohibitive, and the vessel will become contaminated with radioactivity.

One may alternatively determine the total growth rate by microscopically monitoring the mitotic index, l_m, and the duration of mitosis, T_m. The total specific growth rate is given by the ratio l_m/T_m (Johnson 1961). The mitotic index technique is generally too labor intensive for practical purposes; many cells must be counted and monitored for every sample.

In general, by monitoring both the viable and nonviable cell concentrations, one can determine the specific growth and death rates. However, if the dead cells lyse, whole cell counts will not fully account for the number of nonviable cells. Accordingly, one may wish to analyze for a stable intracellular component that is released by the disintegrated cells. Two commonly measured components are lactate dehydrogenase (Arathoon and Birch 1986; McQueen et al. 1987; Smith et al. 1987; Petersen et al. 1988; Gardner et al. 1990) and DNA (Croughan and Wang 1989).

When translating cell component measurements into cell number measurements, one must be sure to:

1. Determine the average amount of component per cell over the concurrent time interval. This may vary with the age of the culture, growth rate, or environmental conditions. All calculations should be performed on an incremental basis over each time interval.
2. Determine if the component is stable in the culture and, if not, correct the measurements for component instability.
3. Determine appropriate sample storage conditions and assay protocols, especially if activity assays are used.

One of the most prevalent hydrodynamic effects is the removal of cells from the microcarriers, which generally increases with the level of agitation. With some cells, such as diploid FS-4 cells, hydrodynamic removal is lethal (Croughan and Wang 1989). Only viable cells are left on the microcarriers while only dead cells are found in suspension. This situation simplifies the calculation of the specific death rate, since it is given by the specific rate of removal.

With other cells, such as aneuploid Chinese hamster ovary cells, hydrodynamic removal is often neither lethal nor irreversible (Croughan and Wang 1990). The cells will frequently reattach to the microcarriers after removal. Such reversible removal makes data interpretation more complex, but also apparently renders the cells less susceptible to hydrodynamic death.

Hydrodynamic effects on cells can be assessed in terms of changes in cell metabolism. It may then be imperative to maintain nearly identical chemical environments, since significant changes in metabolism may arise through small differences in chemical environments. If the cells are used to produce a particular protein, hydrodynamic effects on both the rate of production and product quality may be of primary interest. Intracellular components released by lysed cells, such as proteases or other enzymes, may alter the medium constituents or cell-secreted products. These alterations may, in turn, affect the growth and metabolism of the cells.

Subtle and sophisticated measures of hydrodynamic effects have recently been employed, such as two-dimensional polyacrylamide gel electrophoresis (PAGE) maps of intracellular proteins (Passini and Goochee 1989). Future assessment of hydrodynamic effects will undoubtedly involve a mixture of traditional and new experimental techniques.

9.1.3 Mass and Heat Transport
When investigating the effects of hydrodynamic forces or momentum transfer, one must eliminate any other effects due to mass and heat transfer. If a stagnant cell culture is shaken, the subsequent mixing may induce cell growth simply through the elimination of chemical gradients (Stoker 1973; Stoker and Piggott 1974; Maroudas 1974; Dunn and Ireland 1984). In microcarrier cultures, the temperature and chemical gradients between the cells and medium are generally insignificant (Croughan et al. 1987). As long as the microcarriers are suspended, and as long as the bulk chemical concentrations are maintained at suitable levels, the effects of agitation will occur solely through momentum transfer.

9.1.4 Use of Inert Microcarriers
Cell growth in microcarrier cultures can be affected by both the cell concentration and microcarrier concentration (Mered et al. 1980; Hu et al. 1985; Croughan et al. 1988). To investigate the effects of microcarrier concentration, one may want to use an "inert microcarrier." An inert microcarrier would have the same size and density as a normal microcarrier, but would be chemically inert and incapable of supporting cell growth or attachment. Inert microcarriers could be used to change the solids concentration, even during inoculation, without affecting the chemical environment or the number of cells inoculated per "active" microcarrier. Inert microcarriers could be used to study the effects of microcarrier concentration, independent of the effects of cell concentration.

Sephadex G-50 beads (Pharmacia, Uppsala, Sweden) fulfill the requirements of an inert microcarrier (Croughan et al. 1988). These beads have negligible charge and do not support cell growth or attachment. The beads with dry diameters between 90–106 μm have a size distribution, upon hydration, close to that of Cytodex 1 microcarriers (Pharmacia). In subsequent paragraphs, the term "inert microcarrier" will refer to this size fraction of Sephadex beads.

9.2 HYDRODYNAMIC EFFECTS ON CELL GROWTH

9.2.1 Growth under Mild Agitation
Under flow conditions that do not lead to substantial removal of cells from their growth surface, cell growth appears to be unaffected by fluid shear stresses. For endothelial cells growing on glass coverslips, Dewey et al. (1981)

found that cell growth was not affected by fluid shear stresses up to 8 dyne/cm². In more recent experiments with endothelial cells growing on plastic coverslips, Sprague et al. (1987) found that cell growth was not affected by shear stresses up to 30 dyne/cm².

In microcarrier cultures with mild agitation, cell growth appears to be unaffected by hydrodynamic forces. For both FS-4 and CEF cells, uniform growth is observed over a range of nondetrimental levels of agitation that are sufficient to suspend the microcarriers and provide for adequate surface aeration (Hu 1983; Sinskey et al. 1981).

In general, if cell growth or death was influenced by any type of hydrodynamic mechanism, net cell growth should be affected by a change in a fundamental hydrodynamic variable, such as agitation power, fluid viscosity, or volume fraction solids. However, changes in any of these three variables has no effect on the net growth of FS-4 cultures under mild agitation (Sinskey et al. 1981; Hu 1983; Croughan et al. 1987, 1988, and 1989). Thus, for FS-4 cells, it appears that cell growth and death are not significantly enhanced or reduced by hydrodynamic forces with mild agitation. This conclusion provides an important baseline for the analysis of growth under high agitation and will likely hold true for all but the most shear-sensitive cell lines.

9.2.2 Cell Removal Through Hydrodynamic Forces

When anchored cells are exposed to sufficiently high shear stresses, cell removal from the growth surface will generally become significant. In experiments with endothelial cells on glass coverslips, Viggers et al. (1986) observed extensive cell removal from the growth surface for shear stresses of 128 dyne/cm². In experiments with BHK, Vero, and MRC-5 cells grown on plastic or glass slides, Crouch et al. (1985) observed extensive cell removal for shear stresses in the range of 30 dyne/cm² or greater. In experiments with kidney cells on glass slides, Stathopoulos and Hellums (1985) observed cell removal for shear stresses of 7 dyne/cm² or greater.

Under flow conditions that result in substantial removal of cells from their growth surface, the viability of the cells that remain attached is generally reported to be greater than 90% (Crouch et al. 1985; Stathapoulos and Hellums 1985). However, these viabilities are measured in terms of trypan blue exclusion and not in terms of the ability to reproduce and grow.

When animal cells grow on a surface, cells in mitosis generally round-up and assume a less flattened morphology than the interphase cells (Alberts et al. 1983). For some cell lines on certain growth surfaces, mitotic cells can be selectively and viably removed by applying mild shear stresses to the growth surface (Prescott 1976; Pardee et al. 1978; Terasima and Tolmach 1962). This procedure, however, does not work for all cell lines and growth surfaces (Freshney 1983). Cell removal is frequently not selective for mitotic cells if the agitation is too excessive (Terasima and Tolmach 1962).

For FS-4 cells grown on Cytodex 1 microcarriers in 125 ml spinners, Figure 9–1 shows the removal of the whole cells from the microcarriers at different stirring speeds (Croughan and Wang 1989). The removal increases with the level of agitation and thus appears to be due to hydrodynamic forces. However, the specific rate of removal does not correlate with the estimated mitotic index (Croughan and Wang 1989). This indicates that the cell removal was random and not selective for mitotic cells. For cells grown on microcarriers without colcemid treatment, removal that is not selective for mitotic cells is generally observed (Ng et al. 1980; Mitchell and Wray 1979).

The lack of selectivity for mitotic cells is surprising if one believes cell removal occurs primarily through shear stresses. In response to a shear field at the growth surface, a rounded mitotic cell will experience a higher distracting torque than a flattened interphase cell. Furthermore, a rounded mitotic cell will probably have fewer attachment sites and a weaker attachment to the growth surface than a flattened interphase cell. If cell removal

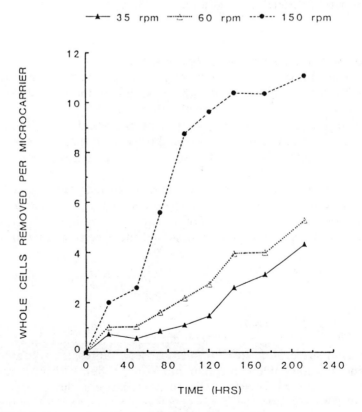

FIGURE 9–1 Removal of whole cells from microcarriers.

occurs through the action of shear stresses, rounded cells in mitosis should be more susceptible to removal than flattened cells in interphase.

However, cell removal can occur not only from shear stresses, but also from normal forces. In microcarrier cultures, normal forces can be generated by pressure fluctuations in the turbulent flow fields. In response to a pressure fluctuation near a microcarrier surface, a cell may be subjected to a distractive normal force. Because the pressure fluctuation will occur on the length scale of a turbulent eddy, which is much larger than a cell, the magnitude of the distractive force will be roughly proportional to the cross-sectional area of the cell on the growth surface. If the cell has a constant number of attachment sites per unit cell surface area, the total attachment force will also be proportional to the cell's cross-sectional area. Overall, both the attachment and distractive forces will be proportional to the cross-sectional area of the cell. The cross-sectional area, or shape, of the cell will then cancel out as a factor. If cell removal occurs through normal forces, a rounded-up cell in mitosis should roughly be no more susceptible to removal than a flattened cell in interphase. For microcarrier cultures, the observed lack of selectivity for removal of mitotic cells may indicate that cell removal occurs primarily through normal forces.

9.2.3 Growth and Death Under High Agitation

When cells are grown on microcarriers in an agitated vessel, a reduction in net growth is frequently observed with an increase in the level of agitation. The reduction could be due to growth inhibition, cell death, or a combination of death and growth inhibition. The biological basis behind this reduction in net growth has recently been investigated through experiments that employ DNA measurements to monitor cell death (Croughan and Wang 1989).

In an overagitated microcarrier culture, cell growth must be occurring if the total number of cells, both attached and removed, is increasing. If cell death and removal are occurring at a specific death rate q, the DNA concentration, D, in the culture fluid should increase according to the relation

$$\frac{dD}{dt} = S_r q C \qquad (9.2)$$

where C is the cell concentration and S_r is the average DNA content per cell. The DNA concentration, D, includes the DNA for both the lysed and whole cells in suspension. In FS-4 microcarrier cultures, the whole cells in suspension are nonviable (Croughan and Wang 1989). If there is random cell removal, such as for FS-4 cells on microcarriers, the average DNA content of the cells removed, S_r, is equal to the average DNA content of the cell population, S_a.

In microcarrier cultures with high agitation, there is extensive cell death and removal. The specific growth rate is strongly dependent on whether secondary growth can occur. Secondary growth, in this chapter, represents growth over microcarrier areas from which cells were previously removed through hydrodynamic forces. In a manner analogous to the stimulation of growth through direct mechanical removal of cells, as described in Alberts et al. (1983), cell growth in a contact-inhibited culture might be stimulated if areas are made available through hydrodynamic removal. If secondary growth can occur, a culture at high agitation may actually grow faster than a culture at low agitation that is more confluent and contact-inhibited.

To investigate the nature of cell growth under excessive agitation, FS-4 microcarrier cultures were grown in spinner flasks at 35 and 150 rpm. At these speeds, the DNA content of the culture fluid, from both lysed and whole cells, was monitored. The relative increase of DNA in the culture fluid at 150 rpm was compared to the predictions of three mechanistic models (Croughan and Wang 1989).

For the first model, it was assumed that the cells in both cultures were growing at the same rate at any given time since inoculation. Cell growth was assumed to be neither inhibited by excessive agitation, nor stimulated by cell removal. The first model represents growth and death with no secondary growth.

For the second model, it was assumed that secondary growth can occur, and that growth in the 150 rpm culture can be corrected for the reduced contact inhibition due to cell removal. For the newly created areas where cells had been removed, secondary growth was assumed to occur at the same rate as normal growth on new microcarriers, with all growth regulated through contact inhibition. The effect of contact inhibition on growth was empirically determined using data from Hu et al. (1985) for FS-4 cells. The second model represents growth and death with secondary growth.

For the third model, it was assumed the reduced net growth at high agitation was solely due to growth inhibition and not caused by cell death or removal. For the third model, the DNA in the culture fluid should not increase at 150 rpm relative to the control at 35 rpm.

Figure 9–2 shows the predictions of the three mechanistic models along with the measured cumulative release of DNA. The predictions of the two growth and death models scatter because the data were analyzed on an interval basis and because the predictions relied on nuclei counts, which are somewhat imprecise ($\pm 10\%$). The measured DNA release clearly does not match the expected release for growth inhibition without cell death and removal, or for growth and death with secondary growth. The measured release data does, however, match the expected release for growth and death without secondary growth (model number one).

To the degree that cell removal was random, as indicated for FS-4 and other cells on microcarriers, the data in Figure 9–2 demonstrate that the cells in the 150 rpm culture were essentially growing at the same rate as the

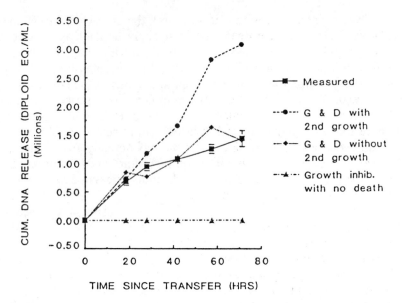

FIGURE 9-2 Measured DNA release versus release expected under three different scenarios.

cells in the 35 rpm culture. If the growth rate at 150 rpm was different than the growth rate at 35 rpm, the amount of DNA released would not have matched the expectation for growth and death without secondary growth.

Thus, for FS-4 microcarrier cultures, hydrodynamic forces appear to affect only cell death and removal but not cell growth. For endothelial cells exposed to mild fluid flow, growth is similarly unaffected by hydrodynamic shear forces (Dewey et al. 1981; Sprague et al. 1987). For these cells, and probably for many animal cells, the decision to replicate does not appear to depend on the hydrodynamic environment, and fluid flow does not appear to influence normal growth regulation.

9.3 HYDRODYNAMIC EFFECTS ON CELL METABOLISM

Hydrodynamic forces may not only lead to cell death and lysis but could also lead to significant changes in cell metabolism. Many of the metabolic events that occur in animal cells are not directly related to cell growth. Whether or not cell growth is inhibited by hydrodynamic forces, metabolic events that are not growth associated could be influenced by agitation.

For endothelial cells, several studies have shown that cell shape, metabolism, and endocytotic activity can be strongly affected by fluid flow. The results are summarized as follows:

1. Cells align and elongate in the direction of flow over one to two days of exposure to shear stresses above 5 dyne/cm^2 (Dewey et al. 1981; Levesque and Nerem 1985).
2. Histadine decarboxylase activity of the cells increases linearly with shear stress for one to two hour exposures to shear stresses above 2.8 dyne/cm^2 (DeForrest and Hollis 1980; Hollis and Ferrone 1974; Rosen et al. 1974).
3. Cell permeability to proteins increases following one hour or longer exposures to shear stresses above 7 dyne/cm^2 (Fry 1968; DeForrest and Hollis 1980).
4. Prostacyclin production increases sixfold upon exposure to a mean shear stress of 10 dyne/cm^2 and increases 16-fold upon a pulsed exposure at 1 Hz between 8 and 12 dyne/cm^2 (Frangos et al. 1985).
5. Exposure to a shear stress of 30 dyne/cm^2 significantly enhances receptor-mediated binding, internalization, and degradation of low-density lipoproteins (Sprague et al. 1987). These results are due, at least in part, to the natural adaptation of endothelial cells to fluid flow in blood vessels.

For other than endothelial cells, there has been a very limited number of direct studies on how cell metabolism is affected by the hydrodynamic environment. For suspended hybridoma cells, Dodge and Hu (1986) found that high levels of agitation resulted in reduced net growth but had no affect on volumetric glucose consumption. This result may indicate that the specific glucose uptake rate of the cells was slightly increased by agitation. However, there may have been conversion of glucose to lactic acid by the enzymes from dead or lysed cells.

For epithelial kidney cells on glass slides, Stathopoulos and Hellums (1985) found that postshear urokinase release was increased by exposure to shear stress levels between 6.5 and 13 dyne/cm^2. For recombinant mouse-L cells on microcarriers, Schulz et al. (1986) found no effects of agitation on specific β-interferon productivity. In addition to these direct studies, there are reports of differences in cell metabolism between agitated and stagnant (or nearly stagnant) cultures (Bryant 1969; Giard et al. 1979). It is unclear, however, whether these differences are due to mass or momentum transfer.

9.4 FLUID-LIFT, AIRLIFT, AND STIRRED-TANK BIOREACTORS

The vast majority of microcarrier cultures are conducted in stirred-tank bioreactors. These versatile vessels provide a relatively homogeneous culture environment that can be readily assessed and controlled. In fact, almost all currently available microcarriers are designed for use in such equipment. Most microcarriers have a specific density in the range of 1.03 to 1.04. This allows for complete suspension with relatively mild agitation. Neutrally buoyant microcarriers are generally not used since they could not be separated from the medium through gravity sedimentation.

When a vessel is agitated with a rotating impeller, the Reynold's number for the bulk flow is given by (Nagata 1975)

$$\text{Re} = ND_i^2/\nu_b \tag{9.3}$$

where ν_b is the kinematic viscosity of the bulk suspension, N is the impeller rotation rate, and D_i is the impeller diameter. If this Reynold's number exceeds approximately 1,000, the flow field becomes turbulent (Nagata 1975). For complete suspension of microcarriers, virtually all stirred-tank bioreactors must be operated in the turbulent regime.

Fluid-life reactors have been employed for microcarrier cultures (Clark and Hirtenstein 1981). When fluid-life reactors are used with microcarriers that have a specific density of 1.04 and a diameter of 185 μm, the maximum linear velocity through the bed is on the order of 5 cm/min. The cells are exposed only to very weak hydrodynamic forces (less than 1 dyne/cm²). The cell density and reactor height, however, are limited by the uptake of oxygen from the medium as it flows past the cells. If one wishes to operate a fluidized bed with high cell densities on an industrial scale, microcarriers with a higher specific density will have to be used. The energy dissipation rates, and fluid forces, will then be comparable to the values in a stirred-tank reactor with microcarriers of a specific density near 1.04.

Airlift reactors have been used for cultures of freely suspended animal and insect cells (Tramper et al. 1986; Boraston et al. 1984; Handa et al. 1987). No published reports have been found that document their use for microcarrier cultures. Direct sparging often causes damage to cells on microcarriers and will be later discussed. With bubbles in the typical size range of 3 mm, the flow around the bubbles will be turbulent.

9.5 MECHANISMS OF HYDRODYNAMIC CELL DEATH

9.5.1 First-, Second-, and Higher-Order Mechanisms of Hydrodynamic Death

In general, the mechanisms of hydrodynamic death can be grouped into first-, second-, and higher-order mechanisms. The first-order mechanisms represent the interaction of a single microcarrier with the surrounding flow field. The second-order mechanisms represent the simultaneous interaction of two microcarriers with the surrounding flow field. The higher-order mechanisms represent the simultaneous interaction of multiple microcarriers with the surrounding flow field.

The first-order mechanisms should be predominant in dilute cultures. The second-order mechanisms should become more significant as the microcarrier concentration is increased. The higher-order mechanisms will dominate only in very concentrated cultures. The higher-order mechanisms have not been investigated.

To provide clear and unique information regarding first-order mechanism(s), experiments should be performed with dilute cultures. These experiments can elucidate the nature of the first-order mechanisms without interference from the second-order mechanisms. Subsequent experiments can be performed with moderately concentrated cultures. The higher rates of damage in the concentrated cultures can be analyzed to investigate the nature of the second- or higher-order mechanisms.

9.5.2 Cell Damage from Turbulence

Independent of whether microcarriers are suspended in an airlift, fluid-lift, or stirred-tank reactor, the fluid-mechanical environment will frequently be turbulent in high-density, large-scale cultures. In turbulent flow fields, short-term hydrodynamic forces arise through the motion of turbulent eddies. In conjunction with the cascade in energy transfer from large to small eddies, there exists a spectrum of eddy sizes down to the viscous dissipation regime (Hinze 1975; Tennekes and Lumley, 1985).

The relative size of an eddy to a microcarrier should play a strong role in the hydrodynamic damage of cells on surface of the microcarriers. If a relatively large eddy formed in a region occupied by a microcarrier, the microcarrier would be entrained and would rotate and translate in a manner that would reduce the net torques and forces on its surface. If a relatively small eddy formed adjacent to the microcarrier, the motion of the microcarrier would be more limited, and the cells on the microcarrier would experience more of the full force of the eddy. Accordingly, cells on microcarriers will be the most readily damaged by small intense eddies of a size and velocity large enough to affect individual cells, but too small to readily entrain entire microcarriers. This hypothesis was first proposed in 1984 and presented in 1985 (Croughan et al. 1985).

In microcarrier cultures, turbulent eddies in the viscous dissipation regime are often intermediate in size between the cells and the microcarriers. The nature of these eddies depends on whether they exist in a state of isotropic equilibrium. There are several results that indicate isotropic equilibrium roughly exists in the viscous dissipation regime in stirred-tank microcarrier bioreactors:

1. For stirred tanks, Nagata (1975) found that the lateral and longitudinal energy spectra became nearly superimposable at high wave numbers. This indicates that local isotropy exists at the viscous dissipation scale.
2. For stirred tanks, Komasawa et al. (1974) found that the energy-containing eddies, or turbulent macroscales, have sizes roughly given by one-fifth the impeller width. Translating these results to most microcarrier cultures, one can estimate that the energy-containing eddies are more

than an order of magnitude larger than eddies in the viscous dissipation regime. Eddies in the viscous dissipation regime can be isotropic if they are much smaller than the energy-containing eddies, which tend to be nonisotropic (Nagata 1975).

3. In analyzing several sets of results, including those of Sato et al. (1967) for stirred tanks, Wadia (1975) found that energy spectra in the viscous dissipation regime followed the prediction of Heisenberg (1948) for isotropic turbulence in statistical equilibrium. The observation held for agitation in an unbaffled stirred tank with an impeller Reynold's number of 8,400, typical of the conditions in a stirred-tank microcarrier reactor.

4. As shown by the tracer-particle studies of Komasawa et al. (1974), the presence of particles in the turbulent flow of a stirred tank does not preclude the existence of isotropic equilibrium. If the bulk-flow Reynold's numbers are high, if the particles are nearly neutrally-buoyant, and if the particles are comparable in size to the Kolmogorov eddy length scale, isotropic equilibrium should exist in the viscous dissipation regime. Microcarriers are nearly neutrally buoyant and are comparable in size to the Kolmogorov eddy length scale.

In light of the evidence presented, it appears that a condition approaching isotropic equilibrium exists in the viscous dissipation regime for many microcarrier cultures.

Under conditions of isotropic equilibrium in the viscous dissipation regime, the size of the smallest eddies is roughly given by the Kolmogorov length scale, L, for the eddies in the viscous dissipation regime (Hinze 1975):

$$L = (\nu^3/\epsilon)^{1/4} \qquad (9.4)$$

where ϵ is the power dissipation per unit mass and ν is the kinematic viscosity. The size of the smallest eddies decreases with an increase in power or a decrease in kinematic viscosity. For sufficiently high power inputs at a given viscosity, the turbulence should generate eddies that are smaller than microcarriers. Cell damage from turbulence should then become evident if the proposed role of eddy length is correct.

The role of turbulent eddies in hydrodynamic damage has been recently investigated for FS-4 microcarrier cultures in spinner vessels (Croughan et al. 1989). Experiments were performed with very dilute cultures, so as to eliminate damage from microcarrier collisions. The power input per unit mass and medium viscosity were both varied to determine if hydrodynamic death correlates with the Kolmogorov length scale for the smallest eddies. Cell damage from time-average velocity fields was purposely avoided and will be discussed in a subsequent section.

Power inputs were originally estimated from the correlations of Nagata (1975). Subsequent measurements (Aunins et al. 1989) indicate the estimates were accurate to within 20%. A 20% error in power input translates to less than a 5% error in calculated eddy length.

Figure 9–3 shows the plot of relative observed growth rate versus average Kolmogorov eddy length for a number of different fluid viscosities. Cell death through hydrodynamic forces, as indicated by a decrease in relative specific growth rate, becomes apparent when the average Kolmogorov length scale falls below about 130 μm, or about two-thirds of the microcarrier diameter of 185 μm. As expected, hydrodynamic death occurs when the turbulence generates eddies that are smaller than the microcarriers. For a number of different fluid viscosities, hydrodynamic death in dilute cultures correlates well with Kolmogorov length scale. This correlation describes the combined effect of both viscosity and power in the unique length scale group, $(\nu^3/\epsilon)^{1/4}$. The existence of this correlation provides more evidence that isotropic equilibrium exists in the viscous dissipation regime in overagitated microcarrier cultures.

Hu (1983) and Sinskey et al. (1981) studied the effects of agitation on net growth of cells in microcarrier cultures. Experiments were performed on the effect of vessel geometry but not on the effect of fluid viscosity. Data from both reports show a good correlation between hydrodynamic damage and Kolmogorov length scale, as determined in a more recent analysis by Croughan et al. (1987). Furthermore, each data set was reasonably well described by a single correlation over a range of different vessel geometries. This is further indication that cell death arises through eddies in the viscous

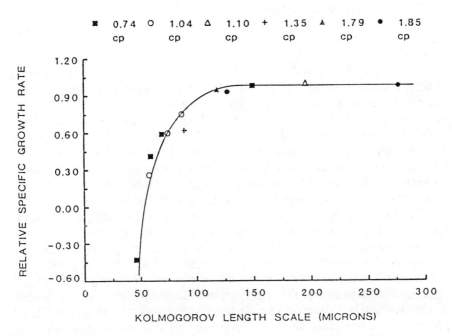

FIGURE 9–3 Relative net growth rate versus Kolmogorov eddy length scale for FS-4 cultures with 0.2 g/l microcarriers.

dissipation regime. Such eddies exist in a state of isotropic equilibrium, which depends only on the local power dissipation rate and kinematic viscosity and not on the vessel geometry. Energy-containing eddies, in contrast, are highly dependent on reactor geometry. If cell death arose through the action of the energy-containing eddies, a strong effect of geometry should have been observed.

9.5.3 Hydrodynamic Forces in Microcarrier-Eddy Interactions

To determine whether the relationship between cell death and eddy length is fundamentally based, one should determine whether the eddies can produce hydrodynamic forces sufficiently strong to damage cells. Currently, however, there is no thorough description of the complex hydrodynamics near a particle in a turbulent flow field. Accordingly, only approximate estimates of the forces can be made.

If viscous forces predominate, Matsuo and Unno (1981) suggest that the shear stress on the surface of a sphere in a turbulent flow field is given by

$$\tau = 2\eta \, (2/15)^{1/2} (\epsilon/\nu)^{1/2} \tag{9.5}$$

where η is the fluid viscosity. If Reynold's stresses are important, Matsuo and Unno (1981) suggest that

$$\tau = 0.37\rho_f \, (\epsilon/\nu) d_p^2 \tag{9.6}$$

where ρ_f is the fluid density and d_p is the microcarrier diameter.

For the conditions under which hydrodynamic cell death was observed in FS-4 microcarrier cultures by Croughan et al. (1989), the average power input per unit mass can be used to initially estimate the shear stresses from equations 9.5 and 9.6. From equation 9.5, shear stress estimates in the range of 1–3 dyne/cm² are obtained. These values are somewhat less than 7 dyne/cm², the minimum shear stress for which significant cell damage has been reported (Stathopoulos and Hellums 1985). From equation 9.6, shear stress estimates in the range of 2–16 dyne/cm² are obtained. The upper half of these values are in the range known to cause cell damage.

The shear stress estimates can alternatively be performed with local instead of average power dissipation rates. In a stirred tank, the power dissipation rates in the impeller discharge stream are much higher than the average dissipation rate. For Rushton turbines, Okamoto et al. (1981) and Placek and Tavlarides (1985) report that the power dissipation rates in the impeller discharge stream are approximately sixfold higher than the average dissipation rates. This ratio can be used to estimate the maximum local power dissipation rates and the corresponding maximum shear stresses. From equation 9.5, the maximum shear stress estimates are in the range of 2–6 dyne/cm², again still less than the values known to cause damage. From equation 9.6, the maximum shear stress estimates are in the range of 10–100 dyne/cm². These values are well within the range known to cause cell death and removal (Crouch et al. 1985; Stathopoulos and Hellums 1985).

As already mentioned with regard to selectivity of cell removal, cells on microcarriers could be damaged or killed not only by shear stresses but also by normal forces. Normal forces will be generated by velocity and pressure fluctuations in the turbulent flow field of a microcarrier culture. In terms of cell death, the critical fluctuations will be those that occur on a scale which is intermediate in size between cells and microcarriers. These fluctuations will involve the eddies in the viscous dissipation regime. Thus, the normal force per unit area on the microcarrier surface, F_n, might be estimated from the magnitude of the pressure fluctuations that occur on the scale of the viscous dissipation regime:

$$F_n \sim P'_{vdr} \tag{9.7}$$

where P'_{vdr} represents root mean square turbulent pressure fluctuation, or pressure intensity, due to the eddies in the viscous dissipation regime.

Turbulence in the viscous dissipation regime is essentially isotropic and has a characteristic Reynold's number near unity (Hinze 1975). The pressure fluctuations due to the viscous dissipation eddies might thus be estimated from an extension of the result presented in Hinze (1975) for isotropic turbulence of low Reynold's number:

$$P' = \rho_f u'^2 \tag{9.8}$$

where u' and P' represent the velocity and pressure intensity, respectively.

In terms of cell death, the relevant pressure and velocity fluctuations are those that occur on the viscous dissipation scale. The effective velocity intensity can thus be estimated by Kolmogorov velocity scale for the viscous dissipation eddies. The normal force per unit area on the microcarrier surface, F_n, is then given by

$$F_n = \rho_f (\epsilon/\nu)^{1/2} \tag{9.9}$$

where ρ_f is the fluid density.

Using average power dissipation rates in equation 9.9, one can estimate normal forces to be on the order of 1–4 dyne/cm² for the cultures that exhibited hydrodynamic death. If the local power dissipation rates in the impeller discharge stream are used, the normal forces estimates are increased to 2–10 dyne/cm². It is not known whether normal forces of this magnitude can damage or remove cells from a growth surface. In fact, a literature review indicates no published data with regard to the effects of normal forces on animal cells. Future research will hopefully be performed in this area.

9.5.4 Kinetics of Cell Growth and Death in Dilute Cultures

Although the hydrodynamics near a microcarrier surface must be investigated more thoroughly, the comparison of experimental data with simplified models can help to elucidate the mechanisms of death from excessive agitation. The data in Figure 9–3, along with other data (Croughan et al. 1987; Cherry and Papoutsakis 1988), indicate that cell death occurs when the turbulence generates eddies that are smaller than the microcarriers. Based upon the correlation of cell death with Kolmogorov length scale, a model was developed to describe the kinetics of hydrodynamic damage in dilute microcarrier cultures. This "eddy-length" model is based upon the following assumptions:

1. Hydrodynamic damage occurs through microcarrier-eddy encounters.
2. Damage will occur only if a microcarrier encounters an eddy smaller than a critical length, L_c, of approximately 130 μm. This assumption is based upon the correlations already discussed, such as those shown in Figure 9–3.
3. Cell death and removal, and not growth inhibition, are the only forms of hydrodynamic damage. This assumption is based upon the experimental results previously presented and discussed.
4. A constant fraction of the culture volume is filled with randomly appearing eddies in the Kolmogorov regime. This assumption is implicit in the derivation of the Kolmogorov scales from an energy balance.

Under the fourth assumption, the effective eddy concentration in the Kolmogorov regime is proportional to the inverse of the average eddy volume, or $(\bar{\epsilon}/\nu^3)^{3/4}$. Because cell death occurs through cell-eddy encounters, the total rate of cell death is proportional to the product of the cell concentration times the eddy concentration. The specific death rate is therefore proportional to the average eddy concentration, or $(\bar{\epsilon}/\nu^3)^{3/4}$. Expressed in mathematical form, the kinetic eddy-length model becomes

$$\frac{dC}{dt} = \mu C \qquad L > L_c \tag{9.10}$$

$$\frac{dC}{dt} = \mu C - q_1 C \qquad L \leq L_c \tag{9.11}$$

$$q_1 = K_e(\bar{\epsilon}/\nu^3)^{3/4} \tag{9.12}$$

where μ is the intrinsic specific growth rate and is independent of the level of agitation, q_1 is the specific death rate due to microcarrier-eddy interactions, $\bar{\epsilon}$ is the average power input per unit mass, ν is the kinematic fluid viscosity, and K_e is a function of the cell and microcarrier properties, and possibly the reactor geometry.

To test the kinetic model presented above, data for the dilute microcarrier cultures were analyzed (Croughan et al. 1989). Figure 9–4 shows a plot of specific death rate versus eddy concentration group, $(\bar{\epsilon}/\nu^3)^{3/4}$, for two different vessels with the same impeller. Each set of data shows a linear correlation, as predicted by the eddy-length model. The intercept values

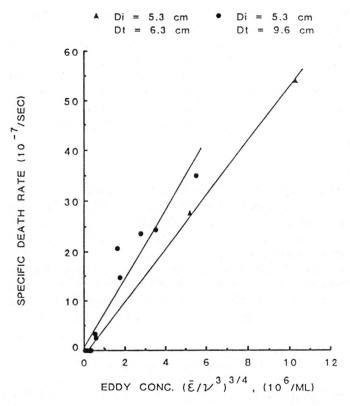

FIGURE 9–4 Specific death rate versus eddy concentration for dilute FS-4 cultures.

near zero follow the assumption of insignificant cell death with mild agitation.

For the data from the 500 ml (9.6 cm) vessel, the value of K_e from linear regression is 6.9×10^{-13} cm³/sec with a 95% confidence interval of $\pm 1.7 \times 10^{-13}$ cm³/sec. For the data from the 125 ml vessel, the value of K_e from linear regression is 5.4×10^{-13} cm³/sec with a 95% confidence interval of $\pm 1.3 \times 10^{-13}$ cm³/sec. The difference between the two values of K_e is not statistically significant within a 95% confidence interval, as determined from the statistical methods presented in Kleinbaum and Kupper (1979). The limited data therefore do not indicate a clear effect of vessel geometry. In fact, a single correlation fits both vessel geometries reasonably well.

The kinetic eddy-length model was found to describe the effect of power input on the net growth of Vero and FS-4 cells on microcarriers, and on the secondary disruption of protozoa in baffled stirred tanks (Croughan et al. 1987). In the protozoa disruption experiments (Midler and Finn 1966), the ratios of impeller-to-tank diameter varied between 0.24 and 0.71. Over this wide range of geometries, the data on secondary disruption can be reasonably described by a single correlation based on average power input.

For BEK cells on microcarriers, Cherry and Papoutsakis (1988) found that damage from two different size impellers was not described well by a single correlation based on average power input. Their data was better described by a correlation based on power dissipation in the discharge region. These results contrast with those above; all sets of data are quite limited. There is clearly a need for more extensive investigations on the effects of reactor geometry.

9.5.5 Cell Damage from Time-Average Flow Fields

In a turbulent flow field, there are both time-average and time-fluctuating pressure and velocity components. If the time-average velocity components change greatly over small intervals in position, strong hydrodynamic forces could arise and damage cells.

To evaluate the role of time-average velocity components in a rigorous fashion, one should determine the position-dependent time-average flow profile around a microcarrier as it circulates through various time-average flow fields in a stirred tank. Because this approach is very difficult and currently impossible, one might instead employ a simplified approach and ignore the dynamic effects due to microcarrier circulation. When the time-average flow field is broken down into small regions, and if turbulent velocity fluctuations are ignored, the problem can be simplified to the situation depicted in Figure 9–5. This schematic of a sphere in a shear field represents a small region in the reactor where the gradient in the time-average velocity field is approximately constant. Such gradients will subsequently be referred to as time-average shear rates. If the flow field was not turbulent, the gradients would represent the undisturbed laminar shear rates.

Lin et al. (1970) developed an analytical steady-state solution to the flow profile near a neutrally buoyant sphere, as depicted in Figure 9–5. The relevant Reynold's number is given by

$$\text{Re} = Yr_m^2/\nu \qquad (9.13)$$

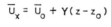

$$\bar{U}_x = \bar{U}_o + Y(z - z_o)$$

FIGURE 9–5 Schematic of a sphere caught in a simple shear field.

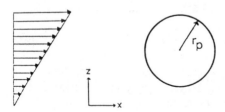

where r_m is the microcarrier radius and Y is the undisturbed shear rate. For microcarrier bioreactors, the Reynold's numbers calculated from equation 9.13 are generally below 1, and the flow can be approximated as in the creeping region. The maximum shear stress, τ, on the sphere surface is then given by

$$\tau = 3\eta Y \qquad (9.14)$$

where η is the fluid viscosity.

As shown in equation 9.14, the maximum shear stress depends upon the time-average shear rate and fluid viscosity. In most vessels, the maximum time-average shear rate is proportional to the impeller tip speed (ITS) and is given, in the reference frame of the rotating impeller, by (Oldshue 1983):

$$Y = K_1 \text{ ITS} = K_1 \pi N D_i \qquad (9.15)$$

and where K_1 is a constant. If a radial flow impeller is used, the maximum shear rate occurs in the radial jet from the impeller. Analyzing the flow profiles developed by Nagata (1975), one estimates that K_1 takes on a value near 0.4/cm for a typical paddle impeller.

Both Hu (1983) and Sinskey et al. (1981) found that cell growth does not correlate with impeller tip speed. In the experiments of Hu (1983), the maximum time-average shear rate was approximately 16/sec, as calculated from equation 9.15. The maximum shear stress was approximately 0.4 dyne/cm^2, as calculated from equation 9.14. This maximum shear stress is much less than 7 dyne/cm^2, the minimum value reported to cause cell damage or removal (Stathopoulos and Hellums 1985). Thus, it appears that cell damage from time-average shear fields was not occurring in the experiments performed by Hu (1983). This probably accounts for the lack of correlation between cell damage and maximum time-average shear rate.

To determine whether cell damage results from time-average shear fields, one should not only evaluate shear stresses, but also investigate the effects of viscosity. If cell damage from time-average flow fields is occurring, an increase in viscosity will increase the amount of damage, since this will lead to higher shear stresses. If cell damage from turbulence is occurring, an increase in viscosity will reduce the amount of damage, since this will dampen the turbulence. Thus, viscosity can be used to distinguish between damage from turbulence and from time-average flow fields.

An increase in viscosity will cause cell damage only if a reactor has sufficiently high time-average shear rates. As was mentioned earlier, the maximum time-average shear rate for most reactors occurs in the discharge stream from the impeller. In some reactors, the maximum time-average shear rate can occur in a region where there is a close clearance between the rotating impeller and a stationary vessel component. For instance, if there is a close clearance between the impeller and vessel wall, the tangential flow profile in this region may periodically assume a character similar to the flow between concentric rotating cylinders. The maximum tangential shear rate can then be estimated from the solution of Lee (1966) for the flow profile between concentric rotating cylinders, and is thus given by

$$Y = \frac{4\pi ND_t^2}{D_t^2 - D_i^2} \qquad (9.16)$$

where D_t is the vessel diameter, N is the impeller rotation rate, and D_i is the impeller diameter.

To investigate the conditions under which damage from time-average shear fields might occur, experiments were performed with dilute microcarrier cultures in spinner vessels with strong increases in viscosity (Croughan et al. 1989). Spinners were used with a clearance of only 5 mm between the impeller tip and vessel wall. Operation at an impeller speed of 220 rpm generated time-average shear rates of approximately 160/sec, as estimated from equation 9.16. A viscosity of 1.5 centipoise (cp) or higher might therefore induce damage from time-average shear fields, since this will lead to shear stresses greater than 7 dyne/cm² (equation 9.14). Shear stresses of this value are the minimum reported to cause cell damage (Stathopoulos and Hellums 1985).

Accordingly, the onset of damage from time-average shear fields was investigated by increasing the fluid viscosity above 1.5 cp at 220 rpm. The maximum shear stress from time-average flow fields, or τ_{max}, was less than 5 dyne/cm² in all cultures except one, where τ_{max} was approximately 9 dyne/cm². If cell death from time-average flow fields was occurring in this culture, but not in the other cultures, the specific death rate should be higher than would be expected for the given viscosity and power. Figure 9–6 shows that this hypothesis was correct. The specific death rates from Figure 9–4, and the best-fit linear correlation, are shown for cultures with $\tau_{max} < 5$ dyne/cm². The specific death rate for the culture with $\tau_{max} = 9$ dyne/cm² is far above the correlation for the other cultures. The difference is statistically significant within a 99% confidence interval and is probably indicative of cell damage from time-average shear fields. However, more experiments are necessary to test reproducibility. Nonetheless, it appears that cell damage from time-average shear fields can be roughly predicted through fluid-mechanical modeling and through knowledge of a critical shear stress for cell damage.

9.5.6 Cell Death Through Microcarrier Collisions

For dilute cultures under high agitation, cell death occurs primarily through microcarrier-eddy encounters and, in unusual circumstances, time-average flow fields. For concentrated cultures under high agitation, cell death might also occur through hydrodynamic interactions between microcarriers. These interactions will be referred to as collisions, even though they may primarily

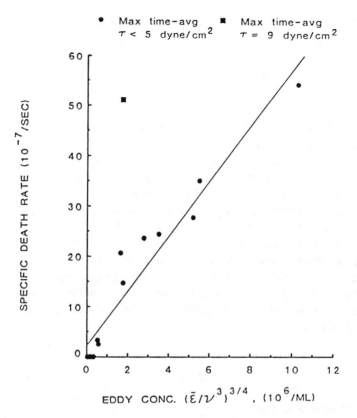

FIGURE 9-6 Specific death rate versus eddy concentration at different levels of maximum shear stress from time-average shear fields.

involve the fluid flow between microcarriers that come in close but not actual contact.

Because microcarriers are very small and nearly neutrally buoyant, they will primarily follow fluid streamlines. Almost all relative motion between two neighboring microcarriers will arise from the action of turbulent eddies. The viscous forces or pressure fluctuations from a single eddy might throw two neighboring microcarriers against each other. The volumetric frequency of such collisions, f_c, would be second order in microcarrier concentration, C_m, and would thus be given by:

$$f_c = K_2 C_m^2 \tag{9.17}$$

where K_2 is a collision frequency constant. The amount of cell death from each collision will be proportional to the cells per unit surface area, ψ_m, and the effective surface area exposed to each collision, K_3. If a cell in an exposed area had a probability K_4 of death from each collision, the volumetric rate of cell death from collisions, $(dC/dt)_{coll}$, would be then given by

$$\left(\frac{dC}{dt}\right)_{coll} = -K_2 K_3 K_4 \psi_m C_m^2 = -q_2 C_m C \tag{9.18}$$

where C is volumetric cell concentration. The specific death constant, q_2, represents the combined product of the parameters, K_2, K_3, and K_4 divided by the total surface area per microcarrier, K_5, or $4\pi r_p^2$.

Since all three parameters K_2, K_3, and K_4 will vary with the level of agitation and the strength of the hydrodynamic forces involved with the collision, the combined term q_2 should be a function of the level of agitation and fluid viscosity. For mild agitation with FS-4 cultures, there is no effect of adding inert microcarriers, and thus the value of q_2 is zero (Croughan et al. 1988). In general, the value of q_2 will depend not only on hydrodynamic variables, but also on the cell and microcarrier properties.

If the eddy-length model is now extended to include cell death from microcarrier collisions, the attached (or viable) cell concentrations should follow the equations

$$\frac{dC}{dt} = \mu C \tag{9.19}$$

with mild agitation, and

$$\frac{dC}{dt} = \mu C - q_1 C - q_2 C_m C \tag{9.20}$$

with strong agitation, where the criteria for cell death involves the level of agitation and may not be a simple function of Kolmogorov eddy length.

Let us assume that cell damage occurred from a microcarrier collision. If one of the microcarriers was an inert microcarrier with no cells, the resulting damage would be, on average, half of what would occur if both microcarriers had cells. Thus, if inert microcarriers are added at high levels of agitation, equation 9.20 should be extended to

$$\frac{dC}{dt} = \mu C - q_1 C - q_2 C_m C - (q_2/2)C_i C \tag{9.21}$$

where C_i is the concentration of inert microcarriers.

To investigate whether cell death occurs through microcarrier collisions, inert microcarriers (Sephadex beads) were added to microcarrier cultures agitated at 150 rpm in spinner vessels (Croughan et al. 1988). If the model presented in equation 9.21 is correct, the data should follow the relation

$$\mu_{obs} = b - (q_2/2)Ci \tag{9.22}$$

where μ_{obs} is the average net growth rate as given by

$$\mu_{obs} = \left(\frac{1}{C} \frac{dC}{dt} \right)_{avg} \tag{9.23}$$

and with constant q_1, C_m, and average total growth rate, μ,

$$b = constant = \mu - q_1 - q_2 C_m \tag{9.24}$$

If there is a second-order damage mechanism, and if all first-order mechanisms are relatively independent of microcarrier concentration, the observed growth rate should decrease in a linear fashion with the inert microcarrier concentration (equation 9.22). The rate of damage from the second-order mechanism (involving the inert microcarriers) should be proportional to the slope. The rate of damage from the first-order mechanism should be proportional to the difference between the intercept b and the value of $(\mu - q_2 C_m)$.

Figure 9–7 shows the average net growth rate observed after attachment versus the inert microcarrier concentration. The data follow the predictions of equations 9.22 to 9.24 and clearly indicate that there are at least two

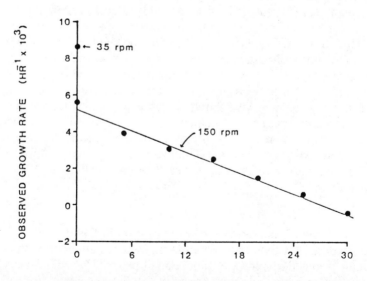

FIGURE 9–7 Specific net growth rate versus inert microcarrier concentration.

distinct death mechanisms. One mechanism is second order in microcarrier concentration and accounts for the decrease in net growth with an increase in inert microcarrier concentration. The other mechanism is first order in microcarrier concentration and accounts, in part, for the decrease in net growth from the 35 rpm culture.

The linearity of the data in Figure 9–7 does not strictly prove that the first-order mechanism was independent of microcarrier concentration. Nonetheless, the data follow a model that assumes this independence over the range of concentrations investigated. The pertinent criterion is the rate of energy dissipation on the microcarrier surface. If microcarriers are added at constant speed, the power draw will go up, but so will the amount of energy dissipated by other microcarriers. The average energy dissipation rate on each microcarrier may remain relatively unchanged. Thus, the increased power draw from increased solids may be offset by the increased dissipation on the other microcarriers. This effect may account for the linearity of the data in Figure 9–7.

The average spacing between the microcarrier surfaces ranged from 50–380 μm , while the average Kolmogorov length scales ranged from 50–60 μm. The average microcarrier spacing was greater than the average Kolmogorov length scale for all but the most concentrated culture. In order for an eddy-microcarrier mechanism to be independent of microcarrier concentration, one might think that average spacing between the microcarrier surfaces would have to be greater than the smallest eddy diameter, or somewhat loosely, the Kolmogorov length scale. Strictly, however, the Kolmogorov length scale is not defined as an eddy diameter, but rather as the typical distance over which viscous dissipation occurs in the absence of solids. This subtle difference does not invalidate the logic behind the eddy-length model, but it should make one wary of comparing eddy lengths with the distance between microcarriers. For microcarrier-eddy interactions to be independent of microcarrier concentration, it is not required that the microcarrier spacing be greater than the Kolmogorov length scale, but rather that the microcarrier spacing is greater than the distance over which viscous dissipation occurs when an eddy interacts with a microcarrier. This distance should be somewhat less than the Kolmogorov length scale, since the presence of solids is known to increase power dissipation (Hiemenz 1977; Monin and Yaglom 1971).

9.5.7 Alternative Mechanisms of Hydrodynamic Cell Death
In the turbulent flow field in a stirred tank, dynamic reductions in pressure will occur as the fluid flows around the impeller. For sufficiently strong agitation, the reduction in pressure may lead to the formation of vapor cavities, a phenomena known as cavitation. Cavitation will arise only if the local pressure falls below the fluid vapor pressure at the given operating temperature. If cavitation were to occur in a microcarrier bioreactor, it could

potentially lead to cell damage or lysis. The role of cavitation in hydro-dynamic cell death has not been adequately analyzed.

Cherry and Papoutsakis (1986) proposed that cells could be damaged by collisions between microcarriers and the solid components of the vessel, such as the impeller, but later found no evidence to support this mechanism (Cherry and Papoutsakis 1988). Because microcarriers are very small and almost neutrally buoyant, they probably do not rapidly penetrate the boundary layers surrounding the solid vessel components.

9.6 CELL DAMAGE FROM DIRECT SPARGING

9.6.1 Effect of Antifoam Addition and Direct Sparging on Cell Growth

Direct sparging of air or oxygen potentially represents a simple and inexpensive method of supplying oxygen to large-scale cultures. The effect of direct sparging has been examined for FS-4 microcarrier cultures in spinners (Aunins et al. 1986). One culture was sparged at a superficial gas velocity of 0.01 cm/sec with a 90:10 air-CO_2 mixture. Foaming was eliminated through the daily addition of 20–40 ppm Medical Emulsion AF antifoam (Dow Corning, Midland, MI). A second culture was grown with antifoam in an identical vessel with no sparging. A third culture was grown with no antifoam and no sparging. All cultures contained 0.2 g/l microcarriers in Dulbecco's Modification of Eagles Medium (DMEM) with 5% fetal calf serum (FCS). They were replenished with nutrients on an interval basis (Croughan et al. 1988) and were essentially at saturation with the 90:10 air-CO_2 gas mixture.

Figure 9–8 shows the results of the experiments. In the cultures that were not sparged, cell growth with antifoam was essentially identical to cell growth without antifoam. This indicates that antifoam AF, at levels between 40 and 180 ppm, had no effect on cell growth. For tPA-CHO cells on microcarriers, Aunins et al. (1986) similarly found no effect on cell growth from antifoam AF at 100 ppm.

For the culture that was directly sparged, the viable cell concentrations were lower than in the controls. This is evidence of significant cell damage from the sparging. The damage was clearly evident even though the superficial gas velocity was only 0.01 cm/sec and there was no significant foam formation.

Subsequent to the experiment described above, which covered the growth period up to 180 hours after inoculation, antifoam was no longer added to the sparged culture. A foam layer began to form at a rate of approximately 0.5 cm/day. After four days, nearly all the microcarriers had collected on the vessel wall above the foam layer. Many microcarriers were spattered through bubble bursts into distant regions of the vessel. If one

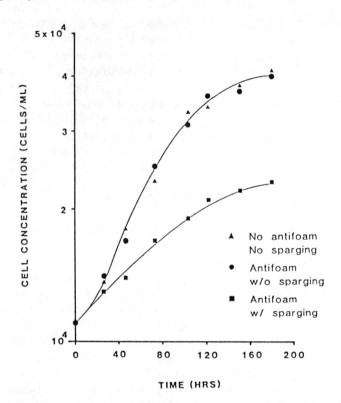

FIGURE 9–8 Effect of antifoam and sparging on net cell growth.

wishes to keep the microcarriers in the liquid phase, excessive foam formation must be strictly avoided.

9.6.2 Mechanisms of Cell Damage from Sparging in Microcarrier Cultures

For suspension cultures, there has been a considerable amount of research on cell damage from direct sparging (Kilburn and Webb, 1968; Tramper et al. 1986 and 1988; Handa et al. 1987; Murhammer and Goochee, 1988 and 1990; Oh et al. 1989; Passini and Goochee, 1989; Gardner et al. 1990; Kunas and Papoutsakis, 1990a and 1990b). For microcarrier cultures, to our knowledge, there is only one investigation reported in the published literature (Aunins et al. 1986). Several investigators have made significant advances in deciphering the complex mechanisms of damage for suspension cultures. These mechanisms will likely play a role in microcarrier cultures.

In the absence of bubbles, suspension cells are relatively resistant to damage from hydrodynamic forces (Augenstein et al. 1971; Dodge and Hu 1986; McQueen et al. 1987; Smith et al. 1987; Petersen et al. 1988; Oh et al. 1989; Kunas and Papoutsakis 1990a). Furthermore, damage of suspension cells from sparging appears to be essentially unaffected by an increase in fluid viscosity through dextran supplementation (Handa et al. 1987). Overall, it appears unlikely that suspended cells are killed or damaged by simple hydrodynamic forces in the fluid adjacent to rising bubbles. If this mechanism were significant, an increase in viscosity should have reduced the damage, as it would have slowed the bubble rise velocities and dampened the turbulence near the bubbles. Furthermore, the presence of bubbles without a vortex and headspace does not appear to cause damage in itself (Kunas and Papoutsakis, 1990a).

Because cells on microcarriers are much more sensitive to hydrodynamic forces than freely-suspended cells, mechanisms of damage may come into play for microcarriers cultures which are insignificant for suspension cultures. Simple calculation illustrates that the damage observed from sparging, shown in Figure 9–8 could have been hydrodynamic in nature. In the sparged culture, the bubbles rose in a column that occupied approximately 10% of the culture volume. As discussed in Aunins et al. (1986), the power input to the liquid near the bubbles may be approximated by the power required to match the buoyancy forces. Thus, for the sparged column region, the power input from sparging per unit mass of fluid, ϵ_s, may be roughly calculated by:

$$\epsilon_s = \frac{Q(\rho_f - \rho_b)gH}{\rho_f(H(\pi D_t^2/4)\, 0.1 - H_o)} \tag{9.25}$$

where ρ_b is the gas density, Q is the volumetric flow rate of gas in milliliters per second, g is the gravitational acceleration, H is the reactor height, and H_o is the gas hold-up in the sparged region. The gas hold-up was estimated to be 20 ml from the increase in culture volume upon sparging.

Equation 9.25 was used to estimate the value of 100 cm²/sec³ for the local power input per unit mass in the sparged region. This is well into the range for which death occurs from impeller-generated power, even with the mild viscosity increase brought about by the antifoam supplementation. Thus, the damage from sparging could have been due, at least in part, to power dissipation in the fluid adjacent to rising bubbles. Future research will hopefully elucidate the mechanisms and kinetics of cell damage from sparging in microcarrier cultures. The damage depicted in Figure 9–8 might have been reduced or eliminated if Pluronic F68 had been employed.

9.7 PROTECTIVE POLYMERS

To eliminate or reduce cell damage from hydrodynamic or interfacial forces, many investigators have added polymers to their medium. The most commonly used polymers are pluronic polyols (Swim and Parker 1960; Runyan and Geyer 1963; Kilburn and Webb 1968; Mizrahi 1984; Handa et al. 1987;

Murhammer and Goochee 1988 and 1990). Methylcellulose has also been used (Kuchler et al. 1960; Bryant 1966, 1969; Holmstrom 1968; Birch and Pirt 1969; Tramper et al. 1986), along with sodium carboxymethylcellulose (Mizrahi 1984), polysucrose (Pharmacia 1981), dextran (Schulz et al. 1986), and hydroxyethyl starch (Mizrahi 1984). In general, metabolic uptake of these polymers is insignificant (Mizrahi 1984). The polymers appear to have no nutritive value (Bryant 1986 and 1969; Mizrahi 1984), although they can provide trace metals in deficient medium (Thomas and Johnson 1967) and can eliminate protein precipitation from medium with poor serum (Swim and Parker 1960).

It is widely believed that these polymers protect animal cells from mechanical damage. In agitated systems, cell disruption is frequently observed if the medium is not supplemented with either serum or one of these protective polymers (Swim and Parker 1960; Kuchler et al. 1960; Runyan and Geyer 1963; Bryant 1966; Thomas and Johnson 1967; Holmstrom 1968; Kilburn and Webb 1968; Birch and Pirt 1969; Mizrahi 1984; Murhammer and Goochee 1988; Kunas and Papoutsakis 1990b). Because the known protective polymers and serum can at least partially substitute for each other with regard to mechanical protection, it appears that serum contains a protective polymer, although it has not been identified.

In light of the recent results presented in this chapter, one might think that the polymers may protect cells from turbulent damage by increasing the medium viscosity. Although viscous reduction of turbulent damage may account for some of the protection, it does not appear to fully account for all the results presented in the literature.

In many instances, a great degree of protection is observed, qualitatively, with very low polymer concentrations or viscosity increases. For instance, serum supplementation apparently provides a great deal of protection and yet increases the viscosity only a few percent. Methylcellulose supplementation, at a typical concentration of approximately 1 g/l and with a typical molecular weight of approximately 15,000, also apparently provides a great deal of protection (Kuchler et al. 1960; Bryant 1966; Holmstrom 1968; Birch and Pirt 1969) and yet leads to a viscosity increase of only 18% (Croughan et al. 1989). For a dilute FS-4 microcarrier culture, an 18% viscosity increase would provide only a 31% reduction in hydrodynamic death. It is difficult to say whether a 31% reduction accurately represents the qualitative results presented in the literature. It appears, however, that the protection provided by some polymers is more than would be expected through only viscous reduction of turbulent damage.

The mechanism of the protective effect is under investigation. Kuchler et al. (1960); Bryant (1966); Mizrahi (1984); and Murhammer and Goochee (1988, 1990) have hypothesized that the polymers adsorb onto the cells and form a protective shell. This adsorption would occur along with a concurrent reduction in polymer concentration in the culture supernatant. For methylcellulose, Bryant (1966) observed a steady decrease in the culture viscosity, potentially indicating polymer attachment onto the cells. For carboxymethylcellulose, hydroxyethyl starch, and pluronic polyols, Mizrahi (1984) found a clear protective effect even though over 99% of the added polymer remained in the supernatant. To our knowledge, no one has directly measured whether there is any adsorption of the polymers onto the cells, although work is being performed in this area. From the data currently available, the protective shell hypothesis is neither directly supported nor refuted, although it appears to be correct (Murhammer and Goochee, 1990).

Another mechanism of protection may involve viscoelastic interactions between the polymers and the turbulent flow fields. These interactions can occur even with very low polymer concentrations. In the experiments of Croughan et al. (1989), low molecular weight dextrans were used as the thickening agents. Viscoelastic interactions were purposefully avoided to isolate the effects of viscosity. For some of the protective polymers, such as methylcellulose and carboxymethylcellulose, viscoelastic behavior has been reported (Amari and Nakamura 1973 and 1974; Hoyt 1985).

9.8 RECOMMENDATIONS FOR FUTURE RESEARCH

The results in this chapter point to several interesting areas for future research. The topics have been discussed throughout the chapter and include the following.

Effects of Reactor Geometry on the Kinetics of Hydrodynamic Death. As shown by many of the results presented in this chapter, hydrodynamic cell death often correlates well with average power input per unit mass (Croughan et al. 1987 and 1989). Nonetheless, only a limited range of geometries have been tested. Furthermore, some have found a better correlation with power dissipation in the impeller discharge region (Cherry and Papoutsakis 1988).

In a stirred tank reactor, cells are only periodically exposed to the impeller discharge flow and for only short time intervals. In general, the smaller the volume where dissipation rates are high, the lower fraction of the culture this volume contains. There will be a trade-off between high local dissipation rates and frequency of exposure to such rates. The net effect may often result in a good correlation between average death rate and average power dissipation rate.

Future research should investigate a wider range of reactor geometries and should incorporate quantitative information on circulation velocities and local rates of mass transfer. Empirical results might be reported in terms of kinetic constants, as was done for the data in Figures 9–4 and 9–6. If possible, the effect of scale on flow regimes should be addressed, as recommended by Aunins et al. (1989). The kinetics of mass transfer should be considered along with the kinetics of hydrodynamic death. For microcarrier cultures, hydrodynamic death appears to involve only the smallest eddies, which are not strongly dependent on geometry. In contrast, mass transfer involves not only small eddies, but also large eddies, which are strongly dependent on geometry. Suspension of the microcarrier also involves large eddies and is again strongly dependent on reactor geometry (Nienow 1985). If the effects of geometry on the large eddies are manipulated, high rates of mass transfer should be obtainable without high rates of hydrodynamic death.

Effects of Normal Forces on Animal Cells.   Normal forces may account for the primary hydrodynamic effects on animal cells in turbulent stirred tanks. For cell removal from microcarriers, the observed lack of selectivity for mitotic cells may indicate that cell removal occurs primarily through normal forces. The role of normal forces has not been investigated and deserves attention.

Microscopic Fluid Flow Fields.   The hydrodynamic forces experienced by cells in microcarrier cultures can currently be calculated in only an approximate manner. Future research should more precisely determine the magnitude and direction of the forces experienced by the cells. This will require extensive modeling of the microscopic flow fields that arise in microcarrier-eddy interactions and microcarrier collisions. The most thorough models might incorporate the protrusion of the cells into the flow field along with the mechanics of cell deformation. The results of these models will allow for a more precise and thorough approach to bioreactor design and optimization.

Reversible Cell Removal and Secondary Growth.   For CHO microcarrier cultures, cell removal from excessive agitation is frequently reversible and nonlethal (Croughan and Wang 1990). Secondary growth readily occurs over areas from which cells have been previously removed. For FS-4 microcarrier cultures, in contrast, cell removal is irreversible and lethal, and secondary growth does not occur. At a moderate level of agitation, the secondary growth in the CHO microcarrier cultures is often fast enough to overcome any reduction in attached cell concentrations from hydrodynamic death and removal. Secondary growth can lead to reduced hydrodynamic sensitivity and may occur because the cells are removed whole. Such reversible removal may be related to the attachment properties of the cells. Future research should be performed to more thoroughly elucidate the nature of cell removal and its relation to secondary growth.

Hydrodynamic Effects on Cell Metabolism.   Industrial animal cell culture will frequently be used for the production of proteins. The performance of these industrial processes will depend not only on the cell concentration and growth rate, but also on the cell metabolism. It is therefore important to understand the hydrodynamic effects on both cell metabolism and cell growth. Future research will hopefully extend the currently very limited information on the hydrodynamic effects of cell metabolism.

Mechanisms of Cell Damage from Direct Sparging. For microcarrier cultures, direct sparging potentially represents a simple and inexpensive method for oxygenation of large-scale bioreactors. If the mechanisms of cell damage from sparging are elucidated, a viable sparging technique may become apparent. In general, the use of sparging should be considered in light of the trade-off between oxygen transfer and cell death. The kinetics of cell damage from sparging should be quantitatively described along the kinetics of oxygen transfer. The kinetic expressions can be used to determine the sparging conditions and reactor design that lead to optimum performance, as illustrated by Tramper et al. (1988). Performance should be evaluated both in terms of productivity and product quality. If cell damage from bubbles involves turbulence, the use of thickening agents or viscoelastic polymers may reduce the amount of damage. Protective agents such as pluronic polyols might also reduce the amount of damage.

Viscoelastic Reduction of Hydrodynamic Death. The protective mechanism of many polymers may involve viscoelastic interactions between the polymers and turbulent flow fields. In general, there is great potential behind the use of viscoelastic agents for the reduction of hydrodynamic death from turbulence. If viscoelastic polymers are added that interfere with the smallest eddies, a strong decrease in hydrodynamic death should be attainable. The viscoelastic polymers should have a characteristic relaxation time equivalent to the burst duration for the smallest eddies. Such polymers might not strongly interfere with the mass transfer processes, as indicated by the results of Quraishi et al. (1977). New experiments should be performed to investigate the use of viscoelasticity in protecting cells from turbulent damage.

9.9 NOMENCLATURE

Roman

b	Intercept value in equation 9.22, per sec
C	Volumetric concentration of attached cells, cells/cm^3
C_i	Inert microcarrier concentration, microcarriers/cm^3
C_m	Microcarrier concentration, microcarriers/cm^3
D	Concentration of DNA in culture fluid, diploid equivalents/cm^3
D_i	Impeller diameter, cm
D_t	Vessel diameter, cm
d_p	Microcarrier diameter, cm
f_c	Frequency of microcarrier collisions, number of collision/cm^3-sec
F_n	Normal force per unit area on microcarrier surface, dyne/cm^2

g	Acceleration due to gravity, 980 cm/sec^2
h	Cell height above growth surface, cm
H	Liquid height in reactor, cm
H_o	Gas hold-up in reactor, cm^3
ITS	Impeller tip speed, cm/sec
K_1	Ratio of time-average shear rate to impeller tip speed, per cm
K_2	Collision frequency constant, number of collisions-cm^3/ microcarrier^2-sec
K_3	Average effective surface area exposed to each collision, cm^2
K_4	Probability of death for a cell in the collision-exposed area
K_e	First-order specific death rate constant, cm^3/sec
L	Length scale for eddies in viscous dissipation regime, cm
L_c	Critical eddy length scale for cell death, cm
N	Rotation rate of impeller or viscometer, rotations/sec
P′	Intensity of turbulent pressure fluctuation, dyne/cm^2
P'_{vdr}	P′ due solely to eddies in viscous dissipation regime, dyne/cm^2
q	Total specific death rate, per sec
q_1	Specific death rate due to microcarrier-eddy interactions, per sec
q_2	Second-order death rate constant, cm^3/microcarrier-sec
Q ·	Flow rate of gas for sparging, cm^3/sec
r_p	Microcarrier radius, cm
Re	Reynolds number for flow, dimensionless
S_a	Average DNA content of cell population, diploid equivalents/cell
S_r	Average DNA content of cells removed, diploid equivalents/cell
u′	Root mean square turbulent velocity component, cm/sec
u'_{max}	Maximum root mean square turbulent velocity component, cm/ sec
v	Velocity scale for eddies in viscous dissipation regime, cm/sec
Y	Shear rate in flow field undisturbed by cell protrusion or microcarrier, per sec

Greek

ϵ	Power input or dissipation per unit mass, cm^2/sec^3
$\bar{\epsilon}$	Average power input or dissipation per unit mass, cm^2/sec^3
ϵ_s	Power input from sparging per unit mass of fluid, cm^2/sec^3
η	Viscosity of fluid, gm/cm-sec
μ	Specific total (or actual) growth rate, per sec
μ_{obs}	Difference between specific total growth and death rate, per sec
ν	Kinematic viscosity of fluid, cm^2/sec
ν_b	Kinematic viscosity of suspension, cm^2/sec
ρ_b	Density of bubbles, gm/cm^3
ρ_f	Density of fluid, gm/cm^3
τ	Shear stress on microcarrier surface, dyne/cm^2
τ_{max}	Maximum τ from time-average flow fields, dyne/cm^2
ψ_m	Surface coverage of microcarriers, cells/cm^2

REFERENCES

Aherne, W.A., Camplejohn, R.S., and Wright, N.A. (1977) *An Introduction to Cell Population Kinetics*, p. 58, Edward Arnold, London.

Alberts, B., Bray, D., Lewis, J., et al. (1983) *Molecular Biology of the Cell*, pp. 611–621, Garland Publishing, New York.

Amari, T., and Nakamura, M. (1973) *J. Appl. Polymer Sci.* 17, 589–603.

Amari, T., and Nakamura, M. (1974) *J. Appl. Polymer Sci.* 18, 3329–3344.

Arathoon, W.R., and Birch, J.R. (1986) *Science* 232, 1390.

Augenstein, D.C., Sinskey, A.J., and Wang, D.I.C. (1971) *Biotechnol. Bioeng.* 13, 409–418.

Aunins, J.G., Croughan, M.S., Wang, D.I.C., and Goldstein, J.M. (1986) *Biotechnol. Bioeng. Symp. Ser.* 17, 699–723.

Aunins, J.G., Woodson, B.A., Hale, T.K., and Wang, D.I.C. (1989) *Biotechnol. Bioeng.* 34, 1127–1132.

Birch, J.R., and Pirt, S.J. (1969) *J. Cell Sci.* 5, 135–142.

Boraston, R., Thompson, P.W., Garland, S., and Birch, J.R. (1984) *Develop. Biol. Stand.* 55, 103–111.

Bryant, J.C. (1966) *Ann. N.Y. Acad. Sci.* 139, 143–161.

Bryant, J.C. (1969) *Biotechnol. Bioeng.* 11, 155–179.

Cheng, L.Y. (1987) *J. Biomech. Eng.* 109, 10–24.

Cherry, R.S., and Papoutsakis, E.T. (1986) *Bioprocess Eng.* 1, 29–41.

Cherry, R.S., and Papoutsakis, E.T. (1988) *Biotechnol. Bioeng.* 32, 1001–1014.

Chittur, K.K., McIntire, L.V., and Rich, R.R. (1988) *Biotechnol. Prog.* 4, 89–96.

Clark, J.M., and Hirtenstein, M.D. (1981). *Ann. N.Y. Acad. Sci.* 369, 33–46.

Crouch, C.F., Fowler, H.W., and Spier, R.E. (1985) *J. Chem. Technol. Biotechnol.* 35B, 273–281.

Croughan, M.S. (1988) *Hydrodynamic Effects on Animal Cells in Microcarrier Bioreactors,* Doctoral Dissertation, MIT, Cambridge, MA.

Croughan, M.S., Hu W-S., and Wang, D.I.C. (1985) Presented at the New England Biotechnology Association Worcester Colloquium 2, Worcester, MA, March 21, 1985.

Croughan, M.S., Hamel, J-F., and Wang, D.I.C. (1987) *Biotechnol. Bioeng.* 29, 130–141.

Croughan, M.S., Hamel, J-F.P., and Wang, D.I.C. (1988) *Biotechnol. Bioeng.* 32, 975–982.

Croughan, M.S., and Wang, D.I.C. (1989) *Biotechnol. Bioeng.* 33, 731–744.

Croughan, M.S., Sayre, E.S., and Wang, D.I.C. (1989) *Biotechnol. Bioeng.* 33, 862–872.

Croughan, M.S., and Wang, D.I.C. (1990) *Biotechnol. Bioeng.* 36, 316–319.

DeForrest, J.M., and Hollis, T.M. (1980) *Exp. Mol. Pathol.* 32, 217–225.

Dewey, C.F., Bussolari, S.R., Gimbrone, M.A., and Davies, P.F. (1981) *J. Biomech. Eng.,* 103, 177–185.

Dodge, T.C., and Hu, W.S. (1986) *Biotechnol. Lett.* 8, 683–686.

Dunn, G.A., and Ireland, G.W. (1984) *Nature* 312, 63–65.

Frangos, J.A., Eskin, S.G., McIntire, L.V., and Ives, C.L. (1985). *Science* 227, 1477–1479.

Freshney, R.I. (1983) *Culture of Animal Cells*, pp. 168, 245, Alan R. Liss, New York.

Fry, D.L. (1968) *Circ. Res.* 22, 165–197.

Gardner, A.R., Gainer, J.L., and Kirwan, D.J. (1990) *Biotechnol. Bioeng.* 35, 940–948.

Giard, D.J., Loeb, D.H., Thilly, W.G., Wang, D.I.C., and Levine, D.W. (1979) *Biotechnol. Bioeng.* 21, 433–442.

Handa, A., Emery, A.N., and Spier, R.E. (1987) *Proc. 4th Eur. Congr. Biotechnol.* 3, 601–604.

Heisenberg, W. (1948) *Z. Phys.* 124, 168.

Hiemenz, P.C. (1977) *Principles of Colloid and Surface Chemistry,* pp. 62–80, Marcel Dekker, New York.

Hinze, J.O. (1975) *Turbulence,* pp. 222–227, 309–310, McGraw-Hill, New York.

Hollis, T.M., and Ferrone, R.A. (1974) *Exp. Molec. Pathol.* 20, 1–10.

Holmstrom, B. (1968) *Biotechnol. Bioeng.* 10, 373–384.

Hoyt, J.W. (1985) *Trends Biotechnol.* 3, 17–21.

Hu, W.S. (1983) Doctoral Dissertation, pp. 204–211, MIT, Cambridge, MA.

Hu, W.S., Meier, J., and Wang, D.I.C. (1985) *Biotechnol. Bioeng.* 27, 585–595.

Hyman, W.A. (1972a) *J. Biomech.* 5, 45–48.

Hyman, W.A. (1972b) *J. Biomech.* 5, 643.

Johnson, H.A. (1961) *Cytologia* 26, 32–41.

Kilburn, D.G., and Webb, F.C. (1968) *Biotechnol. Bioeng.* 10, 801–814.

Kleinbaum, D.G., and Kupper, L.L. (1979) *Applied Regression Analysis and Other Multivariable Methods,* pp. 95–106, Duxbury Press, North Scituate, MA.

Komasawa, I., Kuboi, R., and Otake, T. (1974) *Chem. Eng. Sci.* 29, 641–650.

Kuchler, R.J., Marlowe, M.L., and Merchant, D.J. (1960) *Exp. Cell Res.* 20, 428–437.

Kunas, K.T., and Papoutsakis, E.T. (1990a) *Biotechnol. Bioeng.* 36, 476–483.

Kunas, K.T., and Papoutsakis, E.T. (1990b) *J. Biotech.* in press.

Lee, T.-S. (1966) Doctoral Dissertation, pp. 39–51, MIT, Cambridge, MA.

Levesque, M.J., and Nerem, R.M. (1985) *J. Biomech. Eng.* 107, 341–347.

Lin, C-J., Peery, J.H., and Schowalter, W.R. (1970) *J. Fluid Mech.* 44, 1–17.

Maroudas, N.G. (1974) *Cell* 3, 217–219.

Matsuo, T., and Unno, H. (1981) *J. Environ. Eng. Div. ASCE* 107, 527–545.

McQueen, A., Meilhoc, E., and Bailey, J. (1987) *Biotechnol. Lett.* 9, 831–836.

Mered, B., Albrecht, P., and Hopps, H.E. (1980) *In Vitro* 16, 859–865.

Midler, M., and Finn, R.K. (1966) *Biotechnol. Bioeng.* 8, 71–84.

Mitchell, K.J., and Wray, W. (1979) *Exp. Cell Res.* 123, 452–455.

Mizrahi, A. (1984) *Develop. Biol. Stand.* 55, 93–102.

Monin, A.S., and Yaglom, A.M. (1971) *Statistical Fluid Mechanics: Mechanics of Turbulence* Vol. 1, pp. 412–415, MIT Press, Cambridge, MA.

Murhammer, D.W., and Goochee, C.F. (1988) *Bio/Tech.* 6, 1411–1418.

Murhammer, D.W., and Goochee, C.F. (1990) *BioTech. Prog.* 6, 142–148.

Nagata, S. (1975) *Mixing: Principles and Applications,* pp. 24–32, 126–129, 149–163, Wiley, New York.

Nahapetian, A.T. (1986) in *Mammalian Cell Technology,* (Thilly, W.G., ed.), pp. 151–165, Butterworths, Stoneham, MA.

Ng, J.J.Y., Crespi, C.L., and Thilly, W.G. (1980) *Anal. Biochem.* 109, 231–238.

Nienow, A.W. (1985) in *Mixing of Liquids by Mechanical Agitation* (Ulbrecht, J.J., and Patterson, G.K., eds.), pp. 273–307, Gordon and Breach Science Publishers, New York.

Oh, S.K.W., Nienow, A.W., Al-Rubeai, M., and Emery, A.N. (1989) *J. BioTech.* 12, 45–62.

Okamoto, Y., Nishikawa, M., and Hashimoto, K. (1981) *Int. Chem. Eng.* 21, 88–96.

Oldshue, J.Y. (1983) *Fluid Mixing Technology,* pp. 27–31, 170, 198, 213, McGraw-Hill, New York.

Pardee, A.B., Dubrow, R., Hamlin, J.L., and Kletzein, R.F (1978) *Annu. Rev. Biochem.* 47, 715–750.

Passini, C.A., and Goochee, C.F. (1989) *BioTech. Prog.* 5, 175–188.

Petersen, J.F., McIntire, L.V., and Papoutsakis, E.T. (1988) *J. BioTech.* 7, 229–246.

Pharmacia (1981) *Microcarrier Cell Culture: Principles and Methods,* Technical literature, Pharmacia Fine Chemicals, Uppsala, Sweden.

Placek, J., and Tavlarides, L.L. (1985) *AlChE J.* 31, 1113–1120.

Prescott, D.M. (1976) *Reproduction of Eukaryotic Cells,* pp. 24, 25, 37, 38, Academic Press, New York.

Quraishi, A.Q., Mashelkar, R.A., and Ulbrecht, J.J. (1977) *AlChE J.* 23, 487–492.

Rosenberg, M.Z., Kargi, F., and Dunlop, E.H. (1987) *Biological Responses of Plant Cells to Hydrodynamic Shear Stress.* Presented at the 194th National Meeting of the American Chemical Society, New Orleans, LA.

Rosen, L.A., Hollis, T.M., and Sharma, M.G. (1974) *Exp. Molec. Path.* 20, 329–343.

Runyan, W.S., and Geyer, R.P. (1963) *Proc. Soc. Exp. Biol. Med.* 112, 1027–1030.

Sato, Y., Horie, Y., Kamiwano, M., Yamamoto, K., and Ishii, K. (1967) *Kagaku-Kogaku* 31, 275–281.

Sato, M., Levesque, M.J., and Nerem, R.M. (1987a) *J. Biomech. Eng.* 109, 27–34.

Sato, M., Levesque, M., and Nerem, R.M. (1987b) *Arteriosclerosis* 7, 276–286.

Schulz, R., Krafft, H., and Lehmann, J. (1986) *Biotechnol. Lett.* 8, 557–560.

Sinskey, A.J., Fleischaker, R.J., Tyo, M.A., Giard, D.J., and Wang, D.I.C. (1981) *Ann. N.Y. Acad. Sci.* 369, 47–59.

Smith, C.G., Greenfield, P.F., and Randerson, D.H. (1987) *Biotechnol. Techniques* 1, 39–44.

Sprague, E.A., Steinbach, B.L., Nerem, R.M., and Schwartz, C.J. (1987) *Circulation* 76, 648–656.

Stathopoulos, N.A., and Hellums, J.D. (1985) *Biotechnol. Bioeng.* 27, 1021–1026.

Stoker, M.G.P. (1973) *Nature* 246, 200–203.

Stoker, M., and Piggott, D. (1974) *Cell* 3, 207–215.

Swim, H.F., and Parker, R.F. (1960) *Proc. Soc. Exp. Biol. Med.* 103, 252–254.

Tennekes, H., and Lumley, J.L. (1985) *A First Course in Turbulence,* pp. 19–20, 262–264, MIT Press, Cambridge, MA.

Terasima, T., and Tolmach, L.J. (1962) *Exp. Cell Res.* 30, 344–362.

Thomas, J.A., and Johnson, M.J. (1967) *J. Natl. Cancer Inst.* 39, 337–345.

Tramper, J., Williams, J.B., and Joustra, D. (1986) *Enzyme Microbiol. Technol.* 8, 33–36.

Tramper, J., Smit, D., Straatman, J., and Vlak, J.M. (1988) *Bioprocess Eng.* 3, 37–41.

van Wezel, A.L. (1967) *Nature* 216, 64–65.

Viggers, R.F., Wechezak, A.R., and Sauvage, L.R. (1986) *J. Biomech. Eng.* 108, 332–337.

Wadia, P.H. (1975) Doctoral Dissertation, MIT, Cambridge, MA.

PART
IV

Animal Cell Bioreactor Design, Operation, and Control

Scaleup of Animal Cell Suspension Culture

Malcolm Rhodes
Simon Gardiner
David Broad

The development of monoclonal antibodies and recombinant DNA technology in the mid-1970s has generated a vast number of potential products from mammalian cell culture (Arathoon and Birch 1986). The theme of this chapter is the approach selected at Celltech for the translation of these products from the research laboratory to industrial production. We have chosen, wherever possible, to use suspension culture in a batch or fed-batch mode in airlift bioreactors.

Of course, other views and approaches have been taken and are described elsewhere in this volume. Indeed, there has been controversy over the most basic assumption of our approach; that the development of a unit process capable of scaleup is an important objective. A lot of effort has gone into the development of various techniques for producing animal cell products in small scale, benchtop apparatus. This has often enabled sufficient product for early testing purposes to be prepared in the laboratory, but if large-scale versions cannot be built for commercial production, one may be forced to use large numbers of the original laboratory apparatus. Production of 1 kg of protein might require thousands of mice in the case of monoclonal antibody (MoAb) produced in ascites, or thousands of roller bottles for a recombinant Chinese Hamster Ovary cell (CHO) product from attached culture. Many "high density" cell culture devices are equally difficult to scale up because of mass-transfer or diffusional limitations. High production costs and limited product availability are the likely commercial difficulties experienced when a unit process is not available. However, in some situations, inability to scale up may be a problem that is outweighed by the advantage of being able to launch a new product more quickly than would be possible if the development of a unit process was undertaken.

A particular problem exists with many mammalian cell products intended for in vivo use in humans. Complex proteins such as tissue plasminogen activator, erythropoietin, or immunoglobulins are regarded as biologicals by national regulatory authorities such as the U.S. FDA and the U.K. DHSS. This means that the quality of the final product relies to a great extent on the consistency of the production process. Major changes in the process could potentially alter the biological properties of the product. A change of cell line, for example, or a major process change, would need to be preceded by extensive clinical testing of the product from the new process to establish its equivalence to the original product. This amounts to a significant deterrent to improvements in process technology. By contrast, with a unit process available from the beginning, scaleup to meet increased market demand from a 1,000 l to a 2,000 l bioreactor should require much less product testing because the change in the process would be minimal.

In cases where manufacturing cost is likely to be a deciding factor in determining whether a product is commercially viable, clearly scaleup and economy of scale are going to be more important. In the long run this is likely to be the case for many first-generation animal cell products, where a number of manufacturers compete for a share of the market, or where alternative technologies such as chemical synthesis generate new competing products of lower cost. Major savings can be obtained by scaling up a unit process.

The principles of microbial process technology had already been applied to the large-scale suspension culture of animal cells well before the mid-1970s. Suspension cultures of MBIII mouse lymphoblasts (Owens et al. 1954) and L929 mouse fibroblasts (Earl et al. 1954) were demonstrated in the 1950s, and the use of stirred "spinner" vessels and modified microbial fermenters soon followed (Cherry and Hull 1956; McLimans et al. 1957; Ziegler et al. 1958); quickly demonstrating the potential for scaleup. The first commercial application of this technology was the large-scale growth of BHK 21 cells in suspension for the production of veterinary vaccines, such as foot and mouth disease vaccine (FMDV) (Capstick et al. 1962). This process has been scaled up to 3,000 l working volume bioreactors (Radlett et al. 1972). The same technology was also adapted for the large-scale production of interferon from human lymphoblastoid cells (Namalwa) (Johnson et al. 1979; Pullen et al. 1985; Phillips et al. 1985). In this chapter we will describe the application of these well-established principles and techniques to the scaleup of processes for the manufacture of recombinant proteins and MoAb for diagnostic and therapeutic applications.

10.1 SCALEUP PRINCIPLES

The prime objective in scaling up a mammalian cell unit process from laboratory scale to final production scale is usually to maintain equivalent bioreactor productivity. This ensures that the designed capacity of a given manufacturing facility can be achieved, and manufacturing costs should be similar to those predicted.

In the design of the bioreactor, we must specify the bioreactor geometry, the main reactor and ancillary vessel operating conditions, the utilities, and the instrumentation to control the physicochemical environment. The aim in process scaleup is to attempt to maintain with increasing capacity those key process variables that have a significant effect on cell growth and productivity. This may be a difficult task since the process variables are highly interrelated. An example of this is the control of volumetric oxygen transfer coefficient ($K_L a$) in scaleup for an airlift bioreactor (Figure 10–1).

$$(K_L a_D)_T = f(U_{Gr}, U_{Lr}, D_d, D_r, H_D, \rho_L, \mu, \sigma, D_L)$$

where

$$
\begin{aligned}
U_{Gr} &= \text{superficial gas velocity in the riser,} \\
U_{Lr} &= \text{superficial liquid velocity in the riser,} \\
D_d &= \text{diameter of the downcomer,} \\
D_r &= \text{diameter of the riser,} \\
H_D &= \text{height of the gas liquid dispersion,} \\
\rho_L &= \text{liquid density,} \\
\mu &= \text{liquid viscosity,} \\
\sigma &= \text{interfacial surface tension,} \\
D_L &= \text{oxygen diffusion coefficient in liquid phase, and} \\
(K_L a_D)_T &= \text{overall volumetric mass transfer coefficient based on the} \\
&\quad \text{gas-liquid dispersion volume.}
\end{aligned}
$$

It is usual to scale up reactors by maintaining similarity of the key operating variables such as $K_L a$, impeller tip speed, superficial gas velocity, or power input per unit volume. Fortunately, many of the demands on bioreactor performance of animal cell processes are trivial when compared with microbial systems due to the low biomass concentrations currently attainable. Consequently, culture fluids are usually of low viscosity, and changes in DO_2, pH, nutrient concentration, etc., are slow so that fluid mixing times are not considered to be a significant problem in practice, even in the largest industrial reactors. Similarly, oxygen and carbon dioxide mass transfer requirements are very low, even in high density cultures. At a cell density of

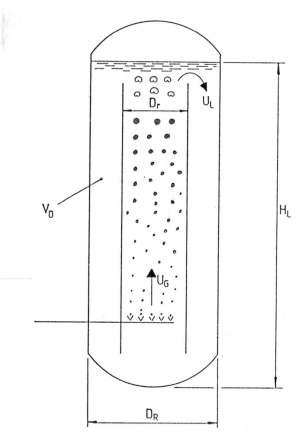

FIGURE 10–1 Airlift bioreactor schematic.

10^8 cells ml^{-1}, the $K_L a$ required is approximately 100 h^{-1}. This is at least threefold lower than is routinely achieved in large-scale microbial fermenters. The low oxygen demand may be attributed to the slower growth and metabolic rates of animal cell cultures, low biomass levels resulting from insufficient nutrient supply, or from the accumulation of toxic metabolites.

Of equal importance to bioreactor scaleup is the design of associated plant systems, which ensure that the successful operation of the reactor is practicable. Cleaning, equipment sterilization, medium preparation and sterilization, aseptic transfers from vessel to vessel, and aseptic operation of the reactor must all be carried out effectively on a large scale. While no difference in principle exists on the larger scale, detailed design based on practical experience is essential to the establishment of a reliable production plant. Some of these aspects will be discussed in more detail in section 10.3.

10.2 BIOREACTOR SCALEUP

Two basic types of bioreactor have been applied to mammalian cell culture. These are the stirred tank reactor (Katinger et al. 1979; Bliem and Katinger 1988) and the airlift reactor. At Celltech we have chosen to concentrate on airlift reactors, for the following reasons. The simple design and construction leads to (1) reduced capital cost, (2) reduced maintenance cost (no shaft bearings seals or drive mechanism to service), and (3) reduced risk of microbial infection, leading to reliable and low cost operation.

Historically, cell culture techniques at Celltech started with roller bottles and spinner vessels and increased to 30 l, 100 l, 1,000 l, and 2,000 l airlift bioreactors. The first airlift reactor at Celltech was a 30 l concentric tube airlift reactor. Systematic studies at this scale established optimum operating conditions for the growth of a wide range of mammalian cell types. The main concerns in perfecting the vessel design and operating procedures were to ensure aseptic operation and to provide adequate mixing to avoid local adverse conditions such as high temperature or low dissolved oxygen.

In addition, physical damage to the cells was reduced by minimizing liquid shear effects. The airlift design was selected partly because of its potential to provide adequate oxygen transfer rates with low shear compared with conventional stirred reactors.

Having established at 30 l scale, a laboratory process at least as effective as that seen in roller bottles or spinners, as judged by maximum specific growth rates, cell viability and antibody production rates, extensive process optimization and development was performed, which yielded increases in productivity of typically 400%. In addition, improved media were developed, eliminating the use of fetal calf serum. More recently, defined serum-free media have been further developed by eliminating more than 90% of the added protein (Rhodes and Birch 1988).

Simultaneously with this rapid advance in process development, the scaleup of the bioreactor and its associated systems was taking place to satisfy rapidly increasing commercial demand for mammalian cell products and increasing quality assurance needs. The simplicity of the airlift design and the predictability of its hydrodynamic performance assisted the design of progressively larger bioreactor units (i.e., 100 l, 200 l, 1,000 l, and 2,000 l). To ensure successful performance at these scales, the effect of aspect ratio (H_L/D_R) and superficial gas velocity (U_G) on reactor performance has been determined. The designs chosen were in all cases able to meet the oxygen transfer rates required by mammalian cell cultures (Wood and Thompson 1986).

10.2.1 Oxygen Transfer

Oxygen mass transfer rates were measured in 10 l, 30 l, 100 l, and 1,000 l airlift bioreactors, and the results compared with the predictions calculated using the empirical equation derived by Bello et al. (1981).

$$(K_L a)_{overall} = 5 \times 10^{-4} \left(1 + \frac{A_d}{A_r}\right)^{-1.2} \left(\frac{P_G}{V_D}\right)^{0.8}$$

$$P_G \simeq G_m \, R \, T \, \ln \left(\frac{P_m}{P_o}\right)$$

Reasonably good agreement with the predicted values was obtained at air-flow rates of 0.01–0.08 l l^{-1} min^{-1} (Wood and Thompson 1986). The maximum $K_L a$ value obtained increased with increasing scale; but sufficient mass transfer to support at least 3×10^6 hybridoma cells ml^{-1} was achieved at all scales. Boraston et al. (1984) reported a maximum oxygen uptake rate of 0.56 mmol O_2 l^{-1} h^{-1} by a culture of NB1 hybridoma cells, which reached 3×10^6 cells ml^{-1} maximum cell density.

This work has been further refined and extended to improve our understanding of scaleup and scaledown of airlift bioreactors. The aspect ratios (H_L/D_R) of the vessels employed by Wood and Thompson (1986) have been correlated with $K_L a$ and U_G. Using this relationship, it is possible to predict the gas flow rate for a given $K_L a$ and reactor geometry. The parameter superficial gas velocity (U_G) is more suitable for scaleup than air flow rate since it takes into account the cross-sectional area of the riser, thus making correlation between different reactor geometries easier (Figure 10–2).

10.2.2 Cell Damage

An environmental factor that is often assumed to be a greater problem in mammalian cell cultures than in microbial cultures is the possibility of mechanical damage to the cells. This is a complex question, which is dealt with in more detail by Petersen et al. (1988). The difficulty is in unambiguously identifying the true cause of cell damage and in deriving empirical relationships that can be used in scaleup.

A potential consequence of the gas and liquid mixing employed to maintain oxygen transfer and chemical homogeneity in the bioreactor is shear damage. This may result from hydrodynamic shear or from damage at the gas-liquid interface. Shear forces generated in cell culture are likely to affect cell shape, plasma membrane integrity (Schürch et al. 1988), specific growth rate, and the capacity for synthesis and secretion of product (Stathopoulos and Hellums 1985). The shear characteristics of agitated vessels have been examined in considerable detail and semiempirical equations have been derived. Far less information is available on shear in airlift reactors, particularly in relation to mammalian cell culture. However, it is clear from the literature that the selection of a variable for scaleup at a constant shear

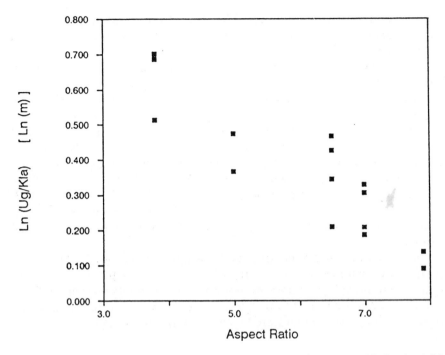

FIGURE 10–2 The effect of airlift fermenter aspect ratio on $K_L a$ (Celltech airlift fermenter scales—10–1000 l).

rate is a complex issue. For example, it has been shown by Hu (1983) that scaleup on the basis of impeller tip speed is not reliable and integral approaches have been adopted. The integrated shear factor (ISF), in addition to tip speed, takes into account the diameter of the impeller in relation to the reactor diameter.

$$ISF = \frac{2 \; \Pi \; N \; Di}{(D_R - D_i)}$$

In evaluating the effects of particular shear regimes in mammalian cells, it is important to consider the time dependence of shear damage. The term "shear work" can be used, which gives an indication of the energy transferred to the system during the course of the fermentation.

Shear work = shear rate × time of application.

The physiological state of the cell population and the chemical composition of the medium have also been identified as factors that influence shear damage (Petersen et al. 1988).

The critical factor in liquid shear damage, from the work of Croughan et al. (1987), appears to be the size and intensity of liquid eddies. Cells attached to microcarriers are only damaged by small eddies of a size and velocity large enough to affect individual cells but too small to move a microcarrier particle. The question to be asked is, when cells are freely suspended under normal agitation conditions, do eddies exist that are sufficiently small and yet sufficiently intense to damage the cells?

When the culture is sparged, strong forces arise adjacent to rising bubbles, and in the region of bubble disengagement. In airlift reactors the shear damage derived from gas sparging is the only concern in contrast to agitated vessels. This may occur when the air is injected into the culture at the sparger, when the bubbles disengage at the liquid surface, and possibly when the bubbles rise through the culture. Handa et al. (1987) have evaluated the effects on cell viability of gas disengagement and protective agents such as polyethylene glycol, serum, and albumin, but little other work has been documented on shear effects through airlift reactors for mammalian cell culture.

In industrial practice, conditions for sparging bioreactors with air and agitation through mechanical stirring or the airlift principle have been determined empirically, and these do not prove to be unduly restrictive. Progress in our understanding of this question would, however, be a major step forward in the reliability and predictability of reactor scaleup.

10.2.3 Hydrostatic Pressure

In large bioreactors, hydrostatic pressure inevitably causes a difference in pressure between the surface of the liquid and the base of the vessel. Cells cycling between these two environments could conceivably grow more slowly, or be less productive than cells experiencing less variable pressures. The effect of high pressure on mammalian cell growth and viability was tested experimentally (Wood and Thompson 1986). Cell cultures were exposed to an atmospheric pressure of up to 20 psi (equivalent water head of 14 m) for 10 cycles of 60 sec each without measurable effect. No problems have been observed in bioreactors up to 5 m tall (i.e., 1,000 l); specific growth rates are identical to those found in small fermenters, spinners, etc. Additional head pressure of up to 10 psi also has no observable adverse effect on cell physiology.

Variations in dissolved oxygen concentration occur between the top of the riser section and the bottom of the downcomer, leading to a cycling between low and high dissolved oxygen conditions. A model has been developed by Merchuk and Stein (1981) to determine the effects of hydrostatic pressure on the dissolved and gas phase oxygen concentrations in airlift bioreactors at varying respiration rates (Figure 10–3). In the riser section, oxygen is utilized at a constant rate along the reactor; however, the driving force for oxygen transfer decreases up the riser due to O_2 depletion and declining hydrostatic pressure. Consequently, the dissolved oxygen concentration (DO_2) reaches a maximum at some point above the sparger and then

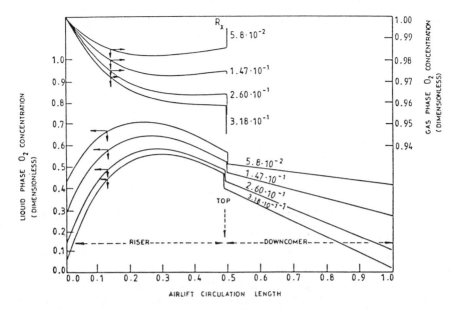

FIGURE 10-3 Profile of dimensionless oxygen concentrations in gas and liquid phases along the liquid flow path in an airlift fermenter (courtesy of Merchuk and Stein, 1981). R_x = Dimensionless respiration rate.

declines. The point of maximum DO_2 depends on the liquid flow rate and the respiration rate. In the downcomer, the DO_2 decreases in a linear fashion. The point of minimum DO_2 is around the sparger, yet measurements are typically taken at the bottom of the downcomer. The difference between maximum and minimum DO_2 can be controlled by varying the amount of gas hold-up, the liquid velocity, or the head pressure.

At constant aspect ratio, as scale increases, the hydrostatic effect tends to increase DO_2 at the bottom of the downcomer, but metabolic utilization tends to decrease DO_2 with the associated increases in circulation time.

Reports on inhibitory effects of high DO_2 on mammalian cells are fragmentary and contradictory, probably because of differences between cell lines, media, growth conditions, etc., used in each study. Our own experience (Boraston et al. 1984) is that dissolved oxygen tension has little effect on the growth rate of hybridoma cells above 10% saturation with oxygen in air at atmospheric pressure. In practice, no evidence for poorer growth or productivity has been seen in large scale airlift bioreactors up to 2,000 l.

10.2.4 Downcomer to Riser Area Ratio

The ratio of downcomer to riser cross sectional areas (A_D/A_R) is critical to mixing and oxygen transfer in an airlift bioreactor because it determines the resistance to flow in liquid circulation. The effects of A_D/A_R have been examined in detail by Bello et al. (1981, 1984) for $0.11 \leq A_D/A_R \leq 0.6$, and the following relationships have been derived.

$$(K_La)_{overall} = 5 \times 10^{-4} \left(1 + \frac{A_d}{A_r}\right)^{-1.2} \left(\frac{P_G}{V_D}\right)^{0.8}$$

$$U_{Lr} = 1.579 \left(\frac{A_d}{A_r}\right)^{0.75} (U_{Gr})^{0.32}$$

for a concentric draught tube airlift reactor.

$$\frac{t_m}{t_c} = \Upsilon \left(\frac{A_d}{A_r}\right)^{0.5}$$

where $\Upsilon = 3.5$ for a concentric draught tube airlift reactor.

As A_d/A_r is increased, K_La, U_L, and t_m all decrease to varying degrees. The optimum A_d/A_r value for minimum resistance is 0.59 whereas that for mixing time is 0.8. The optimum A_d/A_r for mixing and oxygen transfer is 0.68, although values of between 0.59 and 0.75 were not found to be critical. The bioreactors used by Celltech have been designed to give a constant A_d/A_r so A_d/A_r does not require consideration in our scaleup calculation.

10.2.5 Examples of Cell Cultures Scaled Up

Over 200 mouse hybridoma cell lines have been grown in large-scale airlift reactors at Celltech. All cells examined could be grown in a proprietary serum-free medium. Prolonged adaptation from serum-containing medium is not required since all cell lines tested will grow immediately. Defined protein components have been reduced to less than 10 mg/l in recently developed low protein media (Rhodes and Birch 1988). This assists purification and reduces quality control testing needed on the final product.

In addition, a number of human-mouse heterohybridomas and Epstein Barr Virus (EBV) transformed human lymphoblastoid cell lines producing human antibodies have been scaled up successfully. Again, all can be grown in serum-free media.

A large number of recombinant mammalian cell lines have also been grown in suspension in airlift reactors. The majority of these were Chinese hamster ovary lines that produced a variety of human proteins including tissue plasminogen activator, humanized recombinant antibodies, and tissue inhibitor of metalloproteinases (TIMP). (Rhodes and Birch, 1988).

10.3 PROCESS SCALEUP

The two basic types of processes commonly used in mammalian cell bioreactors are batch and continuous. At Celltech we have chosen to concentrate on batch production because of the simplicity, low capital cost, and the reliability of the entire production system including the bioreactor itself, but not forgetting the associated process vessels and pipework. This type of production also suits pharmaceutical production needs in particular, since well-defined batches of product are produced and can be tested and released on meeting acceptance criteria. Batch production also favors the economic production of a series of products, not requiring a dedicated facility for each product.

10.3.1 Batch Process Scaleup

The first requirement for a large-scale batch process is sufficient inoculum culture to start the full-scale process. Since mammalian cell cultures usually require a one-tenth volume inoculum to achieve rapid growth without a lag phase, a series of inoculum bioreactors is normally installed in 10-fold increasing volume; e.g., 10 l, 100 l, and 1,000 l. Each bioreactor is of similar design to the production vessel and is operated in the same basic way as already described in section 10.2. The contents of each inoculum reactor are aseptically transferred to fresh equilibrated medium in the next stage of the process.

The criteria for deciding at what time to make the transfer can vary, but typically involve the attainment of a predetermined amount of growth. This could be a particular cell density, as determined by cell counting carried out on a sample of culture aseptically removed from the reactor through a steam sterilizable sampling valve. On-line measurements that can be related to growth could also be used; e.g., O_2 uptake rate or CO_2 production rate, DO_2, etc.

Each transfer is carried out aseptically through a steam sterilizable transfer line. Transfer typically takes place by gravity feed or by overpressurizing the inoculum donor vessel. Efficient utilization of plant is dependent on predictable, reproducible growth of inoculum stages.

Sterilization of the bioreactor and its associated system is a key step in the process, since ingress of a single bacterium can cause overwhelming infection after only a few hours. The design and construction of the equipment are paramount in the success of this operation. Prior to sterilization the system is pressure tested to detect any leakage. Steam is then injected into the clean, empty vessel via the air line, and air is allowed to escape. Feed and transfer lines are also sterilized by steam, and equipment is designed to avoid cold spots caused by build up of condensate. All pipes must be free-draining and steam traps are used to ensure that condensate is aseptically drained. Connections are welded wherever possible, and valves, filter housings, and other equipment are selected to the usual standards for hygienic and sterile operation. These precautions are standard practice in the pharmaceutical industry, following the principles of good manufacturing practice, but need to be applied especially rigorously for mammalian cell culture compared with, for example, microbial antibiotic fermentation. At the end of the sterilization process the system is allowed to cool without drawing in nonsterile air and is filled with sterile medium, as described below.

Medium is made up from a basal medium powder mix, low pyrogen water, and a series of proprietary components. Neither antibiotics nor serum are normally used. The components are mixed in a stainless steel pressure vessel that has been cleaned and steam sterilized in situ to minimize pyrogen levels. Medium ingredients are selected for low contaminant levels and subjected to stringent quality assurance testing before use. The medium is sterilized by microfiltration directly into the sterile bioreactor. The DO_2 and pH probes are next calibrated, and the conditions in the reactor are then stabilized at the desired initial values. Inoculum is transferred to the main bioreactor at the required time as described below.

During the cell culture process, physiochemical conditions in the reactor are controlled by conventional process control techniques with analog set-point three term controllers for pH, DO_2, temperature, gas mass flow, and are supervised by a computer control system. Samples are taken for growth and nutrient consumption to be determined from off-line assays. Nutrient additions are made from in situ sterilized stainless steel pressure vessels. The prevention of microbial contamination during the process, which may last 14 days, is of prime importance. Process gases are sterilized by micro-filtration and exit gases are also filtered after demisting to prevent contamination. All valves in contact with the sterile plant are either steamed continuously or drained on the nonsterile side. Microbial infection rates of less than 1% have been recorded (Birch et al. 1987) and have been decreasing with increasing experience of operating mammalian cell culture reactors.

At the end of the process the contents of the reactor are clarified by either filtration or by a steam sterilized disc-stack centrifuge and then concentrated by ultrafiltration. Meanwhile, the reactor is cleaned to prepare it for the next batch. An automated clean-in-place system was described by Wilkinson (1987). A key feature of this system is the prevention of cleaning liquids from entering the process using a three valve arrangement.

The effectiveness of the cleaning procedures has been validated to ensure that no batch-to-batch product cross contamination can take place and that no cleaning materials can be carried over.

The validation of all aspects of the process plant is accomplished through the design of an effective process monitoring and control system. With regular calibration, this ensures and documents the effective operation of each batch. Many operations such as plant sterilization, filling the bioreactor, inoculation, medium sterilization, probe calibration, process control, harvesting, and cleaning are carried out automatically. This reduces labor costs and improves batch to batch consistency (Wilkinson 1987).

10.3.2 Perfusion Process Scaleup

Continuous addition of fresh medium to the reactor, with simultaneous removal of cell-free medium, can be achieved by a number of means, such as an internal spin filter inside the bioreactor (Tolbert et al. 1981) or an external loop with a pump and a filter. This type of perfusion culture (or continuous culture with biomass feedback, Pirt 1975) is frequently used with microcarrier cultures but has also been used with suspension culture. The main scaleup problems are the fouling of filters, causing retention of product inside the reactor, and possibly the retention of enzymes, which could destroy or modify the product.

These problems may be avoided by the use of a settling zone device in place of filtration. This process has been operated at 30 l scale (Birch et al. 1987). The potential advantage of perfusion is the ability to obtain high cell densities without having to carefully optimize medium composition. However, this higher cell density does not necessarily result in a useful economic advantage in industrial circumstances. Although a smaller bioreactor may be needed to produce a given quantity of antibody in a continuous settling zone reactor with retention of cells, there is no saving in the quantity of medium required. The capital cost of production plant is very similar for the two systems. The reason for this is that the total volume of sterile plant for the continuous system is very similar to that for the batch system, if sterile medium production and storage and sterile product liquid storage are included. The storage vessels have to be built to similar standards of asepsis, hygiene, cleaning, sterilization, temperature, and pH control and, therefore, are as costly as fermenters.

There is therefore little to choose between batch and continuous production in cost terms. However, the continuous plant is more complex, and therefore less reliable, more prone to microbial contamination, and more difficult to scaleup. The higher cell densities obviously place higher demands on bioreactor performance; e.g., $K_L a$ and O_2 supplementation of the gas stream may become necessary.

10.3.3 Chemostat Process Scaleup

Perfusion culture is not appropriate for processes where production is growth related. This is because with 100% biomass feedback, the specific growth rate of the cells and consequently the production rate tend toward zero (Pirt 1975). However, if cell-containing culture is removed from the bioreactor at the same rate as medium is added, a steady-state chemostat culture can be established and the growth rate of the cells is equal to the dilution rate of the culture. Optimization of the dilution rate and appropriate selection of the limiting nutrient can yield a high productivity system.

In terms of equipment design and operation, chemostat culture is very similar to perfusion culture. The main difference is that a filtration device such as a spin filter is not needed, which is an advantage. Chemostats may be run for a prolonged time at high dilution rates, and the cells divide through many generations. Even low levels of cell line instability could limit the useful run time.

Chemostats do not appear to have been extensively used in industrial production systems with mammalian cells.

10.4   CONCLUSIONS

Suspension culture is the production method chosen by industry for the manufacture of the vast majority of animal cell products. Where adherent cell lines must be used, as in human viral vaccine production, the use of microcarrier cultures grown in suspension is rapidly gaining popularity. In both cases, the main reason is the ease of scaleup of homogeneous processes, leading to economy of scale. Economies result principally from minimizing labor costs. This objective is also assisted by the increased use of automated process control systems. Batch culture is also favored because it is simpler to scale up. It is also well suited to the needs of the pharmaceutical industry, allowing process validation and easier process control. Batch records and traceability of raw materials used are generally simpler to maintain for batch processes than they are for extended continuous processes.

The main engineering issue for scaleup of animal cell cultures is to determine the maximum oxygen transfer rate that can be obtained without causing cell damage. The difficulty is our poor understanding of the mechanisms by which cells may be damaged.

Until this issue is resolved, the choice and design of mammalian cell bioreactors will remain controversial. In practice, this is not a problem for batch suspension cultures, but may be a problem with high cell density processes where the oxygen transfer requirement is higher. Since there is little or no economic advantage to high cell density processes, there seems to be little point in unnecessarily risking mass transfer problems by employing them.

10.5   NOMENCLATURE

$K_L a$	Volumetric mass transfer coefficient
U	Superficial velocity
D	Diameter
H	Height
ρ	Density
μ	Viscosity
σ	Surface tension
d	Oxygen diffusion coefficient
A	Cross-sectional area
P	Power
V	Volume
G_m	Gas molar flow rate
R	Universal gas constant
T	Temperature
P_m	Main line gas pressure
P_o	Gas pressure at liquid surface

N	Stirrer speed
t	Time
P_g	Gassed Power input

Subscripts

G	Gas
L	Liquid
r	Riser
d	Downcomer
D	Gas/liquid dispersion
R	Reactor
T	Total
i	Impeller
m	Mixing
c	Circulation

ISF	Integrated shear reactor
SGV	Superficial gas velocity
DO_2	Dissolved oxygen

REFERENCES

Arathoon, W.R., and Birch, J.R. (1986) *Science* 232, 1390–1395.

Bello, R.A., Robinson, C.W., and Moo-Young, M. (1981) *Adv. Biotechnol.* 1, 547–552.

Bello, R.A., Robinson, C.W., and Moo-Young, M. (1984) *Can. J. Chem. Eng.* 62, 573–577.

Birch, J.R., Lambert, K., Thompson, P.W., Kenney A.C., and Wood, L.A. (1987) in *Culture in Technology* (Lydersen, K., ed.), pp. 1–20, Hanser Publications, Munich, Germany.

Bliem, R., and Katinger, H. (1988) *Trends Biotechnol.* 6, 190–195, 224–230.

Boraston, R., Thompson, P.W., Garland, S., and Birch, J.R. (1984) *Develop. Biol. Stand.* 55, 103–111.

Capstick, P.B., Telling, R.C., Chapman, W.G., and Stewart, D.L. (1962) *Nature* 195, 1163–1164.

Cherry, W.R., and Hull, R.N. (1956) *Anat. Rec.* 124, 483.

Croughan, M.S., Hamel, J.-F, and Wang, D.I.C. (1987) *Biotechnol. Bioeng.* 29, 133–141.

Earle, W.R., Schilling, E.L., Bryant, J.C., and Evans, V.J. (1954) *J. Natl. Cancer Inst.* 14, 1159–1171.

Handa, A., Emery, A.N., and Spier, R.E. (1987) *Develop. Biol. Stand.* 66, 241–254.

Hu, W.S. (1983) Doctoral Dissertation, MIT, Cambridge, MA.

Johnson, M.D., Christofinis G., Ball, G.D., Fantes, K.H., and Finter, N.B. (1979) *Develop. Biol. Stand.* 42, 189–192.

Katinger, H.W.D, Scheirer, W., and Kromer, E. (1979) *Ger. Chem. Eng.* (Engl. Transl.) 2, 31–38.

McLimans, W.F., Davis, E.V., Glover, F.L., and Rake, G.W. (1957) *J. Immunol.* 79, 428–433.

Merchuk, K.C., and Stein, Y. (1981) *Biotechnol. Bioeng.* 23, 1309–1324.

Owens, O., Gey, M.K., and Gey, G.O. (1954) *Ann. N.Y. Acad. Sci.* 58, 1039–1055.

Petersen, J.F., McIntire L.V., and Papoutsakis, E.T. (1988) *J. Biotechnol.* 7, 229–246.

Phillips, A.W., Ball, G.D., Fantes, K.H., Finter, N.B., and Johnston, M.D. (1985) in *Large Scale Mammalian Cell Culture* (Feder, J., and Tolbert, W.R., eds.), pp. 87–96. Academic Press, London, England.

Pirt, S.J. (1975) *Principles of Microbe and Cell Cultivation*, Blackwell, Oxford, England.

Pullen, K.F., Johnson, M.D., Phillips, A.W., Ball, G.D., and Finter, N.B. (1985) *Develop. Biol. Stand.* 60, 175.

Radlett, P.J., Telling, R.C., Whiteside, J.P., and Maskell, M.A. (1972) *Biotechnol. Bioeng.* 14, 437–445.

Rhodes, P.M., and Birch, J.R. (1988) *Biotechnology* 6, 518–523.

Schurch, U., Kramer, H., Einsele, A., Widmer, F., and Eppenberger, H.M. (1988) *J. Biotechnol.* 7, 179–184.

Stathopoulos, N.A., and Hellums, J.D. (1985) *Biotechnol. Bioeng.* 27, 1021–1026.

Tolbert, W.R., Feder, F.J., and Kimes, R.C. (1981) *In Vitro* 17, 885–890.

Wilkinson, P.J. (1987) in *Bioreactors and Biotransformations* (Moody, G.W., and P. Baker, eds.), pp. 111–120, Elsevier, London.

Wood, L.A., and Thompson, P.W. (1986) Proc. Int. Conf. Bioreactor Fluid Dynamics pp. 157–172.

Zeigler, D.W., Davis, E.V., Thomas, W.J., and McLimans, W.F. (1958) *Appl. Microbiol.* 6, 305–310.

Continuous Cell Culture

Mary L. Nicholson
Brian S. Hampson
Gordon G. Pugh
Chester S. Ho

Traditionally, large-scale cell culture has been accomplished in a batch mode using a bioreactor system. Under these circumstances the cells grow and produce protein products until they deplete their nutrient supply, at which point growth slows and the cells proceed to die. Such systems are relatively simple and work very well with bacterial cells. However, skilled labor is needed to scale up cells and turn systems around after each batch. Animal cells grow slower and produce less product per cell, and many require attachment to a substrate for growth or maximum yield. Hence, for animal cells, continuous perfusion culture has proven superior to batch cultures. The four main methods, reviewed in the next section, offer the following advantages:

1. Flexibility in operation and automation of the system since feed, circulation and flow rates, oxygen levels, and pH, etc., can be controlled more precisely (Looby et al. 1987).

We would like to thank Lee Noll for guidance, and Kyle A. Wallace and Nasim G. Memon for their excellent technical assistance. We would also like to thank Lucy D. Phillips for editorial assistance.

2. Continuous steady-state conditions that translate into high cell or product yields (Bodeker 1985; Berg et al. 1988).
3. Retention of the cells in the bioreactor, meaning a higher product yield since the cell's energies go into protein production instead of cell division (Nicholson 1988). The decrease in cell division also lessens the chance for genetic drift in the cell population.
4. Less expensive downstream processing because there are few or no cells to be separated from the product (Murdin et al. 1987).
5. Less expensive media, since cells are maintained at lower concentrations of serum during the nongrowth phase.
6. Decreased degradation of the protein products, as they are rapidly isolated from proteases and elevated temperatures.

A continuous perfusion technology in which cells are immobilized using a ceramic matrix offers additional advantages. These include a homogeneous environment for all of the cells; immobilization of suspension as well as anchorage dependent lines; linear scaleup; and a sufficient oxygen supply to the cells. An example of such a system, trade-named Opticell, will be described after the following background section.

11.1   METHODS OF CONTINUOUS CELL CULTURE

Continuous cell culture can be simply achieved if a portion of the cells are not retained in the bioreactor but are harvested with the soluble product. However, as constant cell division is then needed to replenish the harvested cells, many advantages of continuous culture are lost. In this chapter we will deal only with continuous culture in which cells are retained in the bioreactor. This can be achieved using a filtered cell recycle to a stirred tank. Other methods include retention in the extracellular space of a hollow fiber, immobilization in a gel matrix, or adsorption to a support such as stainless steel, glass, or ceramic.

In the first case, a stirred tank can be used with a cell return such as a tangential flow filter. Media can be fed into the system, and cell-free product extracted from the opposite side of the filter. In a variation on this principle, systems designed by Himmelfarb et al. (1969) and Feder and Tolbert (1983) use a tangentially washed spinning filter in the center of the reactor to remove protein products. As such systems are based upon a standard bioreactor, gas is introduced via sparging or diffusion. Such systems can range from 2–1,500 l in size and may be used with suspension cells or microcarriers. Stirred tank systems that recycle are more cost-efficient because the culture can be maintained for longer periods in a suppressed growth state, and the product is harvested free of cells. The difficulties of these systems usually result from the fragility of the cells and the tendency of filters to clog. Also, engineering problems remain, such as foaming from sparging oxygen into media containing serum.

Another method for cell retention is a shell and tube bioreactor such as a hollow fiber (Knazek et al. 1972; Ku et al. 1981; Weisman et al. 1985). Cells are grown on the shell side while media flows through the tubes. Nutrients, oxygen, and waste products diffuse across the membrane. Such membranes come with 10,000, 50,000, and 100,000 molecular weight cut-offs. Whether the product is passed through depends upon the size of the pores. Such systems can feed from 1–5 l of media per day using one hollow fiber cartridge or 3–50 l/day/system using multiple cartridges. They allow cells to reach densities of about 1×10^8 cell/ml in the extracapillary space and can run for long periods. Product retained with cells is highly concentrated. Costly serum components can be retained on the side of the membrane with the cells, yet the product can be removed in a highly concentrated, cell-free form. However, these conditions promote product degradation and feedback inhibition, and problems exist with gradient formation for nutrients, waste products, and especially oxygen, on the cell side of the fiber (extracapillary space). Such problems have been decreased by fluctuating the pressure as well as initiating counter current flow in the extracapillary spaces, thereby increasing the passage of media across the membrane. But then systems are not scalable: flow of cell nutrients across the membrane is dependent upon the length of the fiber being relatively short. Fibers are also prone to breakage. Only indirect environmental control can be maintained, as the flow through the fibers does not reflect the conditions to which the cells are exposed.

Another method of retention is by adsorption of the cells to weighted matrix beads (Dean et al. 1987). Cells are entrapped in a sponge-like collagen matrix with a 3–5 mm bead size. The beads form a packed, fluidized bed in a closed-loop, clean-in-place, sterilize-in-place system ranging in size from 50–500 l/day feed rate. Such a system can be used for either adherent or suspension cells and maintains a cell density of $\sim 1 \times 10^7$–10^8 cell/ml in the bed. It is available only at large scale and has a relatively high capital and substrate cost. Harvest is very difficult from such a substrate. Alternatively, cells can be encapsulated in solid algonate beads, which are formed in the reactor itself. The beads are semipermeable to 1,000,000 molecular weight and run in a 3 l or 6 l airlift fermenter. Cells reach a fairly high density of 1–5×10^7 cells/ml, and the process can be used for anchorage-dependent as well as suspension cells. However, it is available only in a relatively small size, and there are problems with foaming and bead degradation.

Cells can also be adsorbed to a charged surface such as stainless steel, plastic, glass, or ceramic. Recently, rapid progress has been made in understanding the nature of this adsorption (for reviews Ruoslahti and Pierschbacher, 1987; Hynes 1987). Many adhesive proteins in serum and extracellular matrices contain a cell recognition site: the tripeptide sequence arginine–glycine–aspartic acid (RGD). Including fibronectin, the collagens, vitronectin, and fibrinogen, these proteins adsorb to plastic and glass surfaces by charge attractions and hydrophobic interactions. Cellular attachment is then facilitated by specific receptor attachment to these proteins (Kleinman et al. 1981). The RGD sequence is recognized by cells via a family of structurally related receptors called integrins. These adhesion protein/receptor combinations are involved in multiple functions, including anchorage, migration, differentiation, and cell polarity.

Following initial attachment of the spherical cells, the cytoplasm flattens out and the cell membrane forms multiple other attachments. Cells have been grown adsorbed to 1–3 cm stainless steel turnings in a fluidized bed. Available in a 3–6 l configuration using sparging as the oxygenation method and feeding at 1–4 l/day, the system allows cells to reach a moderate density of 3–5 $\times$ 10^6 cells/ml and can be run in an automated continuous process with a reusable substrate. However, it represents relatively new technology and suffers from common fluidized bed problems of channeling and nonuniform growth. Moreover, cells can be dislodged by the sparging action of gassing the media.

Animal cells can also be immobilized by adsorption to a honey-combed ceramic matrix. In the Opticell, this monolithic structure is known as the "core," from the patented name of Opticore. It consists of many square channels running the length of a cylinder (Figure 11–1) and forming a rigid support for cell adsorption. The cross-sectional channel density 90 to 400

FIGURE 11–1 Ceramic matrices. The smooth ceramic honeycombed matrix on the left is used for attachment-dependent cells when the cells must later be removed from the ceramic. The porous ceramic on the right is typically used for either suspension or attachment-dependent cells when the cellular product of interest is secreted into the media.

channels per square inch results in a very high surface area-to-volume ratio and, consequently, to a relatively high density of cells in the ceramic (10^7–10^8 cells/ml). The matrix or core has been incorporated into a closed-loop, direct perfusion system that is highly automated and flexible for a variety of applications (Bognar et al. 1983; Noll 1984; Lyderson et al. 1985a and 1985b). Opticell systems are currently scalable $100\times$ from 1 l/day to 100 l/day typical feed rate, with an industrial production system at $600\times$ under development. This chapter will further describe the advantages of continuous perfusion culture using this system as an example.

Attachment-dependent cells adsorb and grow on the ceramic much as they do on plastic (Lyderson 1987). Figure 11–2 represents Marvin Darby canine kidney cells (MDCK) growing across the surface of the attachment-dependent (AD) ceramic. Figure 11–3 shows a view of this ceramic wall in cross-section. The MDCK cells are the very thin dark lines on the top edge of the nonporous wall. The ceramic material is solid, preventing cells and

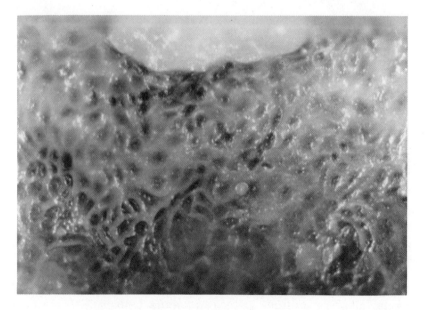

FIGURE 11–2 MDCK cells on AD ceramic. This represents Marvin Darby kidney cells (MDCK) reaching confluence after six days of growth as a monolayer on the AD (smooth, nonporous) ceramic. A small section of the white ceramic can be seen at the top of the photo. All photos of cells on ceramic presented in this chapter were prepared as described below. Cells were fixed to the ceramic with 10% formaldehyde in Delbeccos' phosphate buffered saline for at least 30 min. After washing with distilled water, cells were stained with Harris hematoxylin/eosin or for hybridomas with hematoxylin alone for 7 min and rinsed with water. When possible, these procedures were carried out without draining the core since some loosely attached cells (especially suspension cells) might detach.

FIGURE 11-3 MDCK cells growing on the AD ceramic channel wall. This is a cross-section through the channel wall of the ceramic represented in Figure 11–2. The thin dark line on the top of the channel wall represents MDCK cells (see arrow).

liquid from passing from one channel to another. Traditionally, this approach was limited to anchorage-dependent cell types. More recently, however, it has been used to adsorb suspension cells such as lymphocytes and hybridoma cells to a porous ceramic, the suspension (S) ceramic (Putnam 1987; Lyderson et al. 1987). The mechanism might be explained by our working hypothesis, as follows.

Most hybridomas seem to have a very slight attraction to flat, tissue culture ware. However, they have a small percentage of cytoplasm in comparison to anchorage-dependent cells and since nuclei cannot flatten out, they cannot make multiple tight attachments to a flat support. A tighter bond might be possible if the surface conformed to the spherical shape of these cells, allowing multiple cell/substrate attachments. The porous ceramic, with pores approximately the same size as the cells, could allow for such attachments.

This theory was tested by screening ceramics of the same composition but of varying porosity for their ability to adsorb hybridoma cells. Although anchorage-dependent cells attach well to the smooth ceramic (Opticell AD core), hybridomas do not attach at all. Hybridomas attach only poorly to the ceramic with a smaller pore size (most pores < 0.5 μm). The ceramic to which hybridomas attach best (Opticell S core) has pores that were approximately the size of a spherical cell (10–20 μm) (Figure 11–4). Initial

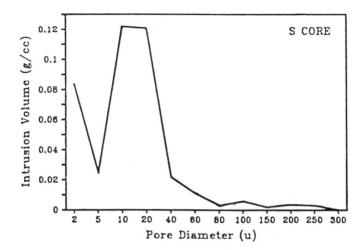

FIGURE 11-4 Pore size distribution of S ceramic. Mercury intrusion volume of the S ceramic as a measure of pore size distribution. Data provided by Corning Glass Works, Corning, NY.

attachment occurs within 10 min. When L243 hybridoma cells that have attached and grown on the S ceramic are examined, there seem to be legitimate attachments to the surface (Figure 11-5). These cells are nevertheless not attached as tightly as the anchorage-dependent cells. The shear force of an air/water interface provided by shaking a ceramic core while half full of media will dislodge most suspension cells but not anchorage-dependent cells. (Note: The porous ceramic is often used for anchorage-dependent cells, when cell harvest is not necessary, because the porosity generates a greater surface area for cell attachment than does the flat AD surface.)

11.2   AUTOMATION

Perfusing medium past an immobilized culture provides the opportunity for improved monitoring and automation. In the Opticell, sensors placed in the flow path both before and after the culture detect changes in the medium with each pass through the culture. These sensors are coupled to on-line autoanalyzers, providing a sensitive measurement of changes in culture dynamics. Diaphragm and membrane sensors used in bioreactors with immobilized cells are less likely to foul since the recirculating fluids have fewer cells or particles in suspension. The Opticell currently monitors both pH and dissolved oxygen (DO_2). Although pH changes vary in importance with the application, the oxygen uptake has proven to be consistently useful. With the aid of a computer, calculated uptake rates can be used to automate

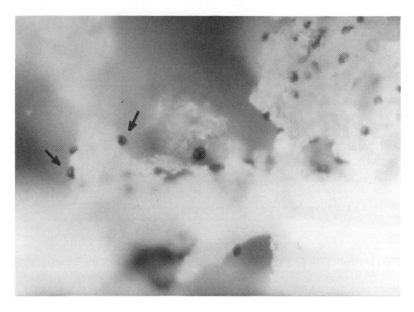

FIGURE 11–5 L243 hybridoma cells on S ceramic. This photo represents L243 hybridoma cells following several days of growth on the S core. Notice the slight flattening of the hybridoma cells that are attached to the outcropping of the S core on the left of the photo (see arrows).

the culture process. As sensor technology advances, other parameters may be utilized to gain a greater control over the production environment. Medium recirculation rates, gas setpoints, and ultimately product harvest rates can be automatically controlled based on this continuous feedback system.

As shown in Figure 11–6, the Opticell ceramic matrix or core is incorporated into a recirculating flow loop that continuously perfuses medium over the immobilized cells. Specifically, medium is drawn from the reservoir by a recirculating pump, it flows through a probe chamber where pH and oxygen are measured, then passes through the core to nourish the cells. It then flows through a second probe chamber where oxygen is measured as the medium travels to a gas exchange device and finally returns to the reservoir.

One of the most difficult problems with high density cultures is their requirement for large amounts of oxygen to maintain the health of the cells. In bacterial fermenters, sparging has been efficiently used to accomplish this. However, animal cell culture media are more complex than bacterial fluids due to the common supplementation with bovine serum. Sparging gas typically results in foaming that can lead to cell death if not controlled. A benefit of an immobilized perfusion bioreactor is the ability to oxygenate the media separately from the cells. Oxygen-depleted medium leaving the

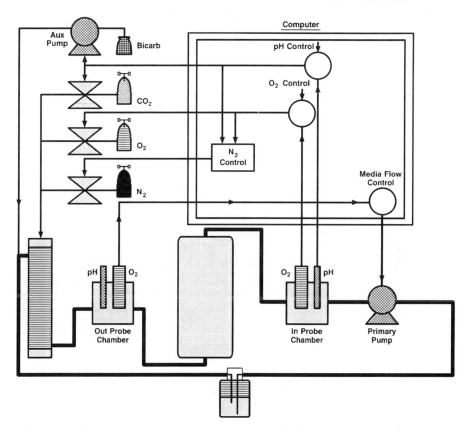

FIGURE 11-6 Opticell process control. Counterclockwise flow around the circulation loop passes through the primary pump, the "IN" probe chamber, the ceramic core, the "OUT" probe chamber, and then the permeator before returning to the media vessel.

culture chamber is passed through a permeator that contains silicone tubing. A mixture of gases enriched in oxygen passes this tubing and diffuses into the culture medium. In this way, foaming is completely eliminated. Also, the medium is well mixed before reentering the cell culture chamber, so a uniform concentration of dissolved gases is delivered to the cells. Finally, the silicone provides a sterile barrier between the gas and the medium, which eliminates the need for sterile gas and the risk of contamination due to filter problems.

In such a system, component selection for the flow loop is very important. The Opticell recirculating pump is a peristaltic type that is completely nonintrusive and eliminates contamination problems that may occur with other pump technologies. Only the detachable tubing, not the pump, need be sterilized. Over many years of use, the peristaltic pump has proven to create no shear problems when seeding the core or with the small numbers of suspension cells that may remain in the circulating loop. The probes (Ingold) are sterilizable and used extensively in the fermentation industry. The oxygen probes are based on the Clark principle and, with proper care, show the excellent long-term stability required by the continuous cultures. pH probes are combined glass electrodes with measurement and reference half-cells in a single housing.

Figure 11–6 also depicts the basic process control features implemented as part of the Opticell life support system. Classic proportional/integral (PI) feedback control loop algorithms are employed for each of three control variables: oxygen IN, oxygen OUT, and pH. Setpoints for each of the variables are entered into a control computer that in turn manipulates various outputs to achieve a match between the setpoint and the actual value measured by each probe. Basic O_2 IN and pH control are achieved by modulating the proportion of oxygen and CO_2 gas, respectively, supplied to the gas exchange device. Nitrogen gas is used as makeup gas and to prevent interaction between the two control loops. Solenoid valves for each gas are automatically sequenced by the computer to achieve the desired proportioning.

The pH control is extended by the use of an auxiliary pump that automatically meters in a base or bicarbonate solution when necessary. Early in the culture, CO_2 must be added to the medium to bring the pH down to the desired value; however, as the culture matures it generates more CO_2 and lactic acid. When CO_2 removal via the permeator (CO_2 addition versus removal is controlled by varying the CO_2 gas percentage to the permeator) and lactic acid removal via feed/harvest procedures are not sufficient to keep the pH high enough, base can be added in a controlled manner. This strategy is implemented by the computer by extending the output range of the control loop to allow calculation of negative CO_2 values. Thus, if the output of the control loop is negative, base or bicarbonate is metered proportionate to the negative value. In addition, as the culture grows, more oxygen will be removed from the medium as it moves past the cells in the ceramic matrix. The O_2 OUT value will drop. The system will then increase the recirculation rate to deliver more media and, hence, more oxygen to the culture per unit time. The net effect is that the O_2 OUT level is held constant, thereby precisely controlling oxygen gradients throughout the matrix.

Temperature control is implemented in some cases by placing the unit within an independently controlled incubator or warm room such that the entire flow loop is surrounded by air at the desired temperature. Due to the temperature gradients within the incubator as well as local heating by the circulation pump, the temperature of the medium delivered to the ceramic in such a configuration may vary. Highest accuracy is achieved by integrating temperature as a fourth control variable. Two measures are taken: the temperature of the medium just before it enters the ceramic and the temperature of the air or liquid used to transfer heat to the medium. The output of the medium temperature control loop is used to modulate the setpoint of the external temperature control loop (via a cascaded Proportional Integral Derivative [PID] feedback control loop). The output of this control loop then modulates an electrical heating element or steam valve to add heat to the air or liquid, respectively. Although the external temperature is allowed to rise above the desired medium temperature, a maximum limit is imposed to ensure that transient conditions do not generate excessively high temperatures, which may degrade medium components.

With these basic automatic control functions in place, the last requirement that needs to be addressed is the replenishment of the medium, both to replace nutrients consumed by the cells and to remove toxic waste products. This can be done by simply performing a batch change of all the medium in the flow loop on a periodic basis. However, the feast-or-famine situation this creates has been observed to cause very cyclical metabolism and, ultimately, lower productivity from a culture (Putnam 1987; Lyderson 1987). In order to eliminate this problem, Opticell uses a continuous feed and harvest system whereby two additional pumps continuously add and remove small and equal amounts of medium, maintaining a constant volume within the loop. Pumps are automatically operated by the control computer to achieve the correct exchange rate.

In addition to the above mentioned control feature, the computer will calculate and plot the oxygen consumption rate (OCR), which is proportional to the oxygen gradient across the ceramic matrix times the flow rate through the matrix. Specifically

$$OCR = (O_2 \text{ In} - O_2 \text{ Out}) \times \text{Flow Rate} \times K$$

where K is a constant of proportionality that takes into account Henry's law concerning the conversion of oxygen partial pressure (measured by the oxygen probes) into an oxygen concentration value. As detailed in the next section, the trend of the OCR has been shown to correlate very well with both the growth curve and productivity of the culture (Berg 1985; Lyderson et al. 1985a; Putnam 1987). Because it provides immediate on-line feedback, the trend of the OCR has proven to be useful in both managing the culture (e.g., setting medium addition rates, infection timing, transition time for conversion of serum containing to serum-free medium) and in optimizing the culture.

11.3 ASSESSING CELL YIELD AND PRODUCTIVITY IN AN IMMOBILIZED SYSTEM

One of the difficulties often encountered in an immobilized system is that the immobilized cells cannot be easily sampled. One must deduce their total numbers, metabolic state, and productivity from the media in which the cells are bathed. Various parameters such as OCR, rate of nutrient utilization, and production of waste products are commonly used to assess the state of the cells in a timely fashion. It is always important, however, to determine how well these parameters correlate to the more difficult-to-determine parameters such as cell number and especially productivity.

In the Opticell, the primary on-line monitor of cell metabolism is the OCR. While OCR is actually an estimate of total cellular metabolism, we have also found correlations to cell number in growth and at plateau (Lyderson et al. 1985a). Even with a 100× scaleup we have found that OCR is proportional to the number of live L243 hybridoma cells on an S core (Figure 11–7). It should be noted, however, that when OCR falls dramatically in a healthy culture due to a problem such as feed and/or harvest, oxygen, or pH regulation, it does not usually indicate a change in viable cell number. The OCR recovers too rapidly after eliminating the cause of the problem to be explained by an increase in the cell number. Instead, most of the cells seem to drop back to a reduced metabolic state when faced with severe environmental conditions. It has been shown in thymocytes, for example, that glucose starvation leads to a reduction in protein synthesis, RNA synthesis, cellular ATP levels, ability of cells to transport amino acids,

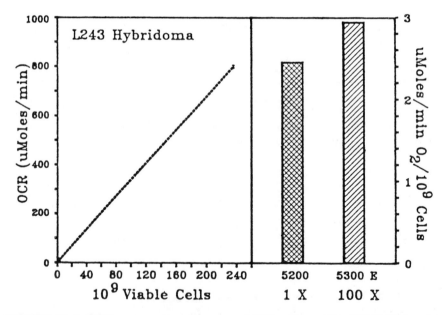

FIGURE 11–7 OCR versus cell number over 100× scaleup. The 5200R (1× scale) was run as described in Figure 11–11. The 5300E (100×) was run as described in Figure 11–21. Cells were removed from the S cores by shaking the cores while half full of Delbecco's phosphate buffered saline followed by draining. This was repeated about five times. Using this method, less than 5% of the cells remained on the core as determined by fixing and staining the cores after cell removal.

and, later, to nuclear fragility and cell lysis (Young et al. 1981 and 1979). A slight change in energy charge ([ATP] + 1/2 [ADP] / [ATP] − [ADP] − [AMP]), no matter what the cause, will dramatically reduce the rates of protein synthesis (Mendelsohn et al. 1977). The rate of protein synthesis, however, will begin to recover very quickly after reexposure to glucose (Young et al. 1979). In order to continue to supply the energy needs of processes necessary for survival (i.e., ion transport) in a stressful environmental condition, cells may suppress less critical cellular functions, such as protein synthesis.

OCR will change dramatically if the cells find themselves in less than optimal conditions. If the temperature, pH, feed rates, or media components drift out of the optimal range for the cells, OCR will plummet (Pugh, 1988). For example, a 2 °C drop in temperature will cause an 8% drop in OCR. If OCR is monitored on-line, this will be noted and can usually be corrected in time to allow recovery of the culture. As animal cell culture systems increase in size and cost to run, monitors become critical to successful operation. The fact that OCR is such a sensitive and quickly responding indicator of a problem is the real essence of its value in a bioprocessor. If there were no such monitor of the metabolic state, the culture could be

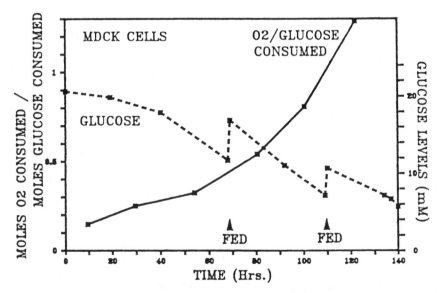

FIGURE 11–8 Oxygen consumed versus glucose consumed by MDCK cells. MDCK cells were grown in the Opticell and fed in a batch mode with Delbecco's minimal essential medium (DMEM), 4.5 g/l glucose, plus 10% FCS. Samples taken from the feed, media, and harvest vessels were assayed for glucose. The dashed line represents the level of glucose in the media vessel (in the circulation loop with the cells). The solid line represents the moles of oxygen consumed as calculated from on-line OCR measurements (μM/min) divided by moles of glucose consumed.

irreversibly damaged before the problem was noticed. It goes without saying that in diagnosing and attempting to correct the situation, OCR is an equally responsive indicator of a positive result.

OCR will also fluctuate with the glucose level in the system. As can be seen in Figure 11–8 and Figure 11–9, as the glucose drops, the cells become more efficient at its utilization, i.e., the moles of oxygen consumed per mole of glucose consumed increases. In practical terms, a detrimental effect on growth or productivity is rarely found if the glucose level remains above 1 mg/ml glucose (5.5 mM). If, however, the glucose level in a high density culture immobilized on the ceramic drops below 1 mg/ml, then productivity and/or growth often drops.

Experience shows that when bioprocessor parameters are kept constant from run to run, the rate of culture respiration as measured by OCR and yield of product do not vary by more than 10–15%. This reproducibility between runs becomes significant when there is a need to develop production schedules and deliver quantities of product on time. In many cases, outstandingly high production is of little value unless it can be simply and predictably reproduced in the production facility. Over the years, many cell

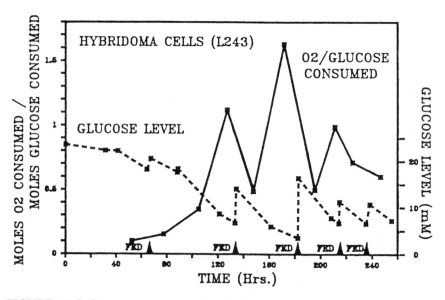

FIGURE 11–9 Oxygen consumed versus glucose consumed by L243 hybridoma cells. L243 hybridoma cells were grown in the Opticell and fed in a batch mode with DMEM (4.5 g/l glucose) plus 10% FCS. Samples taken from the feed, media, and harvest vessels were assayed for glucose. The dashed line represents the level of glucose in the media vessel. The solid line represents the moles of oxygen consumed as calculated from on-line OCR measurements (μM/min) divided by moles of glucose consumed. Note that the moles O_2 consumed/mole glucose consumed spikes up each time the glucose level drops.

lines have been repeatedly grown in Opticell bioreactors for the production of both cells and secreted proteins. Data from such runs support the predictability possible in these bioreactors.

As an example, attachment dependent MDCK cells were seeded on 4,164 cm² AD cores and 850 cm² plastic roller bottles in two identical tests. All cultures were seeded at the same density and provided with equal volumes of DMEM high glucose medium supplemented with 10% fetal bovine serum (FBS) based on surface area. After five days the cells were harvested and the Opticell 5200's yielded 8.2 and 8.8 × 10⁸ cells, respectively, while the roller bottles yielded 1.62 and 1.36 × 10⁸ cells. This represents yields averaging 2.0 × 10⁵ cells/cm² for the bioreactor and 1.75 × 10⁵ cells/cm² for the roller bottles. More importantly, however, the variability in the Opticell yields was only 7% and the OCR curves were very similar (Figure 11–10). The use of the OCR has been proven not only to be a good predictor of reproducibility between ceramics of the same size, but also of reproducible growth trends when compared to systems containing more surface area (i.e., multiple or larger ceramic matrices). As will be discussed later, the OCR slopes of different scale cultures will be similar if similar run parameters are maintained.

Most important of all, however, OCR can often be a reliable and speedy indicator of productivity (Putnam 1987). This may be due to the steep curve that relates protein synthesis to adenylate energy charge in the cells (Men-

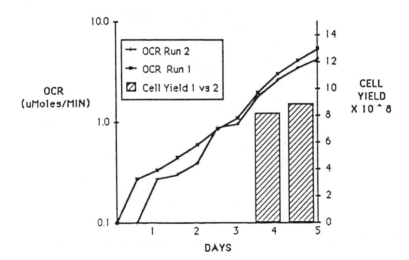

FIGURE 11–10 OCR and cell yields for duplicate runs in an Opticell 5200R. Attachment dependent MDCK cells were seeded at 2 × 10⁴ cells/cm² onto 4,164 cm² AD-51 Opticores. Cultures were fed with DMEM (4.5 g/l glucose) supplemented with 10% fetal bovine serum (FBS). After five days of growth, cells were harvested with 0.25% trypsin containing 0.02% EDTA. The Opticell 5200R yielded 8.2 and 8.8 × 10⁸ cells, which averages 2.0 × 10⁵ cells/cm².

delsohn et al. 1977). We have consistently found that when a protein is produced by healthy cells, if OCR is high, protein production is high. The ratio of antibody production rates versus OCR can be remarkably stable (Figure 11–11 and Figure 11–12). We have also noted this effect with products from genetically engineered mammalian cells (unpublished data). On the other hand, if the product appears in dying cells (i.e., virus-infected cells), a fall in OCR is a good indicator of production rates.

11.4 MAINTAINING STEADY-STATE CONDITIONS

Perfusing media past immobilized cells can achieve a relatively homogeneous micro- as well as macroenvironment around the cells. Difficulties can arise, however, if the flow past the cells is dead-ended or nonuniform, blocked by a cell mass buildup, or separated by a physical barrier such as a membrane (i.e., as with hollow fibers). Some sections of the system may

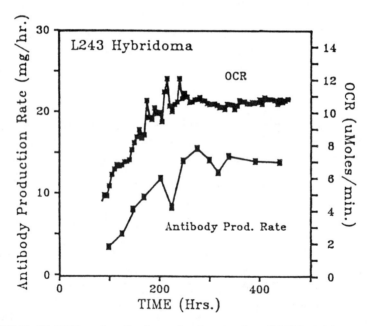

FIGURE 11–11 OCR and antibody production rate for a L243 hybridoma cell line. An Opticell 5200R was seeded at 1×10^8 cells and grown in DMEM with 4.5 g/l glucose supplemented with 10% Serum Plus (Hazelton Biologics, Inc.) and gentamicin (50 mg/l) for 10 days. This media was then exchanged for 1% Serum Plus medium and maintained at a feed rate of 1 l/day for 18 more days. Total cell yield from the core was 2.95×10^9 viable cells. Oxygen consumption rate and antibody production rate as calculated by enzyme linked immunoassay (ELISA) (Kirkgaard and Perry) are plotted as a function of time.

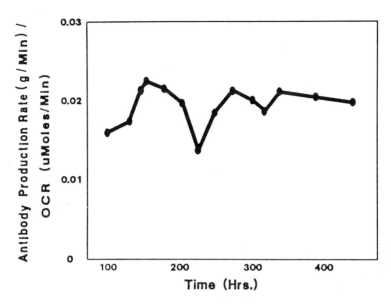

FIGURE 11–12 Antibody production rate/oxygen consumption rate for L243 hybridoma. The ratio of antibody production rate to oxygen consumption rate is plotted from the data in Figure 11–11 as a function of time.

then become starved for oxygen and nutrients and eventually become necrotic. In the Opticore, this has been prevented by: (1) designing a flow diverter to provide uniform flow distribution as the media enters the channels at one end of the core; and (2) forcing flow to follow a straight path down the channels to the other end of the core. This tends to prevent the system from forming pockets of low flow, which often accelerate the formation of the blockage. We have now tested many cores for flow patterns after cell growth. We have also fixed and stained numerous cell-covered cores to examine cell growth patterns. In general, neither suspension cells (in a core run vertically) nor anchorage-dependent cells seem to clog the flow channels. Except under rare circumstances where large clumps of anchorage-dependent cells dislodge, travel around the loop, and are trapped at the incoming end of a channel, cell growth has never lacked uniformity from channel to channel across the core or down its length (Lyderson et al. 1985a). However, if a run of several months is planned with suspension cells, it is best to run the core vertically, as a small number ($\sim$5–10%) of the total cells are always free in the recirculating loop. In a horizontal core, these few cells will settle out on the lowest channel walls, and over an extended period of time (three to five months) can build up, resulting in lower flow to some of the channels. Antibody production will then drop proportionally.

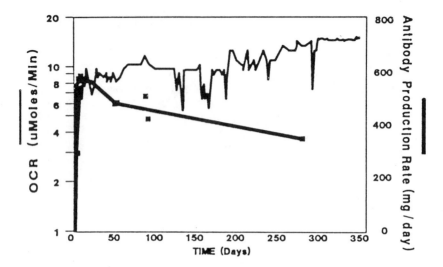

FIGURE 11–13 Hybridoma maintained in the Opticell for over one year. L243 hybridoma cells were grown in DMEM (4.5 g/l glucose), 10% FCS, 0.5 U penicillin G, and 50 μg/ml streptomycin base for 10 days. This media was then exchanged for 1% FCS plus antibiotic and maintained for over a year. Cells were continuously feed and harvested with this media at a rate of $\sim$1.2 l/day. OCR is the thinner line plotted on a log curve and antibody production rate (APR) is the thicker line plotted on a linear curve.

A hybridoma line (L243) was run successfully on the S core in the vertical position for over one year (Figure 11–13). Only four hours per week was needed for maintenance on this system, demonstrating the low labor requirements of continuous perfusion systems. Multiple experiments were performed on this run, and not all of the data have yet been analyzed. Nevertheless, over the period of one year, oxygen consumption rates seemed to remain fairly constant and antibody production rates fell only $\sim$30%. This attests to the potential for continuous culture.

If productivity remains high, it is always more cost effective to keep the bioreactor system running than to restart the system and repeat the growth phase. However, there is some concern that the doublings of a cell line over an extended period will result in genetic drift. For example, a hybridoma clone that has lost the ability to produce antibody might overgrow a culture. We have performed cell counts on the free population of hybridoma cells and compared these numbers to the total cells on the cores (Table 11–1). If these cells are growing in 8 l spinner flasks with 10% Serum Plus (a serum substitute, Hazelton Biologics, Inc.), they have a consistent doubling time of about one day. In the Opticell, following the log growth phase, doubling time for the cells in 1% Serum Plus is lengthened to 7.5–10 days. This means that the turnover rate for cells has decreased an order

TABLE 11-1   Ratio of Immobilized to Nonimmobilized Hybridoma Cells

Scaleup Instrument	% Serum Substitute	Doubling Time
8 l Spinner	10% Serum Plus	1 day
Opticell 5300E	1% Serum Plus	7.5–10 days

Immobilized Cells in Opticell	
Elapsed run time (hours)	Viable cells
575	2.36×10^{11}

Nonattached Cells in Opticell				
Elapsed run time (hours)	Viable cells/l	Liters/day	Cells/day	% of Total on core
384–505	1.73×10^8	100	1.73×10^{10}	7.3
523–573	1.55×10^8	153	2.37×10^{10}	10.0

Doubling times of L243 cells in 8 l spinners in 10% Serum Plus was compared to cells in the Opticell in DMEM (4.5 g/l glucose) plus 1% Serum Plus. (This was following growth for nine days in DMEM (4.5 g/l glucose) plus 10% Serum Plus.) In the 8 l spinner flasks, the cells grew from 2×10^5 to $\sim 8 \times 10^5$ cells/ml over 48 hours. Consistent growth was only maintained in these 8 l spinners by counting cells at every split and continually purging the gas above the cells with a 40% O_2 and 0–5% CO_2 mixture controlled using an Opticell 5200R unit.

of magnitude over a freely dividing population. Hence, the rate of genetic drift should be reduced a similar amount. In other words, maintaining a hybridoma in the Opticell unit for one year at low serum levels should presumably cause no more genetic drift from the original population than does growing the cells for somewhat over one month in high serum conditions. Genetic drift should be even less of a problem with genetically engineered mammalian cells, as these tend to cease division when they run out of surface area for attachment. Moreover, most secreted products made in recombinant mammalian cells immobilized on the ceramic are collected cell-free. Even with hybridomas there are only ~ 1–2×10^5 cells/ml in the product as opposed to 5–10×10^5 cells/ml in a spinner culture. This can prolong the life of filters used to clarify the media during downstream processing.

11.5 MAINTENANCE MEDIA

In the above experiments the serum level is higher in the stirrer culture than it is in the Opticell. However, cells in this system are typically not limited by nutrients but rather by the surface area available for attachment to the ceramic. This is demonstrated by the fact that increasing the serum level or feed rates when a culture's OCR has plateaued in the Opticell does not usually increase OCR or protein production. In fact, the ability to maintain cells at low serum or serum-free levels is one of the advantages of an immobilized culture system. Cells are usually grown in media high in serum or serum substitutes (5–10%) in the Opticell until they reach their maximum OCR. Then they are switched to low-serum or serum-free conditions by exchanging low- for high-serum media over a period of about one to two days (Putnam 1987). This protocol speeds the growth phase, which is usually finished in about seven to 10 days for hybridomas. During the transition period, productivity of the culture usually remains constant or increases slightly (Figure 11–14). Protein products can then be collected in low-serum media.

Serum-free media can also be used from the beginning if the cells have been previously adapted to it. This adaptation typically takes approximately one week or more in spinner flasks, as opposed to the one to two days

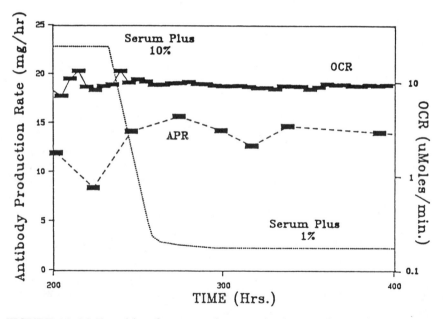

FIGURE 11–14 Transition from growth to production media. An S-51 core was seeded with 1×10^8 L243 hybridoma cells. They were grown for 10 days in DMEM, 10% Serum Plus, and gentamicin (50 mg/l). At hour 231 the transition to production media was accomplished by changing the feed media from 10% to 1% Serum Plus. The system was run in the continuous feed/harvest mode at a rate of 1.2 l/day with a loop volume of ~600 ml; i.e., two exchanges per day. The solid line is OCR plotted on a log curve. The dashed line is antibody production rate as measured by ELISA (Kirkgaard and Perry IgG Kit). The dotted line represents the transition from 10% to 1% Serum Plus in the recirculating loop.

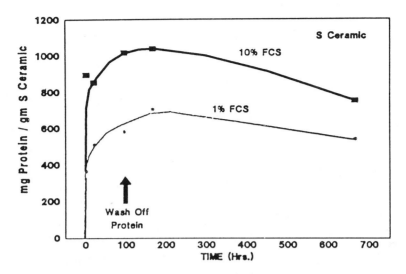

FIGURE 11-15 Protein adsorption and wash-off from the S ceramic. Sections of S core ceramic weighing approximately 0.55–0.65 g were incubated at 37 °C in 5.5 ml DMEM containing either 1% (thin line) or 10% (thick line) FCS. This would be equivalent to ~600 ml/S51 core (~76 g). Samples remained stationary in solutions during this time. At either 1.5, 20, or 96 hours, core pieces were rinsed three times in 5 ml of Delbecco's phosphate buffered saline (DPBS). Protein was then solubolized off the ceramic with 10 ml of 0.5% sodium dodecyl sulfate in DPBS. Protein was assayed by a modification of the Lowry procedure (Lowry et al. 1951; Florini et al. 1977).

Also, with 96 hours remaining, core pieces were rinsed three times in 5 ml DPBS and incubated at 37 °C in DMEM with no serum. During this wash-off period, samples were gently agitated on a platform shaker to simulate flow in the Opticell. At hour 166 and 672 (from the start of the experiment), core pieces were rinsed and assayed for protein as described above.

required in the Opticell. The switch to low-serum conditions in spinners could be related to the outgrowth of a different population. The time needed to convert to low-serum conditions in the Opticell is shorter and the cells grow very little at this time. Adaptation in this unit probably does not represent a shift in the makeup of the cell population. If serum-free media is used from the start of the culture, cores should first be soaked in serum-containing media or bovine serum albumin (BSA), and then rinsed off. As with anchorage-dependent cells, the attachment and growth of the hybridomas seem to be improved by prior adsorption of serum proteins to the core. Evidence also indicates that proteins adsorbed to the core attach tightly and are not removed by repeated washing or by several days in culture (Figure 11–15). If greater than 10% FCS is used, more protein will adsorb but this excess will wash off again. Maximum absorption of BSA will occur with ~7 g/l BSA; however, at this concentration only ~0.3% of the BSA will adsorb to the ceramic. Since this initial protein-ceramic binding appears to be strong, there is little worry that these proteins will later detach, causing problems during product purification.

11.6 CELL DENSITY IN PERFUSION SYSTEMS

Even when a cell can be maintained in low-serum or serum-free conditions in a spinner, an immobilized reactor might be preferred. Unless a gentle, dependable cell return apparatus is utilized, it is necessary to remove portions of the cell population when harvesting product from the stirred system. This reduction in cell density and infusion of fresh medium causes the cell

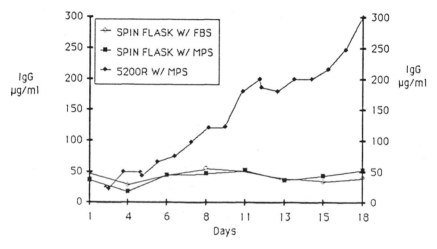

FIGURE 11–16 Eighteen Spinners versus Opticell productivity. In the 250 ml spinner flasks, L243 hybridoma cells were seeded at 2×10^4 viable cells/ml directly into modified Iscove's DMEM/DMEM high glucose (50/50 vol/vol) with 1% FBS or 1% Biotain MPS (Hazelton Biologics, Inc.). Cultures were refed every two to three days by replacing 20% of the spent medium/cell suspension with fresh medium and supplement. Cell density throughout this 18 day test generally ranged from $4–6 \times 10^5$ cells/ml and were similar for cultures with and without serum. Concentrations of IgG in the spinner cultures ranged from 30–60 μg/ml.

L243 cells from the same culture stock were also seeded into an Opticell 5200R; 0.8×10^8 cells were seeded in an S51 Opticore with DMEM high glucose plus 10% FBS. The volume in the ceramic core was 180 ml. The recirculating media volume was changed to 3.5 l of DMEM high glucose plus 1% Biotain MPS after six days of growth. This weaning process was complete by day 10 of the run. A continuous feed and harvest exchange of 1–1.2 l/day was maintained on this 180 ml reactor within the 3.5 l recirculating loop. IgG concentration ranged from 24 μg/ml to over 300 μg/ml.

population to continuously grow and necessitates higher concentrates of nutrients and serum. Additionally, since the cell density remains low, 1×10^6 to 5×10^6 cells/ml, the concentration and daily production of proteins tend to remain low and level. An immobilized cell system allows for a much greater cell density, 5×10^7 to 5×10^8 cells/ml in the media immediately surrounding the cells, which may remain constant or slowly increase with time (Berg 1985; Lyderson et al. 1985a). Thereby, the product concentration and especially daily production can be significantly increased.

To demonstrate this concept, L243 hybridomas secreting a mouse IgG_{2a} at ~ 40 μg/ml in flask cultures were compared as regards production in two 250 ml spinner flasks and the Opticell 5200R. Both types of unit started with cells in the same reactor size and approximately the same number of cells in the reactor. However, in the spinner, cells must be harvested when the system is fed. Both types of unit were run at conditions considered appropriate for maximum continuous production. Figure 11–16 presents the results. The first obvious difference is the higher concentrations of product reached in the 5200R. In addition, Figure 11–17 demonstrates that while the spinner productivity fluctuated, the total amount of IgG produced per day continued to increase in the 5200R. This is probably due to the removal of large amounts of medium and cells from the spinners. Figure 11–18 plots the cumulative amount of IgG harvested from each type of system. This plot indicates cumulative production of ~ 17 mg of IgG from both 250 ml spinners and ~ 2400 mg of IgG from the 3 l Opticell 5200R. If recirculating or final culture volumes are considered, however, the 250 ml spinner that contained 1% FBS produces 25.1 mg IgG, the spinner that contained 1% Biotain MPS™ (Hazelton Biologics, Inc.) produced 27.1 mg IgG, and the

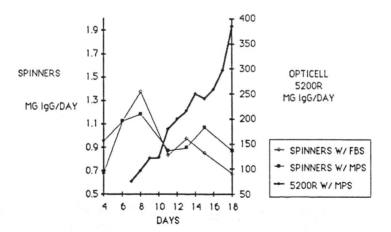

FIGURE 11–17 Spinners vs. Opticell Productivity: Daily IgG Produced. L243 hybridoma cells are grown as described in Figure 11–16. The daily production of IgG is plotted against time in culture.

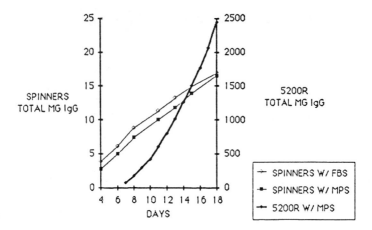

FIGURE 11-18 Spinners vs. Opticell Productivity: Cumulative IgG Produced. L243 hybridoma cells are grown as described in Figure 11-16. The cumulative amount of IgG produced is plotted against time in culture.

Opticell 5200R that contained medium supplemented with 1% Biotain MPS produced greater than 3,400 mg.

For some applications, perfusion bioreactors offer advantages over static environments in maintaining culture viability and attachment. Traditionally, flasks and petri dishes have been used when primary or normal cells were grown. The minimal mixing of media in these systems allows for localized release and binding of autocrine growth factors. Conversely, as cells grow and use the nutrients and dissolved oxygen in the medium, significant boundary layers can be formed. If left undisturbed, these concentration gradients can slow cell division, inhibit further growth, and in some cases affect cell adhesion. Perfusion systems disrupt these diffusion gradients and enable the cells to receive a more even supply of nutrients and at the same time remove waste products more efficiently.

To show how this can affect cell production, a test was performed comparing cell yields between an Opticell 5200R using a smooth AD 51 Opticore and static flasks. A transformed cell line producing a surface-bound protein was seeded at 3.6×10^4 cells/cm² in RPMI-1640 medium with 10% FCS. Over seven days, each system was refed with a media volume proportional to its surface area. On day seven it became evident that the static flasks, which were heavy with cells, would survive no longer. The edges were peeling and holes were forming throughout the cell layer. The ceramic surface, having a perfused medium supply, looked healthy with no disruption evident. While the flasks required harvesting on day seven, the Opticell grew for several more days with no sign of cell loss. The comparative yields represented a 50% increase in cells harvested per square centimeter from the perfusion system. Hence, it can be seen that for some cell types, perfusion bioreactors may have an advantage over static systems in promoting greater culture longevity and total cell yield.

11.7 RAPID PRODUCT ISOLATION AND NUTRIENT MANIPULATION

A perfused immobilized system can offer great flexibility in terms of nutrient manipulation. For example, such systems are more efficient when complete media exchanges are necessary. When reducing serum levels or switching to serum-free conditions, bioreactors such as the Opticell enable a user to easily batch out one media for another. This offers a more complete method of exchange in comparison to dilution. When induction or infection procedures are required, small volumes of reagents can be used at controlled concentrations to uniformly act on a culture. In addition, there are cases when completeness and timeliness of media exchange are improved if the cells are immobilized on a rigid surface. Flexibility in nutrient delivery also enables continuous exchange of media into and out of the system, solving several problems that plague large-scale cell culture today. Two of these problems include glutamine breakdown and product degradation in culture.

Glutamine is required as an energy source for animal cells in culture but spontaneously breaks down in media and at 37 °C. This breakdown is accelerated by phosphate buffers (the most common buffer in animal cell culture media), elevated temperatures, and elevated pH (common in media that has lost CO_2 by exposure to the atmosphere) (Gilbert et al. 1949; Tritsch and Moore 1962). There is also a component of catalyzed deamination caused by enzymes present in the serum component of the media (Griffiths and Pirt 1967). In addition, we have found that with hybridoma cells, glutamine can be the amino acid utilized most rapidly and is therefore present in the culture at a very low concentration. This was also found to be the case with BHK-21 cells (Arathoon and Telling 1982). Ammonia, the product of spontaneous or enzyme-driven deamination, has been shown to be deleterious to many animal cells (Glacken et al. 1986; Ryan and Cardin 1966; Furusawa and Cutting 1962; Eaton et al. 1962). The process of holding large volumes of media in a low-density culture at 37 °C will only accelerate both the disappearance of the nutrient, glutamine and the appearance of the waste product, ammonia.

One of the most common problems with genetically engineered animal cells today is that, unlike monoclonal antibodies, these products may not be stable in media at 37 °C with cells and other cell products (i.e., proteases). Hence, total cellular production levels can be acceptable while the yield of the active product is low. Often the key to the recovery of active product is its removal to a protected environment; i.e., separate from the cells at 4 °C or in the presence of protease inhibitors.

The above problems can be minimized by: 1) increasing the throughput of media per day; 2) maintaining components separately before exposure to the cells; or 3) decreasing the residence time of the media at 37 °C in the circulating loop. The first solution often succeeds but can be expensive. For some products, it is nevertheless cost effective. The second is especially helpful for a glutamine problem because the glutamine can be fed in separately from the media-containing serum and phosphate buffer. This has been implemented on the Opticell by feeding the glutamine (or glutamine plus antibiotic) from a smaller vessel kept cool and dosed in at intervals with an auxiliary pump. Hence, the glutamine remains separate from serum and phosphate buffer until it is in the circulating loop where the cells need it. It is important, even if such a separate dosing system is not assembled, to add the glutamine just before mixing the media (as opposed to ordering media with glutamine already in it). Basal media that originally contained 4 mM of glutamine was found to contain only 2–3 mM of glutamine and 1–2 mM of ammonia when it was used before the expiration date. This is due to spontaneous deamination during storage at 4 °C for several months, even before the media is shipped to the customer.

The last improvement can be made by decreasing the volume in the circulating loop. Originally, a 5200 Opticell system operated with ∼3,300 ml in the loop. We reduced this to ∼1,300 ml and then to 400 ml (of which 175 ml is within the housing for the ceramic). The L243 hybridoma line in Figure 11–13 was maintained successfully at this volume for nine months. We maintained this level by using a shortened harvest line and setting the harvest pump to run 10% faster than the feed pump. This caused the harvest pump to pull out media until the volume dropped below the end of the harvest line, after which it pulled in air. This system maintains a low volume without the risk of draining the core. In developing larger systems, we have decreased the size of the media vessel to take advantage of these positive features. The one risk is that the system has less reserve nutrients if, for example, the feed/harvest ceases to function.

In summary, the major advantages of continuous perfusion cultures result from the fact that the operator can more freely manipulate the system parameters, such as oxygen, pH, flow rates, nutrients, and waste product removal, in order to optimize production based on the needs of the individual culture and the characteristics of the product. These advantages are achieved with low labor requirements.

11.8 SCALABILITY

Immobilized systems have the potential to be linearly scalable, based upon the area available for cell growth. The Opticell units were specifically designed to achieve this linear scalability proportional to the surface area of the ceramic core. The system is not limited by any other feature, such as the ability to deliver media to the cells or provide oxygen in a nondestructive manner, which often limit cell productivity in other systems. At present, the Opticell units are 100-fold scalable from approximately 0.32–32 m^2 in linear surface area for the porous ceramic and from 0.41–42 m^2 for the flat ceramic. Details on the surface area on the various Opticores are provided in Table 11–2. The figures represent *linear* surface area for both the S and the AD cores. In other words, *the porosity of the S core was not taken into account in this table.* We have reason to believe, from protein adsorption studies on the ceramics, that the actual surface area of the S core is ap-

TABLE 11–2 Opticore Specifications

	System			
Core Parameters	5200	5300E	5300E	5500
	Opticore			
	S51	S451	S1251	S7251
Length (in)	5.28	12.00	12.50	24.00
Diameter (in)	1.60	3.66	5.20	11.25
Channels/sq inch	200	200	200	90
Wall thickness (in)	0.012	0.012	0.012	0.017
Channel size (in)	0.059	0.059	0.059	0.088
Channel surface area (sq cm)	8.00	18.18	18.94	54.67
Number of channels	402	2104	4247	8971
Total geometric surface area (sq cm)[1]	3217	38256	80439	490397
Number of cores	1	1	1	1
Ratio A (compare to S451)	0.1	1.0	2.1	12.8
Ratio B (compare to S51)	1.0	11.9	25.0	152.4
Void volume[2]	175 ml	1.7 l	3.7 l	60 l

	Opticore			
	AD51	AD451	AD1251	AD7251
Length (in)	5.28	12.00	12.50	Not available
Diameter (in)	1.60	3.20	5.20	
Channels/sq inch	400	400	400	
Wall thickness (in)	0.012	0.012	0.012	
Channel size (in)	0.038	0.038	0.038	
Channel surface area (sq cm)	5.18	11.77	12.26	
Number of channels	804	3217	8495	
Total geometric surface area (sq cm)	4164	37856	104127	
Number of cores	1	1	1	
Ratio A (compare to AD451)	0.1	1.0	2.8	
Ratio B (compare to AD51)	1.0	9.1	25.0	
Void volume[2]	175 ml	1.4 l	3.7 l	

[1]Geometric surface area only. Actual total surface area will be proportionately larger for each size due to the porosity of the S ceramic (the exact surface area utilized by a cell population will vary with cell type and is difficult to determine). For this reason, geometric surface area comparisons from S to AD Opticores are not valid.

[2]Total volume within housing. Recirculating volume is less, particularly for 5500.

proximately three times that of the AD core. However, the percentage of the extra surface area a cell can use depends to a large degree on the type of cell.

It should be noted that the S cores scale at $1\times$, $12\times$, $25\times$, and $152\times$ (the latter core is being developed for the 5500 unit) whereas the AD cores scale at $1\times$, $9\times$, and $25\times$. In other words, the S451 core and the AD451 core are not exactly the same size, a fact reflected in their respective surface areas. The 5300E unit holds four of either the S or AD1251 cores, making that unit $100\times$ the surface area of an S or AD51 core in a 5200R unit.

Table 11–3 represents a comparison of the Opticell product line. The 5200R (research) will hold one S or AD51 core with a typical seed number of $\sim 1 \times 10^8$ cells and a typical feed rate of ~ 1 l/day at a maximum cell number. This unit is depicted in Figure 11–19 and differs from the larger units primarily because it is temperature controlled and contains a laminar flow unit to allow for sterile manipulations behind the Plexiglas shield. The

FIGURE 11–19 Opticell 5200R (research).

TABLE 11-3 Specifications for Opticell Systems

	System		
	5200	5300E	5500
			(see footnote 1)
Relative scale—S			
Maximum	1× (1 × 51)	100× (4 × 1251)	600× (4 × 7251)
Minimum	1× (1 × 51)	12× (1 × 451)	150× (1 × 7251)
Relative scale—AD			
Maximum	1×′ (1 × 51)	100×′ (4 × 1251)	—
Minimum	1×′ (1 × 51)	9×′ (1 × 451)	—
Maximum recirculation rate	0.48 l/min	24 l/min	120 l/min
Sterilization	Autoclave	Autoclave	CIP/SIP
Medium reservoir size	3 or 5 l	20 l	60 l
Example Performance at Maximum Scale[2]			
Relative scale	1×	100×	600×
Surface area—AD	0.41 m^2	41.6 m^2	Not available
RBE[3]	4.8	489	
Typical seed cell number	1 × 10^8	1 × 10^{10}	6 × 10^{10}
Example of feed/harvest rate	1 l/day	100 l/day	600 l/day
Example of antibody production	0.05 g/day	5 g/day	30 g/day
(at 50 µg/ml, example of feed/ harvest rate, and 300 production days)	15 g/year	1.5 kg/year	9 kg/year
Approximate total medium volume[4]	3.3 or 5.3 l	35 l	300 l
Batch tank equivalent[5]	15 l	1,500 l	9,000 l
Opticore volume	175 ml	14.8 l	240 l

[1]Product currently in development.
[2]Will vary significantly depending on the application and degree of optimization.
[3]RBE = Roller Bottle Equivalent for Roller Bottle with 850 cm^2 surface area.
[4]Based on reservoir filled to capacity.
[5]Based on total liter throughput. Assumptions: Equivalent product concentration, 20 batches per year in tank, 300 production days in Opticell.

cover at the base folds down to reveal four peristaltic pump heads (circulation, auxiliary, feed, and harvest) and two drives for 3.5 inch computer diskettes for data storage from the run.

The 5300 unit can utilize one S451 or AD451 or one S1251 or AD 1251 core. It feeds at approximately ~10 or 25 l/day and is seeded at ~1 or 2.5

$\times$ 10^9 cells. This unit contains the 10$\times$ and 25$\times$ scale. The 5300H (high flow) will support the same cores as a 5300 as well as two S1251 or AD1251 cores. It is seeded at capacity at $\sim$5 $\times$ 10^9 cells and feeds at 50 l/day. Both the 5300 and 5300H range has been built into the newer 5300E (expanded) model pictured in Figure 11–20. This unit can take one S or AD451 core, or one, two, or four S or AD1251 cores, making it adaptable for 10$\times$, 25$\times$, 50$\times$, or 100$\times$ scale. At capacity it will be seeded with $\sim$1 $\times$ 10^{10} cells and

FIGURE 11–20 Opticell 5300E (expanded).

feed at ~100 l/day. The 5300, 5300H, and 5300E units can be temperature controlled by placement in a warm room or specially modified Belco cabinet. A portable laminer flow hood is available to allow for sterile connections near the unit.

When multiple cores are run on the same unit, the separate cores grow cells identically. This is demonstrated in Figure 11–21 where the OCR of the four 5300E cores were measured individually over the length of the run. Maximum variation in OCR among these cores was 10%, but most of the values varied by no more than 1% from the mean. The 5500 system is currently in development and will be 600-fold scale from the 5200. It will be seeded with ~6×10^{10} cells and feed at about 600 l/day. Depicted in Figure 11–22, it is designed to be a clean-in-place, sterilized-in-place modular system that can be linked to standard fermentation vessels to be used for feed and harvest. The unit will contain four S cores, each with a linear surface area of 49 m² for a total linear surface area of over 196 m².

Several studies have shown that the Opticell units are linearly scalable with the surface area of the ceramic (Pugh et al. 1987a and 1987b; Berg

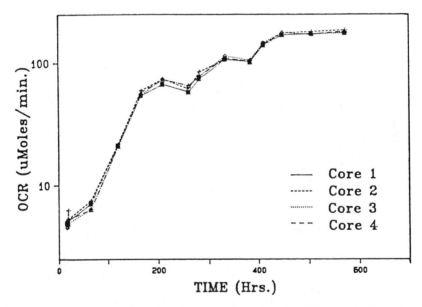

FIGURE 11–21 Oxygen consumption rates from four S-1251 cores independently. The 5300E was seeded at 8×10^9 cells that were grown in DMEM with 4.5 g/l glucose supplemented with 10% Serum Plus (Hazelton Biologics, Inc.) and gentamicin (50 mg/l) for 12 days. They were switched to 2.5% Serum Plus media and fed/harvested at 100 l/day for three days, 1% Serum Plus at 100 l/day for six days, then 158 l/day for three more days. Total cell yield from the four cores was 2.95 $\times$ 10^{11} viable cells. Over the term of this run at specific intervals the OCR of each core was determined independently.

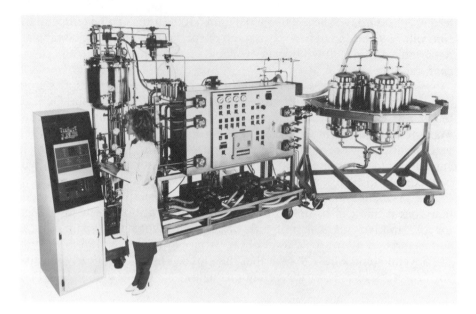

FIGURE 11–22 Opticell 5500.

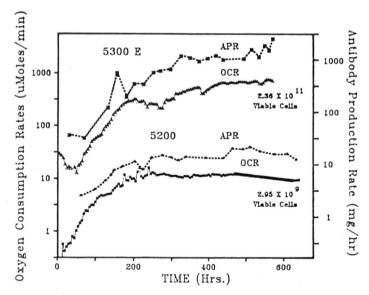

FIGURE 11–23 100× scaleup of OCR and antibody production rate. The 5200R (1× scale) was run as described in Figure 11–11. The 5300E (100×) was run as described in Figure 11–21. Both OCR and antibody production rates are plotted on a log curve against time.

1985; Lyderson et al. 1985a). Experiments were done using L243 hybridomas to compare cell yields, OCR, and productivity as the Opticell unit scales 100-fold from the 5200R to the 5300E. These studies indicate that cell number (see Figure 11–7), OCR, and antibody productivity (Figure 11–23) scale proportionally to the surface area available for cell attachment. Other studies done using BHK-21 and MDCK cells also demonstrate scalability over 25-fold scale, whether the measure is total cell yield (Figure 11–24) or OCR (Figure 11–25) as a function of surface area. While MDCK cells metabolize considerably more oxygen per cell than do BHK-21 cells, the maximum OCR and final cell yield scale linearly in both lines in relation to the surface area of the ceramic.

In conclusion, a perfusion system can hold cells in a stable steady-state condition, thereby extending the life of the culture and increasing productivity with lower labor requirements. If available area for the cells is the limiting factor for cell production, as in the Opticell unit, scalability will be directly proportional to this area. The use of such a system saves infinite time and energy in the scaleup process because each scaleup step ceases to be a new development project. Rather, all factors associated with growth can be scaled proportionally to the available surface area for the cells.

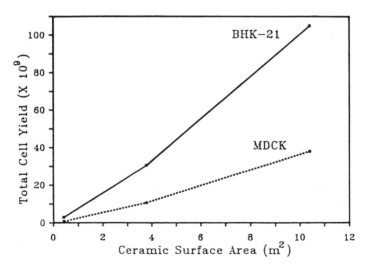

FIGURE 11–24 $25\times$ scaleup of cell yield for MDCK and BHK-21 cells. MDCK cells were grown in DMEM (4.5 g/l glucose) supplemented with 10% FBS. Opticores were seeded with between 2 and 2.5×10^4 cells/cm^2. After five days in culture an average of 1×10^9 (AD51), 10.9×10^9 (AD451), and 38.3×10^9 (AD1251) cells were harvested. BHK-21 cells were grown in DMEM (4.5 g/l glucose) supplemented with 10% FBS. Opticores were seeded with between 2.35 and 2.8×10^4 cells/cm^2. After seven days in culture an average of 2.95×10^9 (S51), 30.6×10^9 (S451), and 105×10^9 (S1251) cells were harvested.

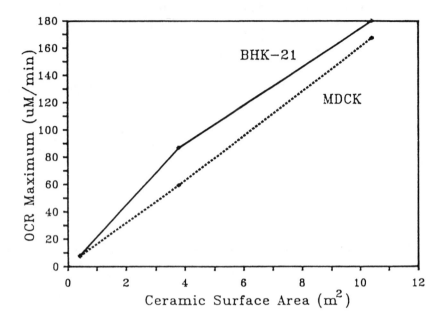

FIGURE 11-25 25× Scaleup of OCR for MDCK and BHK-21 cells. MDCK cells were grown in DMEM (4.5 g/l glucose) supplemented with 10% FBS. Opticores were seeded with between 2 and 2.5 × 10^4 cells/cm². After five days in culture a maximum OCR of 7, 60, and 170 were reached in the three size cores, respectively. BHK-21 cells were grown in DMEM (4.5 g/l glucose) supplemented with 10% FBS. Opticores were seeded with between 2.35 and 2.8 × 10^4 cells/cm². After seven days in culture a maximum OCR of 8, 87, and 180 were reached in the three size cores, respectively.

REFERENCES

Arathoon, W.R., and Telling, R.C. (1982) *Develop. Biol. Stand.* 50, 145–154.

Berg, G.J. (1985) *Develop. Biol. Stand.* 60, 297–303.

Berg, G.J., and Bodeker, B.G.D. (1988) in *Animal Cell Biotechnology* Vol. 3 (Spier, R.E., and Griffiths, J.B., eds.), pp. 321–335, Academic Press Ltd., London, U.K..

Bodeker, B.G.D. (1985) *Labor Praxis* 9, 970–980.

Bognar, E.A., Pugh, G.G., and Lyderson, B.K. (1983) *J. Tissue Culture Methods* 8, 147–154.

Dean, R.C., Karkare, S.B., Phillips, P.G., Ray, N.G., and Rundstadler, Jr., P.N. (1987) in *Large Scale Cell Culture Technology* (Lyderson, B.K., ed.), pp. 169–192, Hanser Publishers, Munich, Germany.

Eaton, M.D., Low, I.E., Scala, A.R., and Utetsky, S. (1962) *Virology* 18, 102–108.

Feder, J., and Tolbert, W.R. (1983). *Sci. Am.* 248(1), 36–43.

Florini, J.R., Nicholson, M.L., and Dulak, N.C. (1977) *Endocrinology* 101, 32–41.

Furusawa, E., and Cutting, W. (1962) *Proc. Soc. Exp. Biol. Med.* 111, 71–75.

Gilbert, J.B., Price, V.E., and Greenstein, J.P. (1949) *J. Biol. Chem.* 180, 209–218.

Glacken, M.W., Fleischaker, R.J., and Sinskey, A.J. (1986) *Biotechnol. Bioeng.* 28, 1376–1389.

Griffiths, J.B., and Pirt, S.J. (1967) *Proc. R. Soc. Biol.* 168, 421–438.

Himmelfarb, P., Thayer, P.S., and Martin, H.E. (1969) *Science* 164, 555–557.

Hynes, R.O. (1987) *Cell* 48, 549–554.

Kleinman, H.K., Klebe, R.J., and Martin, G.R. (1981) *J. Cell Biol.* 88, 473–485.

Knazek, R.A., Gullino, P.M., Kohler, P.O., and Derick, R.L. (1972) *Science* 178, 65–66.

Ku, K., Kwo, M.J., Delente, J., Wildi, B.S., and Feder, J. (1981) *Biotechnol. Bioeng.* 23, 79–95.

Looby, D., and Griffiths, J.B. (1987) in *Modern Approaches to Animal Cell Technology* (Spier, R.E., and Griffiths, J.B., eds.), pp. 342–352, Butterworths, Stoneham, MA.

Lowry, O.H., Rosenbrough, N.J., Farr, A.L., and Randall, R.J. (1951) *J. Biol. Chem.* 193, 265–275.

Lyderson, B.K. (1987) in *Large Scale Cell Culture Technology* (Lyderson, B.K., ed.), pp. 169–192, Hanser Publishers, Munich, Germany.

Lyderson, B.K., Pugh, G.G., Paris, M.S., Sharma, B.P., and Noll, L.A. (1985a) *Bio/Technology* 3, 63–67.

Lyderson, B.K., Putnam, J., Bognar E., et al. (1985b) in *Large Scale Mammalian Cell Culture* (Feder, J., and Tolbert, W., eds.), pp. 39–58, Academic Press Inc., Orlando, FL.

Mendelsohn, S.L., Nordeen, S.K., and Young, D.A. (1977) *Biochem. Biophys. Res. Comm.* 79, 53–60.

Murdin, A.D., Thorpe, J.S., and Spier, R.E. (1987) in *Modern Approaches to Animal Cell Technology* (Spier, R.E., and Griffiths, J.B., eds.), pp. 420–436, Butterworths, Stoneham, MA.

Nicholson, M.L. (1988) in *Proceedings of the 1988 Conference on Commercial Biotechnology*, pp. 164–181, Business Communications Co.

Pugh, G.G. (1988) *Bio/Technology* 6, 524–526.

Pugh, G.G., Berg, G.J., and Sear, C.H.J. (1987a) in *Bioreactors and Biotransformations* (Moody, G.W., and Baker, P.B., eds.), pp. 121–131, National Engineering Laboratory/Elsevier, Amsterdam.

Pugh, G.G., and Bognar, Jr., E.A. (1987b) in *Proceedings of Biotech USA*, 1987, pp. 140–150, On Line International Ltd., London.

Putnam, J.E. (1987) in *Commercial Production of Monoclonal Antibodies* (Seaver, S.S., ed.), pp. 110–138, Marcel Dekker, New York.

Ruoslahti, E., and Pierschbacher, M.D. (1987) *Science* 238, 491–497.

Ryan, W.L., and Cardin, C. (1966) *Exp. Biol. Med.* 123, 27–30.

Tritsch, G.L., and Moore, G.E. (1962) *Exp. Cell Res.* 28, 360–364.

Weisman, M.C., Creswich, B., Calabresi, P., Hopkinson, J., and Tutuyian, R.S. (1985) in *Large Scale Mammalian Cell Culture* (Feder, J., and Tolbert, W.R., eds.), pp. 125–149, Academic Press, Inc., Orlando, FL.

Young, D.A., Voris, B.P., and Nicholson, M.L. (1981) *Environ. Health Perspectives* 38, 89–97.

Young, D.A., Nicholson, M.L., Guyette, W.A., et al. (1979) in *Glucocorticoid Action and Leukemia* (Bell, P.A., and Borthwick, N.M., eds.), pp. 53–68, Alpha Omega Publishing, Cardiff, Wales.

Optimization of the Microenvironment for Mammalian Cell Culture in Flexible Collagen Microspheres in a Fluidized-Bed Bioreactor

John N. Vournakis
Peter W. Runstadler, Jr.

At the current time, seven biotherapeutic drugs, developed using recombinant and hybridoma technology, have reached commercial introduction. These include human insulin, alpha interferon, hepatitis B vaccine, tissue plasminogen activator, human growth hormone, OKT3 monoclonal antibody, and erythropoietin. Many others are in various stages of clinical or preclinical trials and are about to be introduced to the marketplace. Most

We gratefully acknowledge the technical and theoretical assistance of Ed Hayman, Nitya Ray, Amar Tung, Margaret Worden, Sandra Warner, and Mike Young, all of Verax Corporation, in stimulating and executing the preparation of this manuscript; the dedicated staff of the Verax Biosciences Division who contributed to accumulation of much of the information presented; and the Verax Corporation for financial support. In addition, we thank Dr. C. Bisbee of Biogen Corporation, D. Ruzicka of Harvard University, and several members of the Bioprocess Engineering Technology Center of the Massachusetts Institute of Technology for discussions of many of the concepts found in this article.

of these drugs are large, complex proteins having specific carbohydrate structures that require exquisite and precise assembly in order to have the correct biological activity. It is now generally recognized that higher eukaryotic cells are required to make the correct posttranslational modifications needed for the assembly of these biomolecules. As these new drugs begin to enter the marketplace, ever greater emphasis will be placed upon reduced cost of production. This in turn translates to the use of cell culturing processes that are effective at large scale, with high cell densities producing high quality products at high production rates per cell. In addition, the downstream purification process(es) needs to be carried out efficiently at high recovery levels to provide a pure product (often >99%) at the lowest possible cost. A systems approach that makes the culturing and purification processes an integrated process is clearly the direction and objective of process development.

This chapter describes a continuous cell culturing process that has been developed at Verax that focuses on the optimization of the microenvironment for mammalian cell culture. This process is designed to optimize the local microenvironment surrounding genetically engineered cells and hybridomas, leading to high cell density, cell viability and cell-specific productivity. This process facilitates the use of low cost, serum-free media, minimizes product degradation, and thus enhances product quality and potency, and reduces downstream purification and overall production costs.

The Verax Process, illustrated in the flow diagram in Figure 12–1, is the result of synergistic effects derived from the interaction of several unique components including: spherical, three-dimensional, sponge-like, natural collagen Microspheres for the immobilization of cells to achieve high cell densities, and a fluidized-bed perfusion bioreactor designed to deliver the oxygen and other nutrients required and to efficiently remove the secreted biotherapeutic protein products and waste products generated by unusually high cell densities. Figure 12–2 shows that the bioreactor has a recycle loop that includes a gas exchanger, a medium pump, a heater, control parameter probes, and reaction chamber containing the fluidized bed of Microspheres. Fresh medium is added and harvest-containing product is removed at an appropriate perfusion rate.

The key features that derive from the interactions of the elements of the system indicated in Figure 12–1, are:

1. High mass transfer rates caused by the fluid dynamic properties of media flow around and through individual flexible Verax Microspheres in the fluidized-bed system, resulting in efficient intramatrix transport of oxygen and other nutrients and removal of cell products.
2. High viable cell densities of immobilized cells due to the affinity of cells for the natural collagen substrate of the Microspheres, leading to a high productivity.

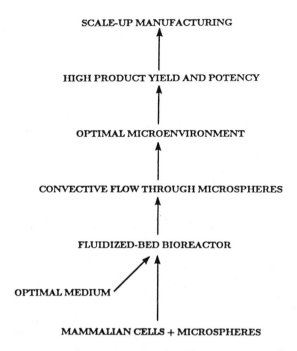

FIGURE 12–1 Elements of the Verax Process. The components listed combine to generate the Verax cell culture system and result in high reactor productivity and product quality.

3. An optimal microenvironment, due to cell-Microsphere and cell-cell contacts and to the formation of a cell-derived extracellular matrix, resulting in local concentrations of growth factors and other important constituents that combine to generate enhanced cell-specific productivity and high product titers.

The densely populated collagen Microsphere is the fundamental unit of the system. When in a fluidized-bed system, which induces a nutrient rich flow around each Microsphere and creates an optimal microenvironment, a Microsphere becomes an independent colony of cells, behaving effectively as an individual microreactor. The nutrient requirements and productivities of the bioreactor are, therefore, the result of the sum of those of the individual microreactors. This concept is illustrated in Figure 12–3, and is the basis for the scalability of the process. Verax systems have an increasing number of Microspheres, going from the bench-top System 10 research bioreactor to the commercial pilot-scale System 200 to the large-scale production System 2000, as indicated in Figure 12–3.

Several features of interest result from this unique cell culture system design. The bioreactor has inherent culture stability and scalability prop-

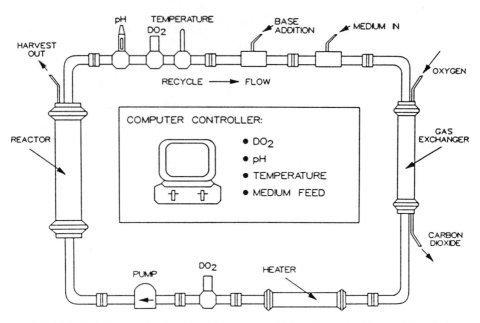

FIGURE 12–2 Verax fluidized-bed bioreactor. This schematic diagram shows the essential features of a Verax bioreactor system including: gas exchanger, recycle pump, fluidized Microspheres, nutrient medium and O_2 input, and CO_2 and harvest liquor output.

erties. Culture stability, in the absence of exogenous selective pressure, derives from the difficulty in propagating a mutation throughout the culture. The short mean residence time (on the order of 2 to 8 hrs) of culture medium in this system facilitates rapid washout of mutant cells and limited exposure of protein product to the culture harvest, which may contain degradative enzymes. Process scalability derives from the additive nature of the basic component of the system, the populated Microsphere. A bioreactor design naturally emerges from this property, allowing reliable scaleup to production level systems while maintaining adequate oxygen and other nutrient supply to the culture (Karkare et al. 1985a and 1985b; Runstadler and Cernek 1988; Young and Dean 1987; Dean et al. 1988; Tung et al. 1988; Runstadler et al. 1989; Hayman et al. 1988).

12.1 VERAX MICROSPHERES AND CELL VIABILITY

The fundamental element of the Verax fluidized-bed, continuous cell-culture system is the three-dimensional, sponge-like, bovine collagen Microsphere, containing small metal weighting particles that allow it to suspend immobilized cells in a fluidized-bed bioreactor. The dimensions and physical

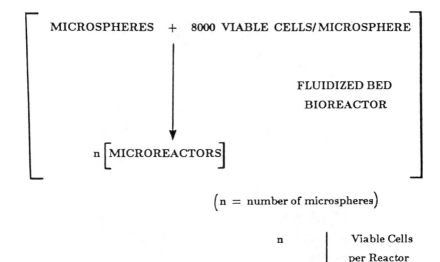

FIGURE 12–3 The Microreactor. Each Microsphere is populated by approximately 8,000 to 12,500 cells to generate individual microreactors, the fundamental scaleup unit in the Verax process.

properties of these Microspheres have been well described elsewhere (Runstadler and Cernek 1988; Tung et al. 1988; Runstadler et al. 1989). Figure 12–4 shows a scanning electron micrograph indicating the structure of a typical Microsphere. Figure 12–4A shows large open pores of approximately 50 μm average diameter, and surrounded by leafy surfaces of collagen, available for the attachment and proliferation of cells. Growth factors and other important molecules adhere to the Microspheres and are, therefore, available in high local concentrations to support cell growth and viability. In addition, as seen in Figure 12–4B, the bovine collagen starting material contains bovine fibronectin. Thus, the material provides a natural matrix available to promote the initial contact of cells having fibronectin surface receptors, provided that bovine fibronectin is sufficiently structurally similar to be recognized by the specific fibronectin receptor found on the surface of the cell type being cultured, as expected based on the high sequence homology among mammalian fibronectin genes.

Two important and unique properties of Verax Microspheres, which allows them to function optimally in fluidized-bed systems, are their porosity and flexibility. These features allow the flow through of medium caused by convective forces. The convective forces are a result of medium

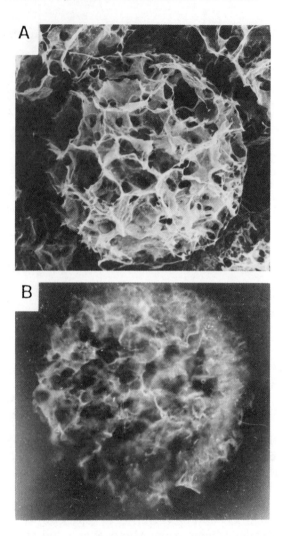

FIGURE 12–4 Verax Microspheres. (A) Scanning electron micrograph (150× magnification) of a standard, weighted Microsphere with 500 μm diameter and 30–50 μm pores. (B) Immunofluorescence staining of Microsphere for fibronectin. The Microsphere sample was incubated with a rabbit antibody monospecific for bovine fibronectin, followed by treatment with a fluorescein isothiocyanate (FITC)-conjugate of goat antirabbit IgG.

flow-induced pressure differences that occur at the surface of the Microspheres suspended in the fluidized-bed bioreactor. The open porous structure of Microspheres, therefore, enhances internal medium flow and nutrient/product exchange. Its physical basis, i.e., the flow dynamics leading

TABLE 12–1   Cell Types Cultured in Collagen Microspheres

Chinese hamster ovary
African Green Monkey kidney
Human embryonic kidney
Baby hamster kidney
Normal rat kidney
Transformed rat kidney
Mouse mammary tumor (C-127)
Human diploid fibroblasts (MRC-5)
Human hepatoma (HEP-G2)
Murine myeloma
Human meloma
Hybridomas

to its existence, is discussed below. In addition, the flexibility property allows cells to populate Microspheres in a manner that optimizes cell/cell and cell/ matrix contacts. As a result, cells differentiate to a natural state that is optimal for the biosynthesis, processing, and secretion of proteins. The freedom of cells to form three-dimensional aggregates similar in structure and cell density to tissues results in part from the flexible nature of the thin collagen sheets arranged in "leafy" arrays throughout the Microsphere.

More than 110 cell lines, including natural secreting cells, attachment-dependent, and suspension cells engineered to produce recombinant proteins, and monoclonal antibody-producing hybridomas, have been cultured in Verax Microspheres and have achieved high cell densities and viabilities. A partial list of the cell types cultured in the process is indicated in Table 12–1.

Table 12–2 summarizes some of the growth and viability data and illustrates the general applicability of the collagen Microspheres for the culture of cell lines of commercial interest. High cell viabilities are usually observed over long periods of time in the fluidized-bed bioreactor system. The majority of cells in such heavily populated Microspheres, e.g., greater than 70%, remain viable and are highly productive.

12.2 THE VERAX FLUIDIZED BED REACTOR

Verax collagen Microspheres, suspended in a fluidized bed, are subjected to medium flow dynamics that result in a continued passage of medium throughout their interior as they are populated to high cell densities. The fluid dynamics of the rapidly recycling culture in Verax bioreactors generates pressure differences that result in the forced flow of nutrients throughout the porous Microspheres. This effect provides convective flow through the Microspheres that leads to a greater uniformity in the distribution of me-

TABLE 12-2   Cell Density and Viability in Collagen Microspheres

Cell Type	Product	Cell Density (× 10⁻⁷ cells/ml)	% Viability
Hybridoma Cell Lines			
Human/Mouse	IgM	5.0	70
Human/Mouse	IgG	8.0	68
Rat/Mouse	IgG	15.0	72
Attachment-Dependent Cell Lines			
CHO[1]	Thrombolytic	18.0	79
CHO	Thrombolytic	20.0	80
CHO	Interleukin	26.0	92
CHO	Thrombolytic	22.0	75
CHO	Cardiovascular Therapeutic	20.0	80
CHO	Hormone	25.0	72
C127	Thrombolytic	9.0	85

[1]CHO, Chinese hamster ovary.

dium components such as oxygen and other cell-culture nutrients than is otherwise possible in other methods for mammalian cell culture. This effect is a property uniquely attributable to the porous, flexible nature of the collagen Microsphere.

The physical basis for the convective flow derives from an understanding of the interaction of the movement of culture liquor around a typical Microsphere in a fluidized bed (Schlichtung 1955; Yuan 1967). The diagram in Figure 12–5 illustrates the principles involved. This sketch shows an individual Microsphere with medium flow approaching with velocity V from the direction indicated. The flow divides and moves around the spherical bead. The symbol P_1 indicates the position on the Microsphere where the flow vector is zero and is referred to as the "stagnation point." At the stagnation point, the momentum of the flow is converted to static pressure. The highest pressure on the bead occurs at position P_1 (the front stagnation point), and a lower pressure occurs at position P_2 opposite P_1 on the bead. The flow, therefore, results in a net pressure difference acting across the diameter of the Microsphere, which is a function of the culture liquor viscosity μ, bead diameter D, and velocity of flow V, as defined by the equation below:

$$P = k \frac{\mu V}{D} \tag{12.1}$$

This pressure differential, generated by the flow of culture medium around the Microsphere, leads to the following effect.

Since the pores and channels in the typical Microsphere follow tortuous paths throughout its interior (as indicated in Figure 12–4), the pressure

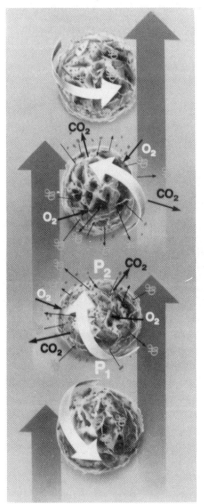

FIGURE 12-5 Forced convective flow. As Microspheres tumble throughout a fluidized-bed, pressure-induced convective forces cause flow of nutrients into and products out of the Microsphere. The tumbling effect continually changes the orientation of the Microsphere in the differential pressure field and enhances mass transfer.

differential described above acts to force fluid through the Microsphere. This forced convective flow moves from higher to lower pressure regions. The result is an enhancement of the movement of culture liquor into and out of the interior of the Microsphere. Since the Microsphere is continually tumbling in the fluidized-bed reactor as shown by the arrows in Figure 12–5, the direction of this movement is continually changing. As this happens, new fluid moves into, and previously trapped fluid, moves out of the Microsphere.

The physical forces described, which lead to the flow through the collagen Microspheres, are optimized by the vertical flow characteristics of the fluidized-bed reactor. The total impact of the fluidized-bed reactor design is, therefore, to help establish the formation of an optimal microenvironment around and within the spherical, densely populated Microspheres.

12.3 CELL-CELL AND CELL-MATRIX INTERACTIONS

It is well documented that tissue-cultured mammalian cells will either retain their morphological and biochemical differentiation and, thus, their protein synthetic and secretory capabilities, or will lose them depending on the chemical composition and structure of the culture substrate (Grobstein 1975; Jackson 1975; Rojkind et al. 1980; Bissell et al. 1982; Bissell and Barcellos-Hoff 1987; Toole 1981; Hay 1981; Li et al. 1987). Typically the sequence of events in the establishment of a productive cell/substrate interaction involves (a) cell attachment, mediated initially by the interaction of a family of structurally related proteins including fibronectin (Hynes 1981), with their cell receptors, the integrins (Hynes 1987; Menko and Boettinger 1987); (b) expression and transport of cellular fibronectin (Obara et al. 1988; Chen and Chen 1987), vitronectin (Gebb et al. 1987), laminin (Graf et al. 1987), proteoglycans (Spray et al. 1987), and other proteins, leading to the formation of an extracellular matrix (ECM) in the presence of appropriate substrate molecules such as native collagen; (c) development of a basal surface in contact with the ECM and an apical layer, coincident with the appearance of a highly organized cytoskeleton (Dabora and Sheetz 1988) in the cell; and (d) establishment of a pattern of gene expression that results in both morphological and biochemical cell-phenotypes that are correlated with high protein synthesis and secretory capacity, as well as with low rates of DNA synthesis and cell replicative activity (Eisenstein and Rosen 1988; Medinia et al. 1987; Ben-Ze'ev et al. 1988; Schuetz et al. 1988). Bissel et al. (1982) has proposed that there is a direct mechanicochemical transduction of information from the extracellular matrix through the cytoskeleton to the cell's nuclear matrix that is responsible for the organization of these events. Grobstein (1975) recognized this interaction in an early review on the subject:

> The wisdom of the matrix, to borrow from Cannon's wisdom of the body, is more likely to be expressed in a language apart from the hereditary one, though interlinked with it.

A series of studies during the past 10 years (Emerman and Pitelka 1977; Bisbee et al. 1979; Turley et al. 1985; Ruzicka 1986) have demonstrated that cells cultured on substrates such as floating collagen gels maintain the kind of morphological differentiation that is optimal for protein productivity and secretion. Cells on such a substrate have been observed to form a continuous epithelial pavement (as in a tissue) on the surface of the floating collagen. In addition, the cells display surface polarization and form microvilli and tight junctions at their apical surfaces. In contrast, it has been clearly demonstrated by Emerman and Pitelka (1977) and Ruzicka (1986) that cells cultured on flat plastic surfaces or on attached two-dimensional collagen sheets will form a confluent epithelial sheet but lose their differentiated ability for protein synthetic and secretory activities. Thus, it is important to consider two factors in developing an optimal method for the large-scale culture of mammalian cells and the production of therapeutic proteins: (1) the correct material composition, and (2) the proper shape (Yang and Nandi 1983) of the substrate.

In view of these results, we have focused our Microsphere development efforts on the use of natural bovine Type I collagen, containing additional components such as bovine fibronectin (see Figure 12–4) as the substrate material, and on the spherical, three-dimensional (Yang and Nandi 1983) configuration as the appropriate morphology for our matrix. Cells populating such Microspheres will find the native collagen/fibronectin mix to be an attractive substrate for the initiation of attachment and subsequent events in establishing a high density culture (Gospodarowicz et al. 1979). The spherical morphology and open leafy structure of the collagen surfaces provides a physical state similar to the floating collagen gels, and thus allows attached cells to make the proper cytoskeletal contacts necessary for achieving and maintaining the appropriate biochemical differentiation for protein synthesis and secretion. Figure 12–6A shows a scanning electron micrograph illustrating the morphology of Chinese hamster ovary (CHO) cells in a Microsphere loaded at low cell density. These cells have invaded the open pores of the Microsphere and have formed substrate (ECM) contacts on the matrix surface. Cell/cell contacts also exist, which in time will result in the population of the Microsphere at near tissue-like cell densities. The shape of the cells in the Microsphere is columnar with basal and apical polarity and has a uniform distribution of surface microvilli (Logsdon et al. 1982). These cells have a much more normal and healthy appearance compared to cells attached to a two-dimensional substrate, as seen in Figure 12–6B.

Newly developed molecular biological methods have made it possible to study the effect of cell-culture substrate on the expression of specific genes in cultured cells. Such studies (Eisenstein and Rosen 1988; Medina et al. 1987; Ben-Ze'ev et al. 1988; Schuetz et al. 1988; Reid et al. 1987) have demonstrated, for example, that hepatocytes cultured on attached, flat, two-dimensional collagen substrates do not express liver-specific protein mRNAs; rather, they synthesize DNA and cytoskeletal protein mRNAs in preparation for cell division. However, when these cells are cultured on hydrated, three-dimensional gels, they form spherical self-aggregates and shift their pattern of gene expression to express low levels of cytoskeletal and high levels of liver-specific protein mRNAs, respectively (Schuetz et al. 1988; Reid et al. 1987). These and similar observations are consistent with the requirement that the cell-culture substrate be three-dimensional in morphology (Yang and Nandi 1983) in order to support high cell densities.

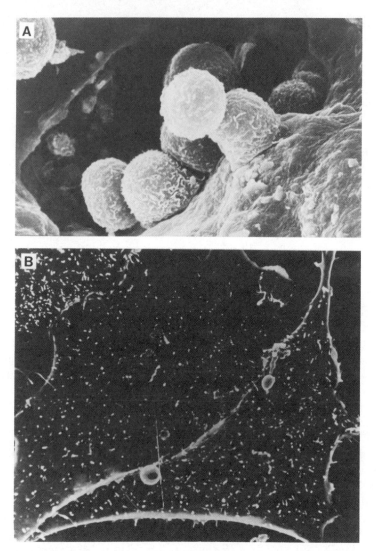

FIGURE 12-6 Scanning electron micrograph of Chinese hamster ovary (CHO) cells attached within a Microsphere. (A) Well-attached cells with an even distribution of microvilli, and having basal to apical polarity, are shown in a microsphere. The leafy-like internal morphology of the Microsphere is available for the attachment of cells. The group of cells in the photograph exhibit cell-matrix and cell-cell contacts. (B) A photomicrograph of a CHO cell attached to a two-dimensional plastic substrate.

12.4 THE MICROENVIRONMENT

In the most general sense, the cell culture microenvironment is the local biochemical, physiological, and physicochemical milieu directly affecting the growth of cells and their metabolism. The behavior of cells, e.g., attachment, spreading, motility, and biosynthetic capacity, are influenced by their immediate surroundings. When considering immobilized mammalian cell culture in a three-dimensional collagen matrix such as in a Verax Microsphere, the microenvironment can be described, using the definitions of Brunner et al. (1982), as consisting of three compartments: the diffusive environment, the contact environment, and cell junctions. Each of these three aspects regulate cell activity by specific molecular mechanisms, acting synergistically to provide a given overall condition for cell behavior. The combined effects of these three components can be optimized by appropriate and judicious choice of cell-culture substrates and bioreactor design and operation to achieve the optimum conditions for mammalian cell growth and productivity. The diagram in Figure 12–7 indicates the complexity of factors that affect the cell microenvironment, and suggests that an integrated approach is required to optimize the productivity of the system.

12.4.1 The Contact Environment: Impact of the Collagen Microsphere

The contact environment of cells in tissues is provided primarily by the extracellular matrix, an insoluble meshwork of protein, carbohydrate, and other matrix-bound molecules synthesized and assembled by the cells them-

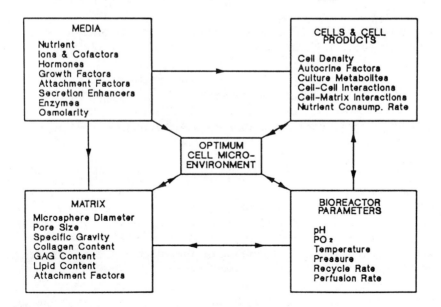

FIGURE 12–7 Schematic diagram indicating factors that affect the cell culture environment, which leads to an optimal microenvironment.

selves. Although the particular concentration distributions of the various ECM components varies among tissues, it consists in general of basement membrane and connective tissue proteins such as collagen and elastin, of adhesive surface glycoproteins such as fibronectin, laminin, and vitronectin, and of proteoglycans, and glycosaminoglycans (Smith et al. 1982; Roberts et al. 1988). These components are organized into complex and specific structures. Nearly every component interacts directly with the cell surface, either through a receptor-mediated interaction such as with the cell surface glycoprotein integrins (Jackson 1975; Rojkind et al. 1980; Bissell et al. 1982; Bissell and Barcellos-Hoff 1987; Toole 1981; Hay 1981; Li et al. 1987; Hynes 1981 and 1987; Menko and Boettiger 1987; Obara et al. 1988; Chen and Chen 1987; Gebb et al. 1987; Graf et al. 1987; Spray et al. 1987; Roberts et al. 1988; Ruoslahti and Pierschbacher 1987; Yamada et al. 1986; Hayman et al. 1985) or via less specific contacts. These adhesion protein receptors appear to connect the external ECM to the intracellular cytoskeleton by direct contact via cytoplasmic structural domains (Ruoslahti and Pierschbacher 1987; Hayman et al. 1985).

Figure 12–8 illustrates some of the known molecular interactions present in the contact component of the microenvironment and describes the interface of the basal surface of cells with the extracellular matrix. Cell adhesion, and its influence on intracellular communication, is affected by the presence of collagen, the fundamental ECM building block, along with other proteins and polysaccharides (Smith et al. 1982; Roberts et al. 1988; Ruoslahti and Pierschbacher 1987; Yamada et al. 1986; Hayman et al. 1985; Gillery et al. 1986; Nagata et al. 1985). Thus, the contact environment, through specific adhesion protein/receptor complexes, provides a transduction pathway for the extracellular regulation of cellular behavior such as adhesion, migration, polarity, differentiation, and quite likely protein synthesis and secretion.

Hydrated collagen gels, as discussed above (Emerman and Pitelka 1977; Bisbee et al. 1979; Turley et al. 1985; Ruzicka 1986; Yang and Nandi 1983), have been employed extensively for animal cell culture in order to closely mimic natural tissue conditions. Collagen is immunologically benign, highly resistant to proteolysis, and is a natural substrate for cell adhesion, in conjunction with the surface adhesion molecules mentioned earlier. In an attempt to reconstruct the natural microenvironment of cells as closely as possible (Brunner et al. 1982), three-dimensional Microspheres are prepared for the culture of mammalian cells at Verax from Type I collagen. These Microspheres are manufactured from bovine hide collagen, which is composed of 80% collagen and 20% other insoluble matrix components. This material serves as a natural scaffolding for the attachment of cultured cells and cell-specific factors either produced by the cells themselves or provided by serum in the media that can be used to precondition the Microsphere. Once attached to the collagen substructure, cells enhance their contact environment by producing and organizing their extracellular matrix on the

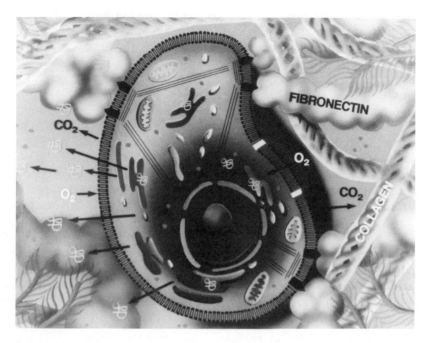

FIGURE 12–8 Cell microenvironment. Schematic diagram of the *contact* and *diffusive* components of the cellular microenvironment within a Microsphere. Fibronectin contacts with cellular fibronectin receptors, the interconnecting network of collagen fibrils, and the diffusion of nutrients and the secretion of products are indicated.

surface of the Microsphere, including the production of cell-specific fibronectin (Yamada et al. 1986; Hayman et al. 1985). Eventually, the Microspheres are coated with a complex spectrum of macromolecules and important minor components such as adherent growth factors (Roberts et al. 1988), which have a very high affinity for matrix proteoglycans. Thus, the collagen Microspheres provide an ideal natural material for the development of the optimum contact environment for cultured cells.

12.4.2 The Diffusive Environment: Impact of the Fluidized Bed Bioreactor

The diffusive environment includes all molecules that are freely diffusible in the extracellular and intracellular tissue space, including hormones and growth factors such as autocrine factors synthesized by the cells and other cytokines, dissolved gases (Storch and Talley 1988) such as oxygen and carbon dioxide, nutrients such as glucose and amino acids, salts and ions such as Na^+, K^+, and Ca^{2+}, and cellular products that include secreted proteins as well as waste products such as lactate and ammonia. This aspect of the microenvironment is also depicted in Figure 12–8. In general, the diffusive substances can access the cell from either the basal/ECM surface or from the apical surface. Many of the molecules in the diffusive compartment are internalized by cells either by active and/or passive transport systems, or by receptor-mediated endocytosis.

In order to optimize the microenvironment, it is necessary to be certain that the diffusive environment, in addition to the contact environment, is considered and that an appropriate system is employed that mimics the natural condition as closely as possible. The fluidized-bed bioreactor was therefore developed in order to function as the diffusive environmental facilitator in harmony with the high specific-gravity, three-dimensional collagen Microsphere cell substrate described earlier. The fluidized-bed bioreactor illustrated in Figure 12–2 optimizes the diffusive environment by two mechanisms; it causes the fluid-flow effects to force culture medium into and through Microspheres and it facilitates the control of nutrient, oxygen, and waste product concentrations by allowing a rapid and controllable recycling of the culture medium and a rapid removal of conditioned medium from the bioreactor (Bailey et al. 1985; Karkare 1986a and 1986b). The tumbling motions superimposed onto the medium flow around the Microspheres, illustrated in Figure 12–5, provides a thorough access of the cycling medium to all cells.

The microenvironment, consisting of diffusive and contact components, can be controlled and optimized in the Verax collagen Microsphere/fluidized-bed system. In addition, it is most significant that cells can be cultured in this system using defined media without supplemental serum, since a central feature of the contact environment is likely to be the production of cell-specific factors that provide the appropriate signals for growth and productivity.

It has been observed with several CHO cell lines that an "adaptation" of the cells occurs on the Microspheres during the culturing process. This is illustrated in Figure 12–9. Several CHO lines were cultured in Verax fluidized-bed bioreactor systems. In each case a significant increase in the *product concentration* is observed following the removal of serum from the medium, in some cases increasing by five times or more. Since the cell densities did not change significantly in these reactors following serum removal, we conclude that *cell*-specific productivity increased sharply. This *cell*-specific productivity reflects an adaptive response of cells to the optimal microenvironment created in the Microspheres during the growth (serum-containing) phase of the culture and is an important effect of the Verax process.

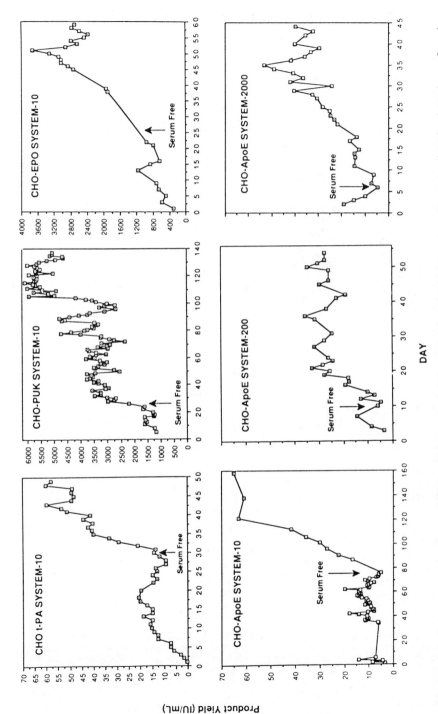

FIGURE 12-9 Chinese hamster ovary (CHO) cell behavior in Verax fluidized-bed bioreactors. Product yield is plotted as a function of days of culture for six Verax bioreactor runs. Product concentration increases dramatically in all cases when serum is removed due to optimization of the microenvironment and adaptation of cells to Verax microspheres.

12.5 BIOREACTOR PRODUCTIVITY AND PRODUCT QUALITY

The Verax Process has been used for the manufacture of more than 40 important recombinant proteins and monoclonal antibodies. A partial list is included in Table 12–3.

The ultimate test of the efficacy of this tissue culture system for the scaleup production of biotherapeutic proteins is its productivity and the *quality* of the product it generates, i.e., the integrity of the structure, the accuracy of the posttranslational modifications, and the specific activity of

TABLE 12–3 Biotherapeutics Produced in Collagen Microspheres

Protein C
Human growth hormones
Tissue plasminogen activator (5)
Pro-urokinase (6)
Urokinase (2)
Factor VIII
Apolipoprotein E (2)
Apolipoprotein Al
Erythropoietin
Interleukins (4)
Soluble Receptors
MIS
Platelet derived growth factor
Monoclonal antibodies: IgM, IgG, IgG2a, IgE
  AntiPMN
  Antiproinsulin
  Antilarge cell carcinoma
  Antinucleoprotein
  Antimurine γ-Inf (3)
  Anti-Ia
  Anti-CSF
  Antihuman IL4
  Antimouse IL4
  Antineuroblastoma
  Antibetagalactosidase
  Antihuman INF
  Antilipid A

The ultimate test of the efficacy of this tissue culture system for the scale-up production of biotherapeutic proteins is its productivity and the *quality* of the product it generates, i.e., the integrity of the structure, the accuracy of the posttranslational modifications, and the specific activity of the protein. A majority of the proteins produced by this process have excellent characteristics based on the above criteria. This is due to the elimination of total bovine serum from the medium, to the short residence time of the product after secretion, and to the impact of the system in creating a microenvironment that enables cells to achieve an optimally productive state (Kivirikko and Myalla 1987; Lee and Chen 1988).

the protein. A majority of the proteins produced by this process have excellent characteristics based on the above criteria. This is due to the elimination of total bovine serum from the medium, the short residence time of the product after secretion, and the impact of the system in creating a microenvironment that enables cells to achieve an optimally productive state (Kivirikko and Myalla 1987; Lee and Chen 1988).

The efficacy of the process is illustrated by the data in Figure 12–10, showing the results of a Verax bioreactor run for a cardiovascular biotherapeutic-producing CHO cell line. The data show that the cell can be cultured for long periods, in this case for six months, in a Verax bioreactor. It becomes most productive when the serum is removed from the growth medium. This is seen by a dramatic rise in the product yield, from an average of 10 mg/l before, to values between 65–80 mg/l after, serum is removed. The product concentration increase results from an increase in the cell-specific productivity, since the cell density did not substantially change after the removal of fetal bovine serum (FBS). The bioreactor productivity was 110–136 mg/day during the final 80 days of the System 10 bioreactor run. This translates to daily productivity values of 30–40 g/day in a Verax System 2000 production-scale system.

In addition and of greater importance, the *protein quality* is improved after serum is removed, with greatly reduced product degradation, as shown by the Western blot analysis in Figure 12–10.

Data like that in Figure 12–10, showing an increase in cell-specific productivity upon removal of FBS from the growth medium, have routinely been obtained for a wide diversity of recombinant CHO cell lines cultured in Verax bioreactors, which is illustrated in Figure 12–9. Other cell types, including mouse mammary C-127 cell lines, exhibit similar, albeit somewhat less dramatic, behavior. It appears that, in general, cells undergo an adaptation to the Verax Microspheres, which is enhanced by the fluidized-bed bioreactor, leading to increased productivity and improved product quality. Such adapted cells may retain their highly productive phenotype and may represent the kind of synergistic transformation described by Rubin (1988), i.e., an epigenetic phenomenon that is *"highly dependent on specific environment conditions"*. The basis for this phenomenon is currently under investigation.

12.6 SUMMARY

Flexible, three-dimensional, collagen Microspheres have been developed to actively promote a natural, optimal microenvironment for large-scale tissue culture of mammalian cells. The transport of nutrients into and cell products out of the Microspheres is enhanced by forced convective flow, which is the result of the tumbling of Microspheres and the dynamic properties of media flow in the fluidized-bed bioreactor.

CARDIOVASCULAR THERAPEUTIC

A. SYSTEM-10 BIOREACTOR PERFORMANCE

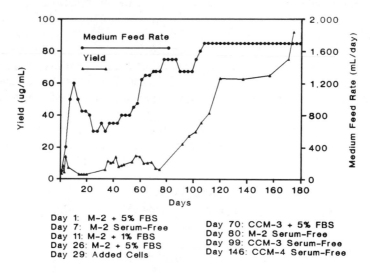

Day 1: M-2 + 5% FBS
Day 7: M-2 Serum-Free
Day 11: M-2 + 1% FBS
Day 26: M-2 + 5% FBS
Day 29: Added Cells

Day 70: CCM-3 + 5% FBS
Day 80: M-2 Serum-Free
Day 99: CCM-3 Serum-Free
Day 146: CCM-4 Serum-Free

B. PRODUCT QUALITY: WESTERN BLOT ANALYSIS

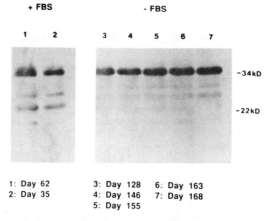

1: Day 62
2: Day 35

3: Day 128
4: Day 146
5: Day 155

6: Day 163
7: Day 168

FIGURE 12–10 Product concentration and product quality during continuous bioreactor operation. (A) The product yield and medium feed rate for a recombinant Chinese hamster ovary (CHO) cell line producing a therapeutic protein in a Verax System 10 bioreactor. Serum-free culture conditions were initiated from day 80. (B) Western blots of harvest samples taken on days during both serum-containing and serum-free phases of the run.

The collagen Microspheres have important characteristics of composition and morphology essential for optimal cell-matrix and cell-cell interactions. These interactions lead to high cell density and productivity through the dynamic modification of the microenvironment by cell-derived extracellular constituents. The collagen and Microsphere/fluidized-bed system provides the means to control and optimize the diffusive and contact components of the cells' microenvironment. Adaptation of cells to this microenvironment often results in dramatic increases in cell-specific productivity. Production of biotherapeutics in this process can be routinely performed in serum-free media, often leading to high productivity and product quality.

REFERENCES

Bailey, K.M., Venkatasubramanian, K., and Karkare, S.B. (1985) *Biotechnol. Bioeng.* 27, 1208–1213.

Ben-Ze'ev, A., Robinson, G., Bucher, N., and Farmer, S. (1988) *Proc. Natl. Acad. Sci. USA* 85, 2161–2165.

Bisbee, C.A., Machen, T.E., and Bern, H.A. (1979) *Proc. Natl. Acad. Sci. USA* 76, 536–540.

Bissell, M., and Barcellos-Hoff, M. (1987) *T. Cell Sci. Suppl.* 8, 327–343.

Bissell, M., Hall, G., and Parry, G. (1982) *J. Theor. Biol.* 99, 31088.

Brunner, G., Nitzgen, B., Weiser, R., and Speth, V. (1982) in *Growth of Cells in Hormonally Defined Media* (Sato, G., Pardee, A., and Sirbasku, D., eds.), pp. 179–201, Cold Spring Harbor Laboratory, Cold Spring Harbor, NY.

Chen, J., and Chen, W. (1987) *Cell* 48, 193–203.

Dabora, S.L., and Sheetz, M. (1988) *Cell* 54, 27–35.

Dean, Jr., R.C., Karkare, S.B., Ray, N.G., Runstadler, Jr., P.W., and Venkatasubramanian, K. (1988) *Ann. N.Y. Acad. Sci.* 506, 129–146.

Eisenstein, R.S., and Rosen, J.M. (1988) *Molec. Cell Biol.* 8, 3183–3190.

Emerman, J.T., and Pitelka, D.R. (1977) *In Vitro* 13, 316–328.

Gebb, C., Hayman, E.G., Engvall, E., and Ruoshlahti, E. (1987) *J. Biol. Chem.* 261, 16698–16703.

Gillery, P., Maquart, F., and Borel, J. (1986) *Exp. Cell Res.* 167, 29–37.

Gospodarowicz, D., Vlodavski, I., Greenburg, G., and Johnson, L.K. (1979) *Cold Spring Harbor Conf. Cell Proliferation* 6, 561–592.

Graf, J., Iwamoto, Y., Sasaki, M., et al. (1987) *Cell* 48, 989–996.

Grobstein, C. (1975) *Extracellular Matrix Influences on Gene Expression*, pp. 9–16, Academic Press, New York.

Hay, E. (1981) in *Cell Biology of Extracellular Matrix*, (Hay, E., ed.), pp. 279–409, Plenum Press, New York.

Hayman, E.G., Pierschbacher, M.D., and Ruoslahti, E. (1985) *J. Cell Biol.* 100, 1948–1954.

Hayman, E.G., Ray, N.G., Tung, A.S., Holland, J.E., and DeLucia, D.E. (1988) *ICSU Short Rep.* 8, 54–55.

Hynes, R.O. (1981) in *Cell Biology of Extracellular Matrix*, (Hay, E., ed.), pp. 295–334, Plenum Press, New York.

Hynes, R.O. (1987) *Cell* 48, 549–554.

Jackson, F.S. (1975) in *Extracellular Matrix Influences on Gene Expression*, pp. 489–496, Academic Press, New York.

Karkare, S.B., Burke, D.H., Dean, Jr., R.C., et al. (1986a) *Ann. N.Y. Acad. Sci.* 469, 91–96.

Karkare, S.B., Dean, Jr., R.C., and Venkatasubramanian, K. (1985a) *Bio/Technology* 3, 247–251.

Karkare, S.B., Phillips, P.G., Burke, D.H., and Dean, Jr., R.C. (1985b) in *Large Scale Mammalian Cell Culture*, pp. 127–149, Academic Press.

Karkare, S.B., Venkatasubramanian, K., and Vieth, W.R. (1986b) *Ann. N.Y. Acad. Sci.* 469, 83–90.

Kivirikko, K.I., and Myalla, R. (1987) *Methods Enzymol.* 144, 96–111.

Lee, C., and Chen, L.B. (1988) *Cell* 54, 37–46.

Li, M., Aggler, J., Farson, D.A., et al. (1987) *Proc. Natl. Acad. Sci. USA* 84, 136–140.

Logsdon, C.D., Bisbee, C.A., Rutten, M.J., and Machen, T.E. (1982) *In Vitro* 18, 233–242.

Medinia, D., Li, M.L., and Bissell, M. (1987) *Exp. Cell Res.* 172, 192–203.

Menko, A.S., and Boettiger, D. (1987) *Cell* 51, 51–57.

Nagata, K., Humphries, M.J., Olden, K., and Yamada, K.M. (1985) *J. Cell Biol.* 101, 386–394.

Obara, M., Kang, M.S., and Yamada, K.M. (1988) *Cell* 53, 649–657.

Reid, L., Abreu, S., and Montgomery, K. (1987) in *The Liver: Biology and Pathobiology*, pp. 717–738, Raven Press, New York.

Roberts, R., Gallagher, J., Spooncer, E., et al. (1988) *Nature* 332, 376–378.

Rojkind, M., Gatmaitan, Z., Mackensen, S., Giambrone, M., Ponce, P., and Reid, L. (1980) *J. Cell Biol.* 87, 255–263.

Rubin, H. (1988) *Nature* 335, 121.

Runstadler, Jr., P.W., and Cernek, S.R. (1988) in *Animal Cell Biotechnology* Vol. 3, pp. 306–320, Academic Press, London.

Runstadler, Jr., P.W., Tung, A.S., Hayman, E.G., et al. (1989) in *Large-Scale Mammalian Cell Culture Technology* Vol. 3, (Lubiniecki, A.S., ed.), Marcel Dekker, New York.

Ruoslahti, E., and Pierschbacher, M.D. (1987) *Science* 238, 491–497.

Ruzicka, D. (1986) Doctoral Dissertation, University of California, Berkeley, CA.

Schlichtung, H. (1955) *Boundary Layer Theory*, pp. 84–85, McGraw-Hill Book Co., New York.

Schuetz, E.G., Li, D., Omiecinski, C., et al. (1988) *J. Cell. Physiol.* 134, 309–323.

Smith, J.C., Singh, J.P., Lillquist, J.S., Goon, D.S., and Stiles, C.D. (1982) *Nature* 296, 154–156.

Spray, D.C., Fujita, M., Saez, J.C., et al. (1987) *J. Cell Biol.* 105, 541–547.

Storch, T.G., and Talley, G.D. (1988) *Exp. Cell Res.* 175, 317–325.

Toole, B.P. (1981) *Cell Biology of Extracellular Matrix* (Hay, E., ed.), pp. 279–295, Plenum Press, New York.

Tung, A.S., Sample, J.vG.S., Brown, T.A., et al. (1988) *BioPharm Manufac.* 1(2), 50–55.

Turley, E.A., Erickson, C.A., and Tucker, R.P. (1985) *Develop. Biol.* 109, 347–369.

Yamada, K.M., Kennedy, D.W., and Hayashi, M. (1986) in *Cell Adhesion; Differentiation and Growth*, pp. 131–143.

Yang, J., and Nandi, S. (1983) *Int. Rev. Cytol.* 81, 249–286.

Young, M.W., and Dean, Jr., R.C. (1987) *Bio/Technology* 5, 835–837.

Yuan, S.W. (1967) *Foundations of Fluid Mechanics*, pp. 306–320, Prentice-Hall, Inc.

High Density Cell Culture

Michiyuki Tokashiki

In the current technical situation, only a few substances of high unit price have been chosen as suitable objects for research on commercial production by mammalian cell culture such as interferon, TPA, or factor VIII. In these cases, the business can be run as long as the production of these substances is ensured no matter how high the costs, and it is not so critical to develop economical processes. In the near future, however, the mass production of low-price substances will be expectedly investigated and the development of economical mass production processes will be needed. For their development, the following problems must be kept in mind: (1) the growth rate of mammalian cells is low; (2) the ability of mammalian cells to produce substances is low; and (3) the cost of culture medium is generally high in the production of substances by mammalian cells.

In order to overcome these problems, it is important to suitably construct the culture process, although the modification of the cells themselves is also needed. Thus, economical culture process should satisfy the following five conditions:

1. High cell density culture can be maintained for a long period of time.
2. The objective substance can reach high concentrations in the culture mixture.
3. The cost of the culture medium is low.

4. The process can be scaled up and remain productive.
5. The operating system is excellent.

What kind of culture processes can meet these conditions?

The low productivity of mammalian cells requires the cultivation of a large amount of cells in order to mass-produce the objective substance. Accordingly, a process that can economically expand its culture volume becomes advantageous. From this point of view, tank cultivation is concluded to be most promising.

In batchwise cultivation of mammalian cells, generally they die several days after the cell density reaches $1-2 \times 10^6$ (cells/ml). Such processes can only keep the cell density low, and the productivity of the objective substance largely depends upon the cell proliferation rate. Accordingly, the productivity of a batchwise cultivator is concluded to be low. Meanwhile, the author's experience has revealed that a cell culture in a cell density as high as 10^7 (cells/ml) is easier to achieve than it is in a low cell density of $1-2 \times 10^6$ (cells/ml). When the same culture medium is employed, this fact produces merits that the consumption of the culture medium is reduced per unit amount of the objective substance produced, and high-density culture can use a more cost-effective medium. For these reasons, the perfusion culture, which can continue cell culture in a high cell density, is regarded to be advantageous.

Additionally, another problem in the mammalian cell culture is the high cost of the culture medium. The problem can be resolved by developing an inexpensive culture medium, but cost reduction will also require the development of improved processes, for example, increased selectivity of harmful metabolite separation from the culture system or recycling of expensive medium components. Perfusion culture is the best choice for introducing these processes. Thus, the development of improved cell culturing systems will be focused in the near future on a culture process where the process for reducing the cost of medium can be incorporated into the perfusion culture in a tank.

Mammalian cells require oxygen to grow. Thus, oxygen must be fed into the culture system in any form. Direct sparging of air or oxygen into the culture system is the simplest and preferred method, but this process causes some cell damage. The direct gas-sparging process can be conducted in many cases for high-density culture, but some kinds of cells are too seriously damaged to continue the cultivation. Accordingly, this aspect of cell culturing needs the development of an oxygen feed process to reduce cell damage.

From the above-mentioned points of view, the author has developed a new type of perfusion culture process using a tank where high molecular components in the culture medium are recycled and the oxygen feeding process uses fluorocarbons. These will be described in the following sections.

13.1 DEVELOPMENT OF A NEW PERFUSION CULTURE PROCESS

Mammarian cells, unlike bacteria or yeast, are very susceptible to growth inhibition by harmful metabolites such as ammonia or lactic acid in general. These harmful metabolites must be removed from the culture system in order to increase the cell density over a certain level. In the current situation where the technique for selective separation of these harmful metabolites has not yet been established, a process for separating the living cells from the culture mixture should be incorporated into the culture system. Since mammalian cells are small, little different in specific gravity from culture media, and susceptible to mechanical shocks, the separation is technically difficult. Thus, the selection of a suitable separation system is one of the critical factors for determining whether the cultivation process is excellent or not.

In most of conventional perfusion culture processes, the cells are separated from the culture mixture by filtration. The biggest problem in the filtration process is clogging. A variety of ideas have been proposed to reduce filter clogging, while some reports have described successful perfusion culture for a long period of time without clogging. Clogging, however, largely depends upon the cell type, the culture medium, and culture conditions. Thus, unstable operations cannot be avoided as long as filters are employed. According to the author's experience, clogging frequently occurs in serum-free culture.

The author and collaborators have developed a cell-settling-type perfusion culture process in which the cells are separated from the culture mixture by gravitational settling as well as a cell-centrifuging-type perfusion process where the cells are separated with centrifugal force.

13.1.1 Perfusion Culture Separating Cells from Culture Mixture by Gravitational Settling

13.1.1.1 Development of Culture Vessels and Cultivation of Mouse-Human Hybridoma X87 Cells Sato et al. (1985) described a perfusion culture in which the cells were separated from the culture mixture using a settling tube. In the gravity field, the settling velocity of cells is generally as low as 2–10 cm/hour. In scaleup, the settling area should be expanded in proportion to the net culture volume. As the culture scale is extended, the sufficient settling area is secured with great difficulty. Therefore, we developed a perfusion culture vessel that easy expands the settling area. This vessel is depicted in Figure 13-1 (Tokashiki et al. 1988).

The culture vessel has the settling zone around the culture zone. In the settling zone, the cells are not agitated and separate from the culture mixture by gravity. The settling area can be increased by expanding the outer wall.

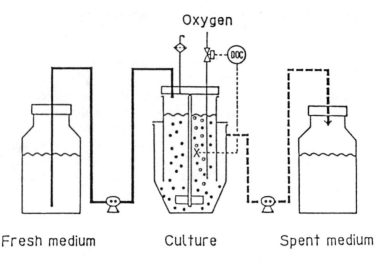

FIGURE 13-1 Perfusion culture apparatus that separates cells from medium by gravitational settling. Reproduced with permission from Tokashiki et al. (1988).

The results of hybridoma cultures according to this method are described below.

Experimental Conditions.   Mouse-human hybridoma X87 cells were cultivated using serum-free medium. X87 cells were established by fusion of human spleen cells with mouse myeloma P3/X63-Ag8-U1 cells that produced a human monoclonal antibody. The serum-free medium contained 9 μg/ml, 10 μg/ml transferrin, 10 μM ethanolamine, and 20 nM Na selenite. Enriched RDF (eRDF) was used as a basal medium (Murakami et al. 1984). Oxygen was fed into the culture mixture and the dissolved oxygen level was automatically maintained at 3 ppm.

Spinner Batch Culture.   Figure 13-2 shows the results of X87 cells on the spinner batch culture. The maximum viable cell density was 2.8×10^5 cells/ml. IgG concentration was 11 μg/ml and the specific productivity of IgG was about 10 μg/10^6 cells/day.

120 ml Perfusion Culture.   Figure 13-3 gives the result of the perfusion culture of X87 cells in a culture vessel of 120 ml net culture volume. The net culture volume is defined as the volume of the space in which the cells can actually exist. The cultivation was continued for about 230 days. With the passage of time, cell debris accumulates in the culture vessel, particularly at the interface between the gas phase and the liquid phase. At this point, almost all of the culture mixture was transferred into another culture vessel and the cultivation was continued. Viable cell density was $1-2 \times 10^7$ cells/

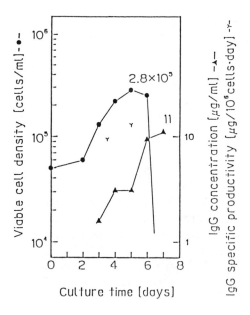

FIGURE 13–2 Spinner culture of mouse-human hybridoma X87 cells. (Medium: ITES + eRDF). Reproduced with permission from Hamamoto et al. (1989).

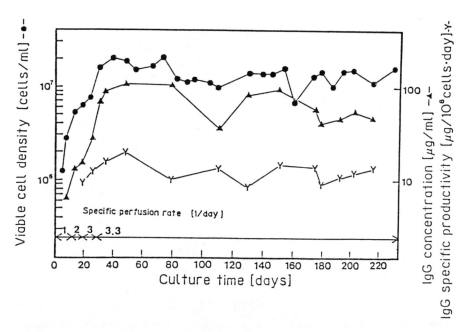

FIGURE 13–3 Perfusion culture of mouse-human hybridoma X87 cells using a perfusion culture vessel that separates cells from medium by gravitational settling. (Net culture volume: 120 ml; medium: ITES + eRDF.)

ml, IgG concentration was 40–100 μg/ml, and IgG specific productivity was about 10 μg/10⁶ cells/day. The viable cell density was 35- to 70-fold, the IgG concentration was 3.5- to 9-fold higher than those in the spinner batch culture, and IgG specific productivity was almost equal to that in the spinner batch culture. From these experimental results, it is obvious that the perfusion culture can be operated for long periods of time without serious problems. Also, the IgG concentration in this culture is severalfold higher than that in the spinner batch culture.

Four Liter Perfusion Culture.   Figure 13–4 shows the result of 4 l perfusion culture with X87 cells. On the tenth day after inoculation, viable density reached 10⁷ cells/ml and this level was maintained, while the specific perfusion rate was kept at 1.8/day. Problems arose on the nineteenth day, and the specific perfusion rate decreased to 1.0/day. After that, viable cell density began decreasing. On the twenty-fourth day, the specific perfusion rate increased again to 1.8/day and the viable cell density rose to 10⁷ cells/ml. IgG concentration was found to be 35–100 μg/ml, the same level found in the 120 ml culture. IgG specific productivity was about 10 μg/10⁶ cells/day.

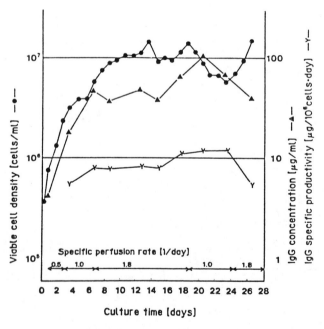

FIGURE 13–4 Perfusion culture of mouse-human hybridoma X87 cells using a perfusion culture vessel that separates cells from medium by gravitational settling. (Net culture volume: 4 l; medium: ITES + eRDF.) Reproduced with permission from Hamamoto et al. (1989).

Perfusion Culture with Multisettling Zones.   In order to facilitate scaleup, we developed a perfusion culture vessel with multisettling zones as shown in Figure 13–5 (Tokashiki and Arai 1989). In this method, the culture vessel had three or more settling zones and the supernatant was taken out of each settling zone. Mouse-human hybridoma X87X cells were cultured in a culture vessel equipped with three settling zones. X87X cells were derived and established from a clone of X87 cells. The net culture volume was 900–1,000 ml. Oxygen was supplied by sparging an oxygen-containing gas directly into the bottom of the vessel.

The result of this process is given in Figure 13–6. While the specific perfusion rate was kept at 2.0–2.3/day, the viable cell density kept the level at 1.8–3.0 $\times$ 10^7 cells/ml.

13.1.1.2 The Culture of Mouse-Human Hybridoma Cells Other Than X87 Cells   At least ten different kinds of hybridoma cells were cultured by this method. Almost all cells cultivatable in quiescent cultures could be culti-

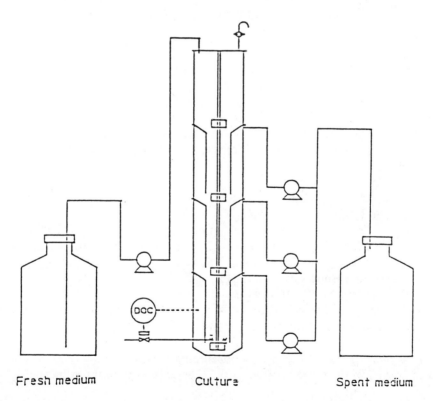

Fresh medium   Culture   Spent medium

FIGURE 13–5 Perfusion culture system with multisettling zones. (Net culture volume: 800 ~ 1000 ml.) Reproduced with permission from Tokashiki and Arai (1989).

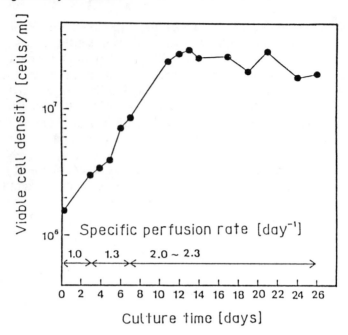

FIGURE 13–6 Perfusion culture of mouse-human hybridoma X87X cells using a culture system with three settling zones. (Net culture volume: 800 ~ 1000 ml, medium: ITES + eRDF.) Reproduced with permission from Tokashiki and Arai (1989).

vated at a high cell density. All these cells could be continuously cultured for long periods of time without any serious operational problems. Two cultures of mouse-human hybridoma cells other than X87 cells are described below.

The Culture of C41 Cells.   Mouse-human hybridoma C41 cells have been established by fusion of human spleen cells with mouse myeloma P3/X63-Ag8-U1 cells and produced human monoclonal antibody IgG_1 against human cytomegalovirus (Masuho et al. 1987). The same serum-free medium that was used in the cultivation of X87 cells was employed.

Figure 13–7 shows the result of the quiescent culture. Maximum viable cell density was 1.0×10^6 cells/ml.

The result of the perfusion culture with the same medium is given in Figure 13–8. Maximum cell density was 9×10^6 cells/ml.

The Culture of C176 Cells.   Mouse-human hybridoma C176 cells were established in the same manner as the C41 cells and produced human IgG_1 against the human cytomegalovirus virus.

Figure 13–9 gives the results of the 120 ml perfusion culture with the same medium that was used in the culture of X87 cells. The cultivation of

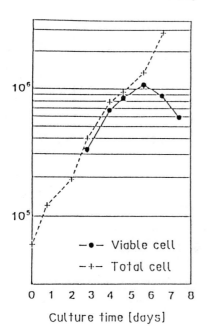

FIGURE 13-7 Serum-free quiescent culture of mouse-human hybridoma C41 cells. (Culture medium: ITES + eRDF.) Reproduced with permission from Toka-shiki et al. (1988).

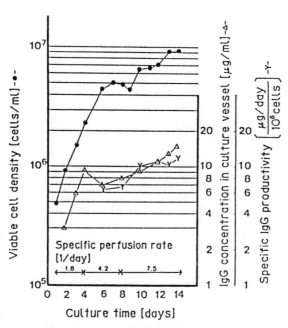

FIGURE 13-8 Perfusion culture of mouse-human hybridoma C41 cells using a perfusion culture vessel that separates cells from medium by gravitational settling. (Net culture volume: 120 ml; medium: ITES + eRDF.) Reproduced with permission from *Hakko Kogaku Kai-shi* 66, 31–35 (1988).

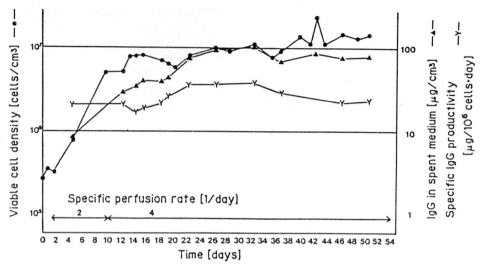

FIGURE 13–9 Cellular growth and IgG production in high density culture of C176 cells using ITES + eRDF medium. (Net culture volume: 120 cm³.) Reproduced with permission from Tokashiki et al. (1988).

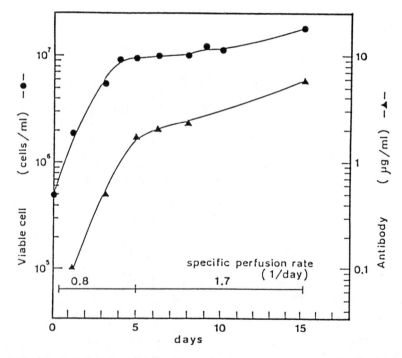

FIGURE 13–10 Growth and antibody production of X63 chimera in high density culture using 10% FCS-eRDF medium. (Net culture volume: 1.2 l.) Reproduced with permission from *Cytotechnology* 2, 95–101 (1989).

the C176 cells was continued for 54 days. During this period, the viable cell density was found to be about 10^7 cells/ml, and IgG specific productivity was almost invariable.

13.1.1.3 The Culture of Genetically Engineered Cells This perfusion culture process can be applied to both hybridoma cells and other anchorage-independent cells. The culture of genetically engineered cells is described below. The cell line used was mouse myeloma X63.Ag8,653, transformed with the chimeric heavy and light chain genes (X63 chimera). This cell line secretes artificially fused immunoglobulin molecules consisting of human heavy and light chain constant domains (γ and κ type) and mouse heavy and light chain variable domains. The antibodies secreted by this cell line specifically bind to common acute lymphocytic leukemia antigen (Nishimura et al. 1987).

Figure 13–10 shows the result of the culture of X63 chimera cells using 10% fetal calf serum (FCS) supplemented with eRDF. On the fifth day after inoculation, viable cell density was about 10^7 cells/ml. On the same day, the specific perfusion rate was increased from 0.8–1.7/day. In the following days, viable cell density and antibody concentration in the spent medium gradually increased with the passage of time.

13.1.2 Perfusion Culture with Separation of the Cells from Culture Mixture by Centrifugation
No centrifuge has yet been used to separate cells from the culture mixture in perfusion culture. In a centrifugal process, operational problems such as filter clogging do not occur, and separation efficiency is high. Therefore, we tried utilizing centrifugation to isolate the cells from the culture mixture (Hamamoto et al. 1989).

13.1.2.1 Influence of Centrifugal Force on Cell Culture Influence of centrifugal force on cell growth was investigated in batchwise culture of hybridoma cells. Two culture systems were compared on the basis of cell growth in quiescent culture. In one system, cells were exposed to a centrifugal force for 10 min three times per day, totalling 30 min. In the other system, no centrifugal force was applied. The cells were mouse-human hybridoma H1 cells, which were established by fusion of mouse myeloma P3/X63-Ag8-U1 with B cells from human tonsil-producing antiherpes virus IgG$_1$ (Fujinaga et al. 1987). Centrifugal forces ranging from 100 g to 500 g were applied in the experiments.

The results are given in Figure 13–11. It shows, in both serum-free and serum-supplemental culture, that the cells could be separated with a centrifugal force of at least 100 g, while 500 g inhibited cell growth.

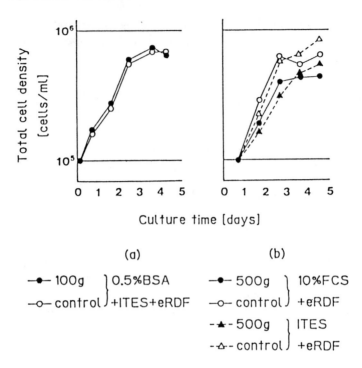

Culture time [days]

(a)   (b)

—●— 100g ⎫ 0.5%BSA   —●— 500g ⎫ 10%FCS
—○— control ⎭ +ITES+eRDF   —○— control ⎭ +eRDF

                              -▲- 500g ⎫ ITES
                              -△-- control ⎭ +eRDF

FIGURE 13–11 Effect of centrifugal force to growth of mouse-human hybridoma H1 cells. (Quiescent culture exposed to centrifugal force for 10 min, three times/ day.) Reproduced with permission from Hamamoto et al. (1989).

13.1.2.2   Perfusion Culture of Mouse-Mouse Hybridoma JTC3 Cells   Figure 13–12 illustrates the experimental system. The culture vessel had a 3 l working volume. In the first step, mouse-mouse hybridoma JTC-3 cells (Wakabayashi et al. 1986) were cultured in a serum-free medium, ITES + eRDF. The hybridoma was established by fusion of mouse myeloma P3/ X63-Ag8-U1 with mouse spleen cells. The results are given in Figure 13–13. On the sixth day after inoculation, viable cell density was found to be 10^7 cells/ml.

13.1.2.3   Perfusion Culture of Mouse-Human Hybridoma X87 Cells   In the second step, mouse-human hybridoma X87 cells were cultured in a serum-free medium, ITES + eRDF.
    The results are shown in Figure 13–14. Perfusion was started by driving the centrifuge just after the inoculation. As the cell density increased, the specific perfusion rate was increased to 2/day. On the seventh day, cell density reached 10^7 cells/ml, and this level was kept almost constant after that time. During the cultivation, cells continued producing antibodies, the

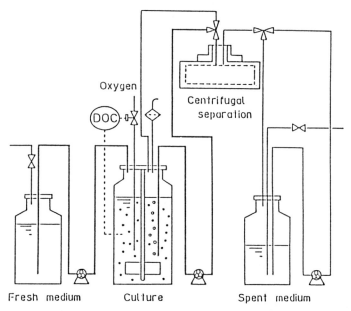

FIGURE 13–12 Experimental perfusion culture apparatus that separates cells from medium by a centrifuge. Reproduced with permission from Hamamoto et al. (1989).

IgG concentration reached 40–80 μg/ml, and IgG specific productivity was found to be about 10 μg/10^6 cells/day.

The centrifugal force was set to 200 g for six days, from day 21 to day 26, to investigate the influence of the centrifugal force. During this period, the viable cell density, IgG concentration, and IgG specific productivity were found to be almost identical to those run in 100 g of centrifugal force. These results were almost equal to those in 4 l perfusion culture with cell separation by gravitational settling shown in Figure 13–4. Thus, perfusion culture combined with centrifugation is applicable if the centrifugal force is suitably selected.

13.2 PERFUSION CULTURE WITH RECYCLING OF HIGH MOLECULAR WEIGHT COMPONENTS

As stated above, perfusion culture seems to perform the best for mammalian cell culture. Even if an excellent perfusion culture process is developed, the high cost of the culture medium remains a problem. Since mammalian cells secrete harmful metabolites that inhibit cellular activities, these substances must be removed from the culture mixture at a low level so the cells can survive for a long period of time. Thus, a certain amount of supernatant must be discarded from the culture system because a technique that can

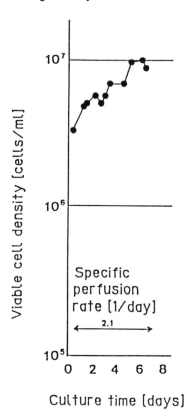

FIGURE 13–13 Perfusion culture of mouse-mouse hybridoma JTC-3 cells using a centrifuge to separate cells from medium. (Medium: ITES + eRDF; net culture volume: 3 l; centrifugal force: 100 *g*.) Reproduced with permission from Hamamoto et al. (1989).

selectively separate these harmful substances from the culture mixture has not yet been developed. In mammalian cell culture, the medium is so expensive that its cost is a substantial portion of the overall cultivation costs, even when a reasonable perfusion culture process has been developed. Hereupon, we experimentally investigated the behavior and the removal of harmful metabolites from the culture mixture.

Buttler et al. (1983) studied a supernatant from a culture of anchorage-dependent cells using microcarriers. They observed that the concentration of ammonia was very high in the medium and suggested that the cell growth might be adversely affected. Iio et al. (1985) cultivated human-human hybridoma cells and observed that ammonia prohibited cells from growing, especially in serum-free cultures. Thus, they performed the cell culture in the presence of a zeolite to remove ammonia from the culture mixture and showed that the ammonia concentration decreased as cell density increased significantly compared to the case with the absence of zeolite.

It has been suggested that lactate is a harmful metabolite, but if the pH of the medium is suitably controlled, inhibition of cell growth by lactate seems to depend upon the cell type and the cultivation conditions. Addi-

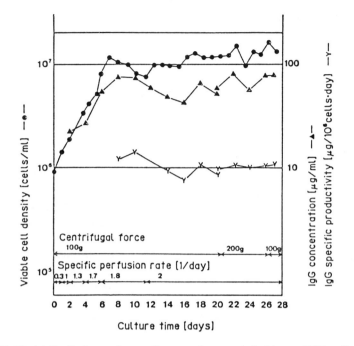

FIGURE 13–14 Perfusion culture of mouse-human hybridoma X87 cells using a centrifuge to separate cells from medium. (Net culture volume: 3 l; medium: ITES + eRDF.) Reproduced with permission from Hamamoto et al. (1989).

tionally, Iio et al. (1985) investigated urea and methyl glyoxal and found that urea was not very harmful, while methyl glyoxal inhibited cell growth. However, methyl glyoxal was not detected in the supernatants. Recently, additional research has been done on harmful metabolites, but it will require some time before the metabolites are identified and separated from the culture mixture economically and selectively.

Thus, we planned to develop a process where only expensive components in the culture medium can be recycled repeatedly (Tokashiki et al. 1988). If such a process is developed, the consumption of the expensive medium components will be largely decreased and the cost of the medium will be reduced. Serum-free media generally contain such proteins as insulin, transferrin, or albumin, which are very expensive compared to basal medium components. These proteins have high molecular weight whereas individual components in basal media have low molecular weight. Although all the harmful metabolites have not yet been identified, some, e.g., ammonia and lactate, are low in molecular weight. Therefore, we tried to subject the spent medium separated from the cells to ultrafiltration in perfusion culture. During this process, the permeate was removed from the culture system and the retentate, including high molecular weight components in the spent medium, was recycled into the culture system (perfusion culture with recycled retentate). This process is illustrated in Figure 13–15.

When the recycled growth factors are still active, they can be decreased in the fresh medium. If the process is applied to the production of a monoclonal antibody by hybridoma cells, the following advantages can be expected. Since the molecular weight of the antibody is about 150×10^3, the molecules of the monoclonal antibody do not permeate through the membrane and accumulate in the culture vessel. In the production of a useful material by mammalian cell culture, the concentration of the target material is generally very low and this low concentration will greatly increase the purification cost. In conventional perfusion culture, this disadvantage cannot be avoided.

Thus it has been said that perfusion culture is inferior to mouse ascites culture, microcapsule culture, and hollow-fiber module culture where this problem is concerned. In any case, as long as high molecular weight substances can be recycled to the culture system, this problem can be solved.

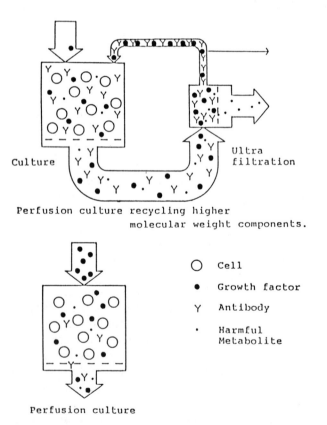

Perfusion culture recycling higher molecular weight components.

O   Cell

●   Growth factor

Y   Antibody

·   Harmful Metabolite

Perfusion culture

FIGURE 13–15 Outline of a perfusion culture that recycles higher molecular weight components compared to a conventional perfusion culture.

13.2.1 Experimental

13.2.1.1 Experimental Apparatus Figure 13–16 illustrates our experimental apparatus. A perfusion culture vessel of 120 ml capacity was used. The ultrafiltration unit was Pellicon Lab Casset XX42 OLCOO (Millipore Inc.) and the membrane was ultrafiltration membrane PTHK OLCOO (nominal molecular weight limit = 10,000).

13.2.1.2 Culture Media Serum-supplemented media and two kinds of serum-free media were employed. eRDF was used as the basal medium. When serum-supplemented media were used, the serum concentration was 10% before high molecular weight substances were recycled and 1% after the start of the recycling process. Serum-free medium A, containing bovine serum albumin, and serum-free medium B were used as the serum-containing media. The composition of serum-free media is shown in Table 13–1.

13.2.1.3 Experimental Procedures Dissolved oxygen was automatically adjusted to 3 ppm by sparging molecular oxygen directly. During the early stages of cultivation, the perfusion culture was conventionally performed without recycling high molecular weight components. After cell density exceeded that in the batch culture, recycling and ultrafiltration was started. Simultaneously, the permeate was taken out of the system and the retentate was recycled into the culture vessel. Moreover, the medium was changed

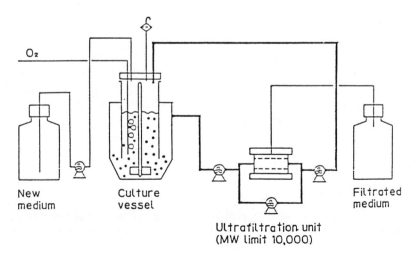

FIGURE 13–16 Experimental perfusion culture system that recycles higher molecular weight components. Reproduced with permission from Takazawa et al. (1988).

TABLE 13–1   Composition of Serum-Free Medium

Serum-Free Medium A: BSA + ITES + eRDF

	Before Recycling	*Under Recycling*
BSA	0.5%	0.025%
Insulin	9 µg/ml	1.8 µg/ml
Transferrin	10 µg/ml	10 µg/ml
Ethanolamine	10 µM	10 µM
Selenite	20 nM	20 nM

Serum-Free Medium B: ITES + eRDF

	Before Recycling	*Under Recycling*
Insulin	9 µg/ml	1.8 µg/ml
Transferrin	10 µg/ml	0.5 µg/ml
Ethanolamine	10 µM	10 µM
Selenite	20 nM	20 nM

Reproduced with permission from *Hakko-Kogaku Kai-shi* 66, 31–35 (1988).

from that for the "before-recycle" to that for the "under-recycle" (see Table 13–1).

13.2.2   Culture of Mouse-Mouse Hybridoma 4C10B6 Cells in Serum-Supplemented Medium

Mouse-mouse hybridoma 4C10B6 cells are a fusion product of mouse myeloma P3/X63-Ag8-U1 cells and mouse spleen cells and secrete IgG2b. The maximum viable cell density attained 1.0–1.5×10^6 cells/ml in quiescent culture.

Figure 13–17 shows the results of perfusion culture with retentate recycling and without recycling. At the early stage, the perfusion culture was carried out without recycling. On the sixth day following inoculation, the viable cell density rose to 6.0×10^6 cells/ml, much higher than that in quiescent culture. The recycling of high molecular weight medium components was then begun. For three days after the start of recycling, the viable cell density was about 1×10^7 cells/ml. After that time, the density gradually decreased. On the eighth day after the start of recycling, namely, the fourteenth day after inoculation, the cell density reached 5.0×10^6 cells/ml. Conversely, the concentration of antibodies in the culture vessel increased with the passage of time and reached 100 µg/ml on the sixth day after recycling began. This value was about fivefold in comparison to that found in quiescent cultures. The concentration of antibodies in the permeate was less than 1 µg/ml on day 2 and day 4. These results show that this process can accumulate antibodies in the culture vessel. The viable cell density,

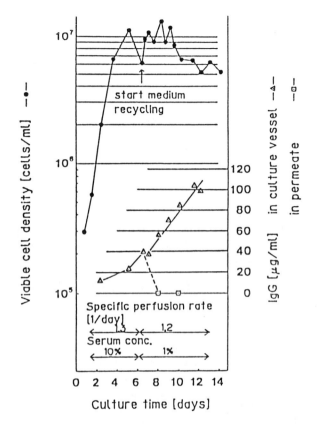

FIGURE 13-17 Perfusion culture of mouse-mouse hybridoma 4C10B6 cells with recycling of culture medium after ultrafiltration. Reproduced with permission from *Hakko Kogaku Kai-shi* 66, 31–35 (1988).

however, was lower than that of conventional perfusion culture under the same cultivation conditions.

13.2.3 Culture of Mouse-Human Hybridoma V6 Cells in Serum-Free Medium

Mouse-human hybridoma V6 cells were established by fusion of human spleen lymphocytes with mouse myeloma P3/X63-Ag8-U1 cells and produced human monoclonal antibody IgG_1 against the varicella-zoster virus (Sugano et al. 1987). The medium used was serum-free medium A shown in Table 13–1. The maximum viable cell density of the quiescent culture of V6 cell was 1×10^6 cells/ml.

Figure 13–18 gives the results of the perfusion culture. In its early stage, the cultivation was carried out without retentate recycling. The volume of

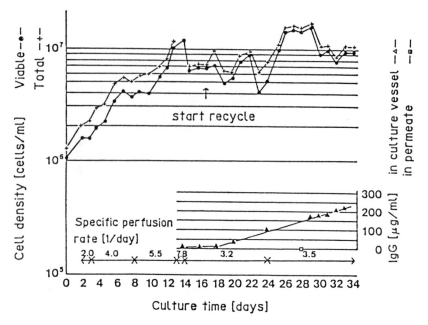

FIGURE 13–18 Perfusion culture of mouse-human hybridoma V-6 cells with re-cycling of culture medium after ultrafiltration. (Culture medium: 0.5% BSA + ITES + eRDF.) Reproduced with permission from *Hakko Kogaku Kai-shi* 66, 31–35 (1988).

insulin, transferrin, and albumin were 9, 10, and 5,000 μg/ml in the medium, respectively. In this case, the maximum viable cell density was 6–12 × 10⁶ cells/ml, and the antibody concentration was found to be about 9 μg/ml in the spent medium. On the sixteenth day after inoculation, recycling of the spent medium supernatant was initiated and the levels of insulin, transfer-rin, and albumin were changed to 1.8, 0.5, and 250 μg/ml in the fresh medium, respectively. The cultivation with recycling was continued for 18 days. The viable cell density reached 1.6 × 10⁷ cells/ml at maximum and it was not less than that in the perfusion culture without recycling. The concentration of antibodies increased with the passage of time in the culture vessel and reached a maximal level of 220 μg/ml on the seventeenth day after the recycling began (i.e., on the thirty-third day after inoculation). This value is 24-fold higher than that in the cultivation without recycling. On the twenty-eighth day after inoculation, antibodies were assayed and IgG was present less than 1 μg/ml in the permeate. As a result, it was concluded that almost all the antibodies accumulated in the culture vessel.

13.2.4 Culture of Mouse-Human Hybridoma C41 Cells in a Serum-Free Medium

Serum-free medium B, shown in Table 13–1, was employed. Figure 13–7 shows the result of the quiescent culture. The maximum viable density was 1.0×10^6 cells/ml. Figure 13–8 gives the result of the perfusion culture without retentate recycling. The maximum cell density was 9.0×10^6 cells/ml, and the antibody concentration reached 6–12 μg/ml at the specific perfusion rate of 7.5/day.

Figure 13–19 gives the result of the perfusion culture with retentate recycling. For the first 10 days, the culture was conducted without recycling. The maximum viable cell density reached 9.5×10^6 cells/ml, and the antibody concentration was found to be 7 μg/ml in the spent medium at a 6.2/day specific perfusion rate. On the tenth day, the retentate recycling was started. At the same time, the insulin and transferrin levels were changed from 9 and 10 μg/ml to 1.8 and 0.5 μg/ml, respectively. Immediately after the recycling started, the viable cell density decreased but increased with time and reached the same level as that without recycling. The concentration of antibody increased with the passage of time in the culture vessel and rose to 150 μg/ml on the tenth day after the start of the recycling process.

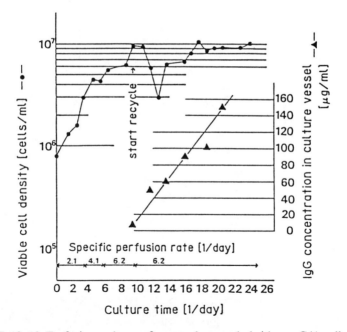

FIGURE 13–19 Perfusion culture of mouse-human hybridoma C41 cells with recycling of culture medium after ultrafiltration (Culture medium: ITES + eRDF.) Reproduced with permission from *Hakko Kogaku Kai-shi* 66, 31–35 (1988).

There were few emergencies in this experiment. Thus, it is worthwhile to discuss the IgG specific productivity in the perfusion culture with retentate recycling. In this culture (Figure 13–19), the average viable cell density was found to be 7.7 × 10⁶ cells/ml, and the increase in IgG concentration was 140 μg/ml in the period from day 10 to day 20. The total volume of the retentate recycled in the overall system, including the tank, ultrafiltration unit, pipes, and pumps, was 350 ml, the net culture volume was 120 ml, and IgG specific productivity was 5.3 μg/10⁶ cells/day. The values were a little smaller than those of the perfusion culture without recycling.

13.2.5 Culture of Mouse-Human Hybridoma H2 Cells

Mouse-human hybridoma H2 cells are a fusion product of mouse myeloma P3/X63-Ag8-U1 cells and B cells from human tonsil and produce antiherpes virus IgG₁ (Fujinaga et al. 1987).

The result of the quiescent culture of H2 cells in 10% FCS-eRDF medium and serum-free medium B is given in Figure 13–20. H2 cells did not exhibit significant growth in ITES-eRDF, but grew well in serum-supplemented medium up to 1.4 × 10⁶ cells/ml density.

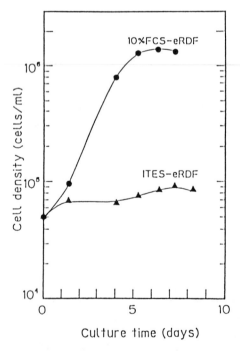

FIGURE 13–20 Cellular growth of H2 in 10% FCS + eRDF and ITES + eRDF. Reproduced with permission from Takazawa et al. (1988).

High cell density perfusion culture was tried in ITES-eRDF and the result is given in Figure 13–21. Perfusion was started two days after inoculation and the rate was increased stepwise as cell growth occurred. Contrary to the result in quiescent culture, H2 cells started to grow after two days lag phase and reached a density of 1.1×10^7 cells/ml by 10 days in ITES-eRDF. The concentration of IgG in spent medium increased with cell growth to a maximum value of 31 μg/ml. The ratio of IgG to the total protein was about 3.4% at this time. The specific productivity of IgG was approximately 20 μg/10^6 cells/day throughout the culture.

Figure 13–22 shows the result of perfusion culture of H2 cells with the recycling of high molecular weight medium components (Takazawa et al. 1988). The medium was serum-free medium B. When cells entered the stationary phase, the recycling process was started. While the recycle was continued, the concentration of transferrin in the medium reduced to 0.5 μg/ml, which corresponded to 5% of the concentration in the original medium. The cultivation was continued for 48 days after inoculation. Cell density was maintained at about 10^7 cells/ml during the recycling phase and the maximum viable cell density was 2.7×10^7 cells/ml. IgG concentration gradually increased after the recycling began, and 48 days later, reached 2.0 mg/ml, over 60-fold more than was obtained in the perfusion culture. The

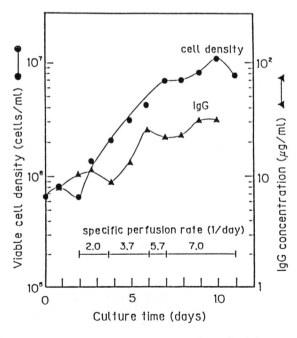

FIGURE 13–21 Cellular growth and IgG production of H2 in a perfusion culture with ITES + eRDF. Reproduced with permission from Takazawa et al. (1988).

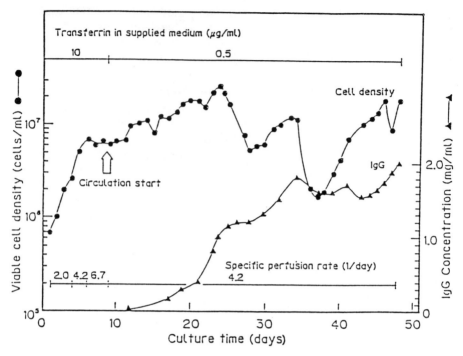

FIGURE 13–22 Cellular growth and IgG production of H2 in a medium circulating culture with ITES + eRDF. Reproduced with permission from Takazawa et al. (1988).

TABLE 13–2   Summary of Two Culture Methods

Indexes	Perfusion Culture	Medium Circulating Culture
Maximum cell density	1.1×10^7 cells/ml	2.6×10^7 cells/ml
Maximum IgG concentration	0.03 mg/ml	2.0 mg/ml
IgG ratio to the total protein	3.4%	61%
Relative affinity of the IgG for the antigen	100	100

ratio of IgG concentration at that time to the total protein in the culture mixture was about 61%, which was approximately 18-fold higher than the perfusion culture without recycling.

The culture of H2 cells is summarized in Table 13–2. It was proven that long-term incubation of monoclonal antibody at high concentration did not affect the specificity since the affinity to the antigen did not decrease after 48-day cultivation.

13.2.6 Discussion

There is a possibility that cell density and IgG specific productivity of the perfusion culture with retentate recycling are different from those of the culture without recycling because the culture environments are different from each other. The cell density of the culture with retentate recycling was found to be almost the same as that without recycling in the V6 cell culture and C41 cells. On the other hand, cell density was obviously higher with recycling than with no recycling in the H2 cell culture, particularly long after recycling was started. This result may be specific to H2 cells, but may occur when the culture with retentate recycling is continued for a long period of time. There is a possibility that some growth factors are secreted by the cells and accumulate in the culture system, thus resulting in high cell density. (see Table 13–2)

In order to determine IgG specific productivity, the material balance of IgG in an experiment must be precisely traced, particularly on a culture mixture that contains a high concentration of the antibody. Strictly speaking, adequate and accurate data to determine this cannot be obtained by these experiments. However, in each culture, the antibody concentration increased with the passage of time and the feedback inhibition of antibody was not observed, even when the concentration was very high. In the serum-free culture of mouse-human hybridoma cells, the concentrations of insulin, transferrin, and albumin were reduced to one-fifth, one-twentieth, and one-twentieth, respectively, of their original concentration when the culture with retentate recycling was compared to the culture without recycling. It does not always mean that the concentration of growth factors can be decreased in the fresh medium in this way. We should more accurately establish the optimal medium condition in both cases and compare the results. Such experiments have not yet been done. H2 cells were subjected to cell culture in a serum-free medium B containing insulin, 1.8 μg/ml, and transferrin, 0.5 μg/ml, but they did not grow at all. Thus, it is possible to decrease the amount of growth factors in the culture mixture in the case of culture using recycling.

13.3 HIGH DENSITY CULTURE USING FLUOROCARBON TO SUPPLY OXYGEN

Conventionally, oxygen has been supplied to a mammalian cell culture system by bringing the culture mixture into direct contact with an oxygen-containing gas or into indirect contact through a hollow fiber module or silicone tubes. If the culture volume is small, the former process can apply a sufficient amount of oxygen from the gas phase to the culture system through a free surface. In large-scale or high-density culture, however, the sufficient supply is not achieved unless an oxygen-containing gas is directly sparged into the culture medium because the free surface is too small for sufficient supply. However, direct sparging forms foam on the free surface and the cells sometimes suffer damage, resulting in lowered activity of the target products. In indirect oxygen transfer through a hollow fiber nodule or silicone tubes, the oxygen transfer resistance is so high that the process is not always suitable for large-scale or high-density culture. Therefore, we tried to use liquid fluorocarbon to supply oxygen into the culture mixture (Hamamoto et al. 1987 and 1988). This process seems to be particularly suitable for large-scale, high-density culture.

13.3.1 Culture of Mouse-Mouse Hybridoma 4C10B6 Cells

Figure 13–23 illustrates an experimental flow sheet. The culture vessel is equipped with a spinner filter for separation of cells from the culture mixture and has a net culture volume of 225 ml. In this system, the oxygen-containing fluorocarbon was fed dropwise from the top, flowed down through the medium, and was trapped on the bottom of the vessel. The fluorocarbon was fed to the oxygen absorption column and the level of boundary surface between the fluorocarbon and the culture medium was kept at a certain level. The absorption column is equipped with an inner shell and the fluorocarbon absorbs oxygen while it rises in the inner column. It then flows over the top of the inner shell and drops to the bottom of the column, is

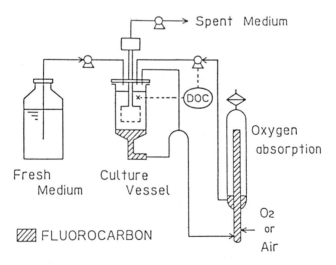

FIGURE 13–23 Perfusion culture system using fluorocarbon to supply oxygen. Reproduced with permission from Hamamoto et al. (1987).

pumped to the top of the culture vessel, and supplies oxygen to the culture mixture as it drops.

Mouse-mouse hybridoma 4C10B6 cells were cultivated. RPMI1640 supplemented with 10% FCS was used as a culture medium.

Cell density and IgG concentration are given in Figure 13–24. For six days after the start of cultivation, oxygen was supplied by sparging into the culture mixture. On the sixth day, when the viable cell density was 5.0×10^6 cells/ml, oxygen supply was switched to the fluorocarbon method. The cell density continued to increase after switching and reached 9.8×10^6 cells/ml. The cells continuously secreted antibody, and IgG specific productivity was about 10 μg per 10^6 cells per day. During the cultivation under supply of oxygen with fluorocarbon, dissolved oxygen was kept almost constant at 3 ppm and no foam was observed in the culture vessel.

13.3.2 Culture of Mouse-Human Hybridoma X87 Cells

Figure 13–25 shows the flow sheet used in the experiment. The perfusion culture vessel was provided with a section for separating the cells from the culture mixture by gravitational settling. The net culture volume was 1.6 l. The fluorocarbon was passed through the culture medium to the bottom while oxygen was transferred into the medium, and pumped up to the ox-

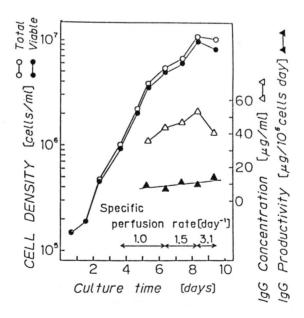

FIGURE 13–24 Perfusion culture of mouse-mouse hybridoma 4C10B6 cells using fluorocarbon to supply oxygen. Reproduced with permission from Hamamoto et al. (1987).

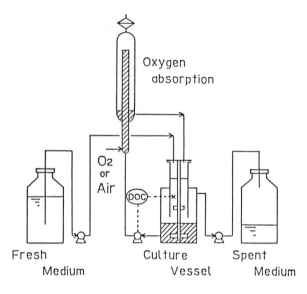

FLUOROCARBON

FIGURE 13–25 Perfusion culture system using fluorocarbon to supply oxygen. Reproduced with permission from Hamamoto et al. (1988).

ygen absorption column. The fluorocarbon was repeatedly recycled in the closed system. Under such conditions, mouse-human hybridoma X87 cells were cultivated in serum-free medium B. The viable cell density is presented in Figure 13–26.

The specific perfusion rate was set to 1.0–1.5/day from the inoculation to the eighth day and cell density seemed to be confluent under such conditions. Beyond that time, the specific rate was increased and the cell density simultaneously increased to $0.9–1.5 \times 10^7$ cells/ml for seven days. Meanwhile, the dissolved oxygen concentration was maintained in the range of 2.0–3.0 ppm.

These results show that the formation of foam can be prevented by using fluorocarbon to supply oxygen in the culture of the hybridoma cells. However, the cell density was not always higher than it was in the culture with direct oxygen sparging. Damage due to foaming depended upon the kind of cells being cultured. This culture process seems to be most effective for cells that are sensitive to foaming.

The high density culture processes developed by the author have been explained by exemplifying the culture of several kinds of hybridomas. In the near future, the useful substances produced by mammalian cells will be used in a wide range of fields, including medicine. In such situations, various kinds of mammalian cells will be cultured to produce a variety of target

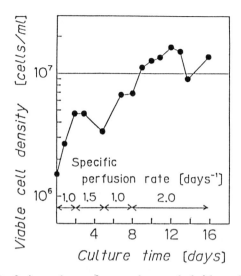

FIGURE 13-26 Perfusion culture of mouse-human hybridoma X87 cells using fluorocarbon to supply oxygen. Reproduced with permission from Hamamoto et al. (1988).

substances under a wide range of cultivation conditions, and the production scale also will be widely varied.

REFERENCES

Butler, M., Imamura, T., Thomas, J., and Thilly, W.G. (1983) *J. Cell Sci.* 61, 351–363.

Fujinaga, S., Sugano, T., Matsumoto, Y., Masuho, Y., and Mori, R. (1987) *J. Infectious Dis.* 155, 45–53.

Hamamoto, K., Ishimaru, K., and Tokashiki, M. (1989) *J. Fermentation & Bioengineering,* 67, 190–194.

Hamamoto, K., Tokashiki, M., Ichikawa, Y., and Murakami, H. (1987) *Agric. Biol. Chem.* 51, 3415–3416.

Hamamoto, K., Tokashiki, M., and Ichikawa, Y. (1988) *Chemical Engineering Symposium Series No. 17,* (Y. Hanano et al., eds.), pp. 143–147, The Society of Chemical Engineering, Tokyo, Japan.

Iio, M., Moriyama, A., and Murakami, H. (1985) *Growth and Differentiation of Cells in Defined Environment,* (H. Murakami et al., eds.), pp. 437–442, Kodansha Ltd. and Springer-Verlag, Tokyo, Japan and Heidelberg, Germany.

Masuho, Y., Matsumoto, Y., Sugano, T., Fujikawa, S., and Minamishima, Y. (1987) *J. Gen. Virol.* 68, 1457–1461.

Murakami, H., Shimomura, T., Nakamura, T., et al. (1984) *J. Agric. Soc. Jpn.* 58, 575–583.

Nishimura, Y., Yokoyama, M., Araki, K., et al. (1987) *Cancer Res.* 47, 999–1005.

Sato, S., Kawamura, K., Hanai, N., and Fujiyoshi, N. (1985) *Growth and Differentiation of Cells in Defined Environment* (H. Murakami et al. eds.), pp. 123–126, Kodansha Ltd. and Springer-Verlag, Tokyo, Japan and Heidelberg, Germany.

Sugano, T., Matsumoto, Y., Miyamoto, C., and Masuho, Y. (1987) *Eur. J. Immunol.* 17, 359–364.

Takazawa, Y., Tokashiki, M., Hamamoto, K., and Murakami, H. (1988) *Cytotechnology* 1, 171–178.

Tokashiki, M., and Arai, T. (1989) *Cytotechnology* 2, 5–8.

Tokashiki, M., Hamanoto, K., and Takazawa, Y. (1988) *Kagaku Kougaku Ronbunshu* 14, 337–341.

Wakabayashi, K., Sakata, Y., and Aoki, N. (1986) *J. Biol. Chem.* 261, 11097–11105.

14

Diffusion and Convection in Membrane Bioreactors

Carole A. Heath
Georges Belfort

The growing need for a cost-effective means of producing large quantities of important biological products from mammalian cells has stimulated the development of numerous types of bioreactors and the modes in which they are run. Initially, bioreactors corresponded only to the conventional batch system in which both cell density and product concentration were very low (generally 10- to 100-fold less than in mouse ascites). Improvements have focused on increasing productivity per unit volume and reducing costs, especially for the expensive downstream processing. Membrane bioreactors, which retain the cells behind a barrier, have proven to be quite successful in both respects perhaps because of their emulation of the in vivo capillary-tissue environment (see the recent review articles on membrane bioreactors by Shuler 1987 and Belfort 1989). Typical system configurations are the hollow fiber, the plate and frame (flat sheet), and the spiral wound modules. The hollow fiber and the flat sheet membrane reactors have been used successfully for culture of mammalian cells (Ku et al. 1981; Seaver and Gabriels 1985a and 1985b; Altshuler et al. 1986; Tharakan and Chau 1986;

The authors would like to acknowledge Millipore Corp. (Bedford, MA) for partial support of this work.

Evans and Miller 1988). Commercial units of both types are available. Culture of animal cells in a spiral wound module is undoubtedly forthcoming.

Membrane systems can provide a much greater surface area per unit volume (important for substrate delivery and waste removal) than typical batch reactors, resulting in increased cell density and product concentration. With recycling, membrane systems can achieve much higher product concentrations than batch reactors and are capable of producing at levels approaching that of living systems. The membrane porosity can be chosen to selectively separate the cells and product using microporous or high molecular weight cutoff (MWCO) ultrafiltration membranes or to concentrate the product with the cells in the extracapillary space using low to medium MWCO ultrafiltration membranes. Both alternatives remove one or more steps in the costly product purification process. In addition, the fragile mammalian cells, which do not have a cell wall like bacteria and yeast, are protected from damaging fluid shear stresses by the membrane barrier. Cells grown in membrane reactors often have lower growth rates. This, in combination with the cell retention, results in lower net consumption of nutrients, lowering medium costs, especially with high recycle rates.

One of the proclaimed disadvantages of membrane systems is the potential for concentration polarization and fouling, which can severely diminish the transmembrane nutrient transport. Fouling can usually be avoided by using hydrophilic membranes with low surface charges, which has the effect of minimizing adsorption on the pore walls and surface of the membrane. Concentration polarization can be held to a minimum with high flow rates through the lumen space, which increases the wall shear rate, and with highly porous membranes, which lowers the protein rejection coefficient.

Another suggested disadvantage is the existence of diffusional limitations. In many systems transport of nutrients and by-products relies primarily on diffusion, thus cells further from the membrane surface may be subject to low substrate and high waste concentrations, leading to cell stagnation and death. The extent to which a necrotic region develops depends on numerous factors, such as medium flow rate, bulk substrate and waste concentrations, membrane porosity and thickness, and intermembrane spacing. In many membrane systems, particularly those using very porous membranes, diffusional limitations may be diminished or even avoided because of the convective leakage flow through the membrane into the cell region, supplementing the diffusive transport. Systems may be classified by their primary mode of mass transfer to and from the cell region, i.e., diffusive, convective, or combined. This distinction is important in assessing the potential for mass transfer or kinetic limitations within a particular operation.

Appropriately, a published set of rules governing the design and operation of membrane devices for all applications does not exist. Instead, system development is generally based on experiment and accumulated experience. Mass transfer models can provide the necessary theoretical background in the quest for optimum use of membrane bioreactors for growing animal cells and harvesting their products. This chapter will review previous work and outline the equations necessary to model diffusive and convective transport in membrane systems. Simple examples of the resulting concentration profiles will then be demonstrated for a flat membrane bioreactor, chosen because of its geometric simplicity and ease of use in modeling. The given equations can be modified easily for application to other membrane systems.

14.1 DIFFUSIVE MASS TRANSFER

Diffusion is the primary mode of mass transfer in membrane bioreactors when there is insignificant convection in the cell region. This situation occurs when medium flow rates are low, membrane porosity is small, and/or the cell region is very densely populated, decreasing its permeability to fluid flow. Mathematical modeling of diffusive mass transfer and uptake of a limiting substrate can indicate bioreactor conditions in which the diffusion of substrate is insufficient to maintain high cell viability and/or productivity. Numerous papers describing theoretical models of substrate diffusion and uptake in hollow fiber bioreactors (the most commonly used type of membrane bioreactor) have already been published.

The first models to be developed were for enzyme reactors (Rony 1971; Horvath et al. 1973; Waterland et al. 1974; Georgakis et al. 1975; Kim and Cooney 1976) followed by those for whole cell systems (Webster and Shuler 1978 and 1981; Webster et al. 1979; Davis and Watson 1985 and 1986; Heath and Belfort 1987; Chresand et al. 1988). Waterland et al. provided both a theoretical model (1974) and supporting experimental data (1975) for diffusion with a first-order enzymatic reaction at steady state using asymmetric hollow fiber membranes. The experimental results, consisting of overall reactor conversions at various flow rates, correlated well with model predictions. A similar approach was taken by Kim and Cooney (1976), who considered the general case in which axial substrate concentration gradients exist in the fiber lumen, essentially representing the system as a plug flow reactor. The slope of the axial concentration gradient, and the subsequent effect on diffusion into the extracapillary region, is dependent on the substrate residence time or solution flow rate. For systems with short residence times and frequent recycling, axial concentration gradients are negligible and the overall system resembles a continuously stirred tank reactor. Using this assumption, Webster and Shuler (1978) derived models for hollow fiber enzyme reactors and whole cell hollow fiber reactors (Webster et al. 1979) utilizing an effectiveness factor (a ratio of the actual reaction rate to the reaction rate without mass transfer limitations) for both first- and zero-order kinetics. Webster and Shuler (1981) also developed a model for simulating transient substrate concentration profiles for whole cell hollow fiber reactors. Davis and Watson (1986) developed a generalized mathematical model and computational solution procedure for describing concentration and temperature profiles in annular reactors for any reaction rate equation. Heath and Belfort (1987) demonstrated axial and radial substrate concentration profiles in a mammalian cell hollow fiber system using zero- and first-order kinetics. All the models indicate the potential occurrence of diffusion limitations.

Many of the papers mentioned above simulate concentration profiles for the reactors as a function of the Thiele modulus or a modified Thiele modulus, allowing the theory to be applied to many systems. The Thiele modulus is a function of the diffusive and kinetic constants, relating substrate uptake to its diffusion resistance. The use of the modulus confers the advantage that the effects of changing design and operating conditions can be seen quickly and easily by observing whether the system is under diffusional or kinetic control. For this reason, the same convention will be used in the following simple mathematical analysis of diffusion.

A mathematical definition of the diffusion/reaction system begins with the equation of continuity for constant density and diffusivity

$$\frac{\partial c}{\partial t} + (\mathbf{V} \cdot \nabla c) = D_e \nabla^2 c + R \tag{14.1}$$

where $c = c(x, y, z)$ is concentration of substrate (usually the limiting one), t is time, $\mathbf{V}$ is the velocity vector (w, u, v), and D_e is the effective diffusivity of substrate in the particular fluid or matrix. The reaction term, R, is generally expressed as a zero- ($c \gg K_m$) or first- ($c \ll K_m$) order limit of the Monod equation

$$R = -\frac{\mu_{max} c}{c + K_m} \tag{14.2}$$

where μ_{max} is the maximum reaction rate and K_m is the Michaelis constant. The following mathematical model description is based on the flat membrane system with z being the axial direction and y the direction perpendicular to the membrane-medium and membrane-cells interfaces. Development of equations to model other membrane systems can be undertaken in a similar fashion. For ease in developing the model equations, assumptions relevant to the particular system are generally made. For the flat membrane system under study, these assumptions include a homogeneous cell suspension in the cell region, an isothermal system, Fickian diffusion characterized by an effective diffusivity D_e, negligible resistance to diffusion by the membrane (i.e., a partition coefficient equal to one), and steady-state substrate conversion. With these assumptions, equation 14.1 in cartesian coordinates reduces to

$$v \frac{\partial c}{\partial z} = D_e \left(\frac{\partial^2 c}{\partial y^2} + \frac{\partial^2 c}{\partial z^2} \right) + R \tag{14.3}$$

In this model, the flat membrane bioreactor consists of three regions, each with different transport mechanisms. Regions 1, 2, and 3 consist of the medium, membrane, and cells, respectively. With consideration of the assumptions, equation 14.3 can be rewritten for each region. The velocity profile in region 1 is generally assumed to be that of laminar flow. In regions 2 and 3, convection is negligible, leaving diffusion as the sole transport mechanism. Again, the membrane is assumed to provide negligible resistance to diffusion compared to that offered by the cells.

Depending on the membrane porosity and whether the cells are anchorage dependent or independent, the region where most of the reaction occurs can be either in the membrane or in the cell region. For purposes of this chapter the cells will be located only in the cell region, which is generally typical of suspension cells such as hybridomas, and the mathematical equations will be developed for that location. Because the module length is much greater than the thickness of the cell region (i.e., $L \gg y_3 - y_2$), an order of magnitude analysis indicates that the concentration gradient in the z-direction is negligible compared to that in the y-direction and thus can be dropped, meaning that diffusion occurs primarily in the direction perpendicular to the membrane surface. The substrate concentration in the cell region can now be described by

$$D_e \frac{d^2c}{dy^2} + R = 0 \qquad (14.4)$$

The boundary conditions necessary to solve equation 14.4 are the definition of the substrate concentration at the membrane-cells region interface

$$c(y = y_2) = c_m \qquad (14.5.1)$$

in which c_m may be a function of z (e.g., long residence time in the reactor), and/or t (e.g., semi-continuous systems), and

$$\frac{dc}{dy} = 0, (y = y_3) \qquad (14.5.2)$$

which indicates no net flux in the y-direction (symmetry) at the midpoint of the cell region. The solution for the zero-order case is found to be

$$c(y) = c_m - \frac{\mu_{max}}{2D_e} [y_2^2 - y^2 + 2y_3(y - y_2)] \qquad (14.6)$$

Similarly, for first-order kinetics the solution is given as

$$c(y) = c_m \left(\frac{\exp[\alpha(y - 2y_3)] + \exp(-\alpha y)}{\exp[\alpha(y_2 - 2y_3)] + \exp(-\alpha y_2)} \right) \qquad (14.7)$$

where $\alpha = (\mu_{max}/D_e K_m)^{1/2}$. Figure 14–1 demonstrates simulated concentration profiles in the cell region of a flat membrane bioreactor as a function of c_m using equation 14.7. Decreasing values of c_m could either reflect a

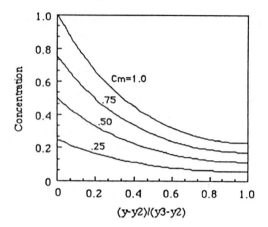

FIGURE 14–1 Simulated substrate concentration profiles in the cell region of a flat membrane bioreactor as a function of C_m with diffusion as the primary mode of mass transfer.

time-related drop in the bulk substrate concentration (e.g., glucose) for a system in which the medium reservoir is changed periodically or it could approximate a situation in which the substrate is significantly depleted along the reactor length (e.g., oxygen), resulting in a different c_m for different values of z. As shown, far from the membrane surface the substrate concentration is very low and would be even lower if the values for uptake and diffusivity used in the simulation were too low or too high, respectively. This type of approach is useful in determining an approximate schedule for reservoir medium exchange and for determining the limiting axial path length of a reactor, both of which may be instrumental in avoiding diffusion limitations.

A first-order Thiele modulus for the flat membrane system can be defined as

$$\phi^2 = y_2^2 \left(\frac{\mu_{max}}{D_e K_m} \right) = (\alpha y_2)^2 \qquad (14.8)$$

and used to plot concentration versus dimensionless distance from the membrane surface as a function of the Thiele modulus. Figure 14–2 shows that the substrate concentration in the cell region is very dependent on this ratio. For a reaction-controlled system ($\phi^2 < 1$) adequate substrate is available throughout the region. However, for a diffusion-controlled system ($\phi^2 > 10$) there is a severe concentration gradient in the cell region, which would lead to low cell viability in the reactor.

To investigate the effect of cell region thickness (which affects both the distance required for diffusion as well as the amount of substrate consumed per membrane area in the cell region) the parameter y_3 is replaced by λy_2. Varying the value of λ in Figure 14–3 demonstrates the effect of changing

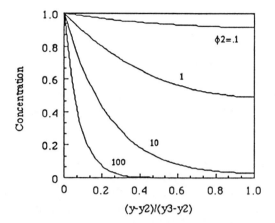

FIGURE 14-2 Substrate concentration in the bioreactor cell region as a function of the Thiele modulus.

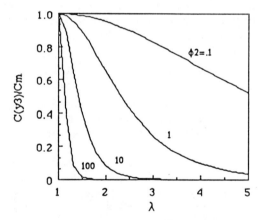

FIGURE 14-3 The influence of the Thiele modulus on substrate concentration in the cell region as a function of cell region thickness.

the thickness of the cell region on the availability of substrate as a function of the Thiele modulus. A value of $\lambda = 1$ corresponds to $y_3 = y_2$, i.e., no cell region, while $\lambda = 5$ corresponds to a cell region thickness of $y_2(\lambda - 1)$ or $4y_2$. The size of the cell region is of utmost importance in situations of diffusional control ($\phi^2 > 10$), as shown by the steep concentration gradients in this regime.

A particularly beneficial technique for assessing the behavior of a membrane bioreactor is to use effectiveness factors similar to the approach of Webster et al. (1979). The effectiveness factor is a ratio of the substrate actually converted in a reactor to that which would have reacted in the absence of mass transfer limitations. In effect, it is a lumped parameter description of efficiency. The variation of the effectiveness factor with different λ and ϕ^2 is shown in Figure 14–4. The effectiveness factor is high only for a kinetically controlled system, especially when the cell region is thick. In a diffusion controlled system, such as one with a large cell region, a longer distance for diffusion is required so many cells may not receive as much substrate as they could potentially consume, resulting in lower effectiveness factors.

The equations described above for modeling diffusional mass transfer in a flat membrane bioreactor cannot fully represent the functioning bioreactor with complete accuracy due to the assumptions made to simplify the governing equations. However, the model can provide a good approximation limited only by the validity of the assumptions for the particular case under study. The assumption of negligible axial diffusion in the cell region may not be justified in systems with a long residence time due to a low flow rate or long axial path length. The error introduced by making this assumption should be assessed prior to making definite conclusions concerning the existence of diffusional limitations. The dependence of c_m on t and/or z can be very important and should be addressed before continuing with model development. The lack of convective flow in the cell region is a requirement of the model, and cases where this does not hold true should be approached either with a convective model or a combined diffusion and convection model. If convective flow contributes only minimally to transport in the cell region, the overall mass transfer will be enhanced and the profiles determined by diffusion analysis will be conservative.

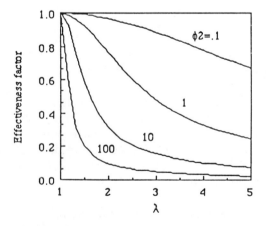

FIGURE 14–4 The effectiveness factor of the flat membrane bioreactor as a function of cell region thickness and the Thiele modulus.

Although not shown, it is very important when using model equations to have accurate values for the model parameters such as the diffusivity and the kinetic constants. In actuality they should be measured for each system in use since the kinetics of one type of cell under given growth and reactor operating conditions may differ by as much as an order of magnitude from those of another cell line. However, to obtain approximate concentration profiles, appropriate values from the many published studies on diffusivity and kinetics of various cell types can be used. Additionally, the cell region concentrations of the "limiting substrate" as well as those of other important nutrients and waste products should be carefully investigated. Buildup of by-products such as ammonia and lactate can be inhibitory to cell growth and productivity. These effects may be apparent even before substrate limitations occur.

When the model equations indicate that diffusional mass transfer limitations exist, steps can be taken to improve the situation either by changing the operating parameters or, more drastically, by changing the bioreactor design. Operating parameters that can be adjusted to improve mass transfer include bulk substrate concentration, medium flow rate, and cell density. Design variables include dimensions of the three regions as well as membrane type and porosity.

14.2   CONVECTIVE MASS TRANSFER

In many membrane systems, diffusion is not the only or primary mode of mass transfer. For systems with high flow rates, open membranes, and porous cell regions, diffusion will be negligible compared to the convective mass transport caused by the transmembrane hydrostatic pressure difference. For a system with an open shell space, fluid in the fiber lumen will flow through the membrane as long as the pressure in the fiber lumen is greater than that in the cell region by a factor that overcomes any membrane resistance. With a closed shell space, fluid will flow through the membrane into the cell region near the reactor entrance and reenter the lumen near the reactor exit. This leakage flow is often called Starling flow (Starling 1896) because of its analogy with fluid transport in the venous and arterial capillary beds. This transmembrane flux is advantageous in most circumstances because of the enhanced mass transfer of substrate to and waste products from the cell region. However, a disadvantage of the leakage flow is that convection in the cell region may cause cell redistribution, dragging the cells toward the far end of the reactor where they may be forced to accumulate at the membrane surface. The heterogeneous cell distribution created by such flows obviously will not result in optimal cell growth and productivity and can generally be avoided by periodic reversal of the axial flow direction. Clearly, there must exist a reasonable range of working pressures to maximize the advantages and minimize the disadvantages of Starling flow for optimal reactor productivity.

Only a few studies have been developed for convective transport (Reach et al. 1984; Heath et al. 1990) or even for combined convective and diffusive transport (Schonberg and Belfort 1987; Salmon et al. 1988) of nutrients in membrane bioreactors primarily because convection in the shell space has been assumed to be negligible compared to diffusion as a result of the low permeabilities in regions of densely packed cells. However, there have been reports that convective mass transport does contribute to the overall rate of mass transfer in some cases. Reach et al. (1984) modeled and measured the quantity of convective flow through the hollow fiber membranes of a bioartificial pancreas and found that convection does add significantly to the diffusive transport of glucose across the membrane, doubling that which occurs by diffusion alone. Furthermore, there have been limited reports of modules designed specifically to utilize convective flow in membrane bioreactors. Robertson and Kim (1985) developed a dual aerobic hollow fiber bioreactor that operates either with convective or diffusive transport of the nutrient medium to the cells in the shell space. Endotronics, Inc. (1988) has incorporated bidirectional convective transport (with cycling) across the fiber membranes of their bioreactor to improve pH control and to reduce substrate gradients in the shell space. Heath et al. (1990), using magnetic resonance imaging (MRI), have directly measured the axial velocity of water (protons) in the shell space of a cell-free hollow fiber reactor due to convective leakage flow and found good comparison with theoretical predictions. Their work is expected to be extended to systems that contain cells.

Modeling the convective fluid flows in a membrane bioreactor directly follows the pioneering work of Apelblat et al. (1974), who modeled fluid flows in capillaries and tissues. The same general approach will be outlined below in the development of equations describing leakage flow in the flat membrane system. Further details can be obtained from Apelblat et al. (1974) or Heath (1988), the latter of which contains equations for both hollow fiber and flat membrane systems.

The convective analysis for any membrane system begins with the simplified equations of continuity and motion (assuming constant density and viscosity). Also necessary is Darcy's Law, which relates fluid permeation velocity and the applied pressure gradient in the porous membrane and cell regions

$$\mathbf{V} = -k\nabla \mathbf{P}/\eta \qquad (14.9)$$

where k is the permeability, $\nabla \mathbf{P}$ is the pressure gradient, and η is fluid viscosity. Although Darcy's Law is generally assumed to be valid for regions of low permeability, as is the case with densely packed cells, recent work by Fowler and Robertson (1988) has demonstrated that the Carmen-Kozeny relation gave a more accurate description of fluid flow through a packed bed of immobilized bacterial cells. Whether this is also true for hybridomas and other mammalian cells has yet to be determined.

The development of equations describing convective leakage flow in the flat membrane system utilizes assumptions of steady laminar flow, time independence, negligible body forces, an isotropic membrane, and a homogeneous, porous cell region that can be characterized by a permeability coefficient, k. Because only dilute solutions are considered, osmotic effects can be ignored. End effects are also neglected because the ratio of medium path length to its thickness is large.

Using the assumptions mentioned above and applicable boundary conditions, the solution in the cell region takes the form of

$$P(y,z) = A + B\left(L/2 + \sum_{1}^{\infty} C_n F_i(\alpha_n y)\cos(\alpha_n z)\right) \quad (14.10)$$

$$u(y,z) = -\frac{k}{\eta}\frac{\partial P}{\partial y} \quad (14.11)$$

$$v(y,z) = -\frac{k}{\eta}\frac{\partial P}{\partial z} \quad (14.12)$$

where A, B, α_n, and C_n are constant coefficients, $F_i(\alpha_n y)$ is an expression containing exponentials, and L is the axial path length. The detailed solution is not presented because of its length. See Heath (1988) for the complete flat membrane system solution.

Equations 14.10 through 14.12 can be used to investigate flow behavior in the cell region for different size reactors and operating conditions. The equations demonstrate that the flow in the cell region is primarily axial (z-direction) with a significant component in the y-direction only near the reactor entrance (and exit) and very near the membrane-cells interface. The inlet and exit axial halves of the reactor are mirror images with flow streamlines curving away from the membrane near the reactor entrance and curving toward the membrane near the exit.

The variation of leakage flow into the packed cell region as a function of membrane thickness and permeability can also be investigated. As expected, increasing membrane thickness decreases the leakage flow to the cells. The convective flow into the cell region also drops when membrane permeability is decreased. Each increase in membrane permeability for a given membrane thickness results in less added leakage flow, especially as membrane permeability approaches the cell region permeability; a case of diminishing returns. It should be noted at this point that if the membrane is anisotropic, different permeabilities will have to be used for the y- and z-directions in the Darcy's law equation for the membrane region. This will result in changed quantities of leakage flow in both the membrane and cell regions.

The effect of changing the cell region permeability on leakage flow in the cell region indicates that a decrease in the permeability, a consequence of increasing cell density during a bioreactor run, yields a decrease in the amount of flow into the cell region. As stated earlier, a homogeneous suspension in the cell region is assumed. An order of magnitude decrease in permeability results in an order of magnitude loss in leakage flow. An increase in membrane thickness with membrane permeability greater than the cell region permeability has only a small effect on decreasing the leakage flow.

Changing the pressure drop from entrance to exit in the medium region also affects leakage flow in the cell region. Smaller pressure drops significantly decrease the leakage flow to the cells. With high pressure drops, the leakage flow is also highly dependent on membrane thickness, while at lower pressure drops the membrane thickness is less important.

The above analysis of convective leakage flow in the flat membrane system supports not only the expected trends (i.e., increased flow with increased pressure drop, permeability, etc.) but also provides quantitative values. The same can be and has been done for other membrane systems as well. Selected aspects of the hollow fiber analysis by Heath (1988) were recently verified experimentally using magnetic resonance flow imaging—a powerful noninvasive technique for high resolution measurement of fluid velocities. With proper experimental corroboration, model equations such as those presented above can be very important in membrane system operation and design for bioreactors as well as for other applications.

Having determined the convective velocities in the cell region, the appropriate expressions can then be inserted into the continuity equation (equation 14.1) and, with the appropriate assumptions, can be solved for concentration. With negligible diffusion (compared to convection) and steady-state substrate conversion, the governing equation is

$$u \frac{\partial c}{\partial y} + v \frac{\partial c}{\partial z} = R \qquad (14.13)$$

where u and v are defined by equations 14.11 and 14.12, respectively. A numerical solution is most appropriate since $u(y,z)$ and $v(y,z)$ are both series expressions of exponentials for the flat membrane case. From this, the concentration profiles of substrate in the cell region can be obtained and used to investigate the existence of mass transfer limitations with convective transport in the cell region.

14.3   SIMULTANEOUS DIFFUSIVE AND CONVECTIVE MASS TRANSFER

Many membrane systems may be operated in the regime where both diffusion and convection contribute to the mass transfer of nutrients and wastes. Solution of the appropriate equations is much more complex. Calculation of a system Peclet number will indicate the relative contributions of diffusion and convection. For the flat membrane system discussed in the previous sections, the Peclet number can be defined as

$$Pe = \frac{\int_0^{L/2} u(y,z)\, dz}{D_e} \tag{14.14}$$

where $u(y,z)$ is the leakage flow in the y-direction through the membrane and is defined in equation 14.11. If the Peclet number indicates a predominance of convective or diffusive mass transfer into the cell region, the easiest approach is to use the appropriate simplified model. If, however, both are important, the simplified models described previously may not provide enough accuracy and thus a more complete model representation is needed. Again, model development begins with the continuity equation in cartesian coordinates. Employing the assumptions of steady-state substrate conversion, negligible mass transfer in the x-direction, and an effective diffusivity D_e in the cell region results in

$$u\frac{\partial c}{\partial y} + v\frac{\partial c}{\partial z} = D_e\left(\frac{\partial^2 c}{\partial y^2} + \frac{\partial^2 c}{\partial z^2}\right) + R \tag{14.15}$$

where u and v are defined by equation 14.11 and equation 14.12, respectively. Since u and v are each functions of both y and z, equation 14.15 remains a complicated second-order, partial differential equation, solvable only by numerical techniques. This is the most complete method of modeling the system and will not be pursued further in this chapter.

Solutions to variations of the problem exist in the literature. Schonberg and Belfort (1987) present a series solution for superimposed diffusive and convective transport in a hollow fiber perfusion bioreactor under the assumption that transport occurs only in the radial direction, not in the axial direction. This approach is appropriate for a membrane reactor with an open cell region; i.e., fluid that flows through the membrane into the cell region is siphoned off rather than forced to return to the medium region as occurs in Starling flow. Salmon et al. (1988) have developed two models (one complete, one approximate) to describe the effects of combined convective and diffusive transport in a hollow fiber bioreactor with axial convection. The approximate model, which omitted radial convection terms because the average radial velocity was zero, was found to adequately describe the cell region transport under certain conditions.

Perhaps a less rigorous but also less complicated method of approaching the problem is to use the Damkohler number, which is a ratio of the maximum reaction rate to the maximum transport rate, the latter consisting of both the convective and diffusive components. The exact definition of the Damkohler number depends on the system under study (i.e., the type of kinetics, etc.). A typical definition for the flat membrane system studied in this chapter is

$$
Da = \frac{W \dfrac{L}{2} \dfrac{\mu_{max}}{K_m}}{D_e + \displaystyle\int_0^{L/2} u(y,z)\, dz}
\tag{14.16}
$$

where W is channel width. Calculation of this number is helpful in determining whether the reactor is reaction or transport controlled. The kinetics and diffusivity are generally inherent in the cell/reactor system used so that a change in the ratio of kinetics to transport is achieved only by altering the convection. This may be done by changing the inlet-outlet pressure drop in the medium region (e.g., varying the velocity), as was shown with the convective model. The Peclet and Damkohler numbers, while not providing as thorough an analysis as a complete model, can be quite helpful and their calculation avoids the complexity of the complete model solution.

Work is currently underway to experimentally verify a combined model of diffusive and convective flow in the cell region of membrane bioreactors by MRI. This is a continuation of recent measurements demonstrating the excellent corroboration between theoretical and experimental convective flow profiles in the shell space of hollow fiber bioreactors (Hammer et al. 1990). Investigation of the resulting substrate concentration profiles is also planned.

14.4 CONCLUSIONS

Despite their minor limitations, model equations such as those presented in this chapter can provide a significant savings in time, effort, and expense by allowing a quick and easy "noninvasive" bioreactor inspection of concentration profiles. Design, operation, and scaleup may also be significantly improved with the insights afforded by the flexibility of theoretical investigations.

REFERENCES

Altshuler, G.L., Dziewulski, D.M., Sowek, J.A., and Belfort, G. (1986) *Biotechnol. Bioeng.* 28, 646–658.

Apelblat, A., Katzir-Katchalsky, A., and Silberberg, A. (1974) *Biorheology* 11, 1–49.

Belfort, G. (1989) *Biotechnol. Bioeng.* 33, 1047–1066.

Chresand, T.J., Gillies, R.J., and Dale, B.E. (1988) *Biotechnol. Bioeng.* 32, 983–992.

Davis, M.E., and Watson, L.T. (1985) *Biotechnol. Bioeng.* 27, 182–186.

Davis, M.E., and Watson, L.T. (1986) *Chem. Eng. J.* 33, 133–142.

Endotronics, Inc. (1988) Presented at the Engineering Foundation Conference on Cell Culture Engineering, Palm Coast, FL.

Evans, T.L., and Miller, R.A. (1988) *BioTechniques* 6, 762–767.

Fowler, J.D., and Robertson, C.R. (1988) Presented at Biochemical Engineering VI, Santa Barbara, CA.

Georgakis, C., Chan, P.C.H., and Aris, R. (1975) *Biotechnol. Bioeng.* 17, 99–106.

Hammer, B.E., Heath, C.A., Mirer, S.D., and Belfort, G. (1990) *Bio/Technology* 8, 327–330.

Heath, C.A. (1988) Doctoral Dissertation, Rensselaer Polytechnic Institute, Troy, NY.

Heath, C.A., and Belfort, G. (1987) *Adv. Biochem. Engr./Biotech.* 34, 1–31.

Heath, C.A., Hammer, B.E., Mirer, S.D., Pimbley, J.M., and Belfort, G. (1990) *AlChE J.* 36, 547–558.

Horvath, C., Shendalwan, L.H., and Light, R.T. (1973) *Chem. Eng. Sci.* 28, 375–388.

Kim, S., and Cooney, D.O. (1976) *Chem. Eng. Sci.* 31, 289–294.

Ku, K., Kuo, M.J., Delente, J., Wildi, B.S., and Feder, J. (1981) *Biotechnol. Bioeng.* 23, 79–95.

Reach, G., Jaffrin, M.Y., Vanhoutte, C., and Desjeux, J.-F. (1984) *ASAIO J.* 7, 85–90.

Robertson, C.R., and Kim, I.H. (1985) *Biotechnol. Bioeng.* 27, 1012–1020.

Rony, P.R. (1971) *Biotechnol. Bioeng.* 13, 431–447.

Salmon, P.M. Libicki, S.B., and Robertson, C.R. (1988) *Chem. Eng. Commun.* 66, 221–248.

Schonberg, J.A., and Belfort, G. (1987) *Biotechnol. Prog.* 3, 80–89.

Seaver, S.S., and Gabriels, J.E. (1985a) *Hybridoma* 4, 63.

Seaver, S.S., and Gabriels, J.E. (1985b) *In Vitro* 21(II), 16A.

Shuler, M.L. (1987) *Methods Enzym.* 135, 372–387.

Starling, E.H. (1896) *J. Physiol.* 19, 312–326.

Tharakan, J.P., and Chau, P.C. (1986). *Biotechnol. Bioeng.* 28, 329–342.

Waterland, L.R., Michaels, A.S., and Robertson, C.R. (1974) *AlChE J.* 20, 50–59.

Waterland, L.R., Robertson, C.R., and Michaels, A.S. (1975) *Chem. Eng. Commun.* 2, 37–47.

Webster, I.A., and Shuler, M.L. (1978) *Biotechnol. Bioeng.* 20, 1541–1556.

Webster, I.A., Shuler, M.L., and Rony, P.R. (1979) *Biotechnol. Bioeng.* 21, 1725–1748.

Webster, I.A., and Shuler, M.L. (1981) *Biotechnol. Bioeng.* 23, 447–450.

Bioreactor Control
and Optimization

Michael W. Glacken

With the increasing variety and demand for products produced by animal cell systems comes the challenge from biochemical engineers to develop principles and strategies for maximizing bioreactor productivity and efficiency. The rate-limiting steps in this process are the complexity and the incomplete understanding of metabolic regulation in mammalian cells, especially as it pertains to responses to changes in the bioreactor state. Since the cellular response to bioreactor perturbations is incompletely known, a necessary step in the rational design of optimal control schemes is the development of quantitative descriptions relating cellular kinetics to the bioreactor environment. This chapter presents a strategy to achieve this, and shows examples of how mathematical descriptions may be utilized to formulate process control strategies to attain the desired result. Direct control of measured process states, such as temperature, pH, and dissolved oxygen will be briefly discussed, as will the indirect control of inferred variables such as lactate production and glutamine concentration. Finally, speculative control strategies based on the limited knowledge of the intercellular regulation of mammalian catabolic pathways will be presented.

15.1 PROCESS GOALS

Cellular metabolism may be regulated via the deliberate manipulation of the bioreactor environment. In turn, the operation of cellular metabolic pathways directly affects bioreactor performance and, consequently, the profitability of a given process. Therefore, before proceeding with a discussion of bioreactor control, generalized statements concerning production goals and process optimization are in order. These goals may be conceptualized with the aid of an objective function that represents production costs per gram of product produced, an example of which is given below:

$$F_{obj} = \frac{t\$_{time} + X\$_{cells} + S_t\$_{materials} + \$(S,P)_{separation}}{P_t t} \qquad (15.1)$$

where

t	= production time, days;
X	= inoculum required, total cells;
S_t	= total amount utilized of the most expensive raw material, grams or moles;
P	= concentration of product, g/l or moles/l;
S	= concentration of contaminant, usually serum proteins;
P_t	= production rate of product, g/day or moles/day;
$\$_{time}$	= fixed costs, \$/day;
$\$_{cells}$	= inoculum costs, \$/cell;
$\$_{materials}$	= costs of raw material, \$/g or \$/mole; and
$\$(S,P)_{separation}$	= separations costs, a function of the product and contaminant concentrations, \$.

Examination of this equation indicates that increases in the volumetric productivity, P_t, will decrease the per gram unit production costs. The volumetric productivity may be maximized for any bioreactor configuration by maximizing the cell concentration (cell/l) and/or the specific cellular productivity (g/cell-day). High rates of productivity may be obtained in bioreactors that provide an adequate supply of nutrients while minimizing the accumulation of toxic waste products. However, some bioreactor configurations that maintain high rates of volumetric productivity (i.e., chemostats) have the disadvantage of utilizing large amounts of raw materials ($S_t\$_{materials}$) while simultaneously diluting the concentration of the product (P). Low product concentrations decrease separation efficiencies, thereby increasing separation costs ($\$(S,P)_{separation}$). High growth rates minimize the costs of time ($t\$_{time}$) as well as the inoculum costs ($X\$_{cells}$). However, these expenses usually represent only a small fraction of the total. Thus, it is clear from this brief discussion, that no one aspect of the process performance should be considered alone; rather, each process characteristic must be examined with respect to the others when attempting to minimize production costs.

To visualize how the bioreactor environment may affect cellular metabolism and, hence, process performance, refer to the overview of mammalian cell culture dynamics illustrated in Figure 15–1. Note the complexity of the environment in which mammalian cells are cultivated. Unlike microbial cells, mammalian cells strictly require both glutamine and a carbohydrate source for catabolic processes that produce ATP and NADPH. In turn, the ATP and NADPH drive the biosynthesis of cellular material and/or the product of interest. If the product of interest results from the expression of an amplified recombinant gene, the cellular energy required to synthesize this protein could be a substantial fraction of the total energy capacity of the cell. In such a case, if ATP synthesis is rate-limiting, high growth rates may retard the rate of product formation since reactions synthesizing cellular material compete for the same ATP pool as reactions creating the product of interest. Since both the rate of growth and the operation of the various catabolic pathways (see Figure 15–2) are regulated by glutamine and glucose levels, nutrient supply often determines bioreactor productivity. This is also true if the availability of biosynthetic intermediates is rate-limiting, since glucose and glutamine are a major source of intermediates (see Figures 15–1 and 15–2).

Note from Figure 15–1 that end-products of glutamine and glucose catabolism, such as ammonium, lactate, and CO_2 are excreted into the medium. Since these end products may be inhibitory to cell growth and product production, their accumulation may also limit culture productivity. Ammonium has been shown to inhibit the growth of mouse L cells (Ryan and Cardin 1966), Madin-Darby canine kidney cells (MDCK) (Glacken et al. 1986), and hybridoma cells (Glacken et al. 1988; Miller et al., 1988; Bree et al. 1988), as well as the production of a number of viruses (Eaton and Scala 1961; Furusawa and Cutting 1962; Jensen and Liu 1961) and interferon (Commoy-Chevalier et al. 1978; Ito and McLimans 1981). Lactic acid accumulation can lower the pH of the culture, subsequently inhibiting growth (Paul 1975), whereas the lactate ion itself has been shown to slightly inhibit growth and antibody formation from hybridoma cells (Glacken et al. 1988).

Nutrient depletion and waste product formation may therefore limit bioreactor productivity by reducing both the maximum achievable cell concentration and the specific productivity (g of product/cell-day). Fed-batch systems, which continually supply nutrients to but do not remove medium from the bioreactor, can eliminate nutrient depletion but not waste product accumulation. On the other hand, continuous systems, which supply "fresh" medium to and withdraw "spent" medium from the bioreactor, are ideal for minimizing waste product accumulation. If the cells can be retained inside the bioreactor, this is termed a "perfusion system," and the waste product levels may be kept as low as desired by simply increasing the perfusion rate. However, this will lower the concentration of desired products and increase serum or growth factor usage. Reviews of various mammalian

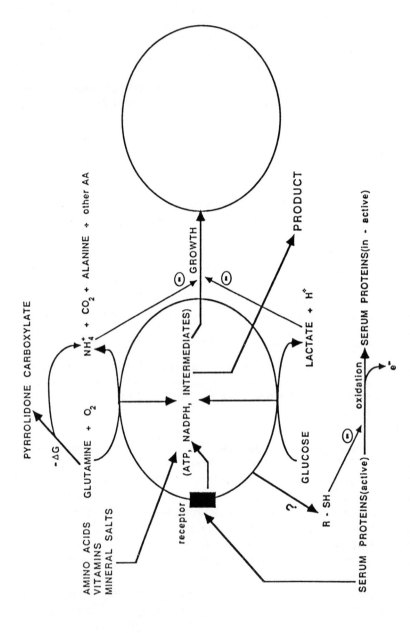

FIGURE 15–1 Overview of mammalian cell bioreactor dynamics.

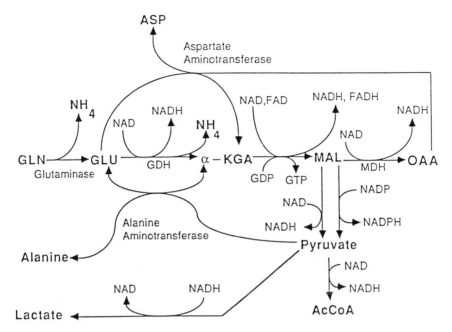

FIGURE 15–2 Pathways of glutamine catabolism (glutaminolysis).

cell production processes should be referred to for more information (Adamson and Schmidli 1986; Glacken et al. 1983).

Additionally, mammalian cells require oxygen in order to synthesize ATP via oxidative phosphorylation, and will not grow or remain viable for extended periods of time under anaerobic conditions. Because the solubility of oxygen in cell culture medium is very low (0.2 mM; Fleischaker and Sinskey 1981), the rate of oxygen transfer to the medium may limit bioreactor productivity. Oxygen transfer will not be discussed here. See the reviews by Van't Riet (1983) for more information on oxygen transfer in general, and Fleischaker and Sinskey (1981) for oxygen transfer in mammalian cell culture.

The final point to be noted from Figure 15–1 is that peptides which interact with specific receptors on the cell surface are required for growth. Often, calf or horse serum is supplemented to the medium to provide the cells with these peptides. Unfortunately, serum is very expensive, ill-defined, and contains large amounts of proteins and their contaminants that may interfere with the purification of the product. Consequently, the addition of serum increases both the raw material and separation costs of a process. Even completely defined serum-free medium usually requires the addition of protein growth factors (e.g., insulin, transferrin) that must eventually be separated from the product.

With this background, some objectives for minimizing production costs by regulating mammalian cell metabolism become evident.

1. Regulate glutamine and carbohydrate metabolism so as to minimize the accumulation of ammonia and lactate;
2. Minimize the oxygen utilization rate, since this would maximize the concentration of cells that can be grown at a given oxygen transfer rate; and
3. Minimize the utilization and concentration of serum proteins or growth factors.

15.2   PROCESS CONTROL: DIRECT CONTROL OF MEASURED VARIABLES

Most commercial bioreactors on the market today are equipped with three standard control loops: temperature, pH, and dissolved oxygen (DO). These three variables must be tightly controlled if more sophisticated strategies are to be successful. Since operation of these control loops are straightforward, only a brief overview of each will be offered. What follows is personal opinion formed from experience with five different commercial laboratory systems. Note that the discussion in this chapter applies only to homogeneous stirred systems.

15.2.1   Temperature Control

The temperature must be maintained within a tight tolerance around 37 °C. Temperatures above 38.0 °C will rapidly kill mammalian cells while temperatures below 37 °C needlessly slow growth. Most commercial units use the outside surface area (A) of the bioreactor to transfer heat. This large surface area minimizes the temperature difference (ΔT) between the heat transfer medium and the culture fluid (Note: heat flux $\propto A\Delta T$). Large temperature gradients may promote high temperature pockets near the bioreactor wall. Since transient exposure of cells to these pockets will accelerate cell death, low surface area immersion heaters inserted through the headplate should be avoided. The three most common heat transfer media are water, air, and a solid electric heating pad, with the heating pad preferred for ease of use. All temperature controllers will overshoot 1 or 2 °C during start-up from cold media, so it is recommended to either: (1) inoculate the culture after a stable temperature of 37 °C is attained, or (2) use an initial set-point of 35 °C, and raise the set point to 37 °C after 35 °C is attained.

15.2.2 pH Control

There are two types of pH control: 1) CO_2 controlling and 2) non-CO_2 controlling. CO_2 controlling schemes are simple and more efficient. CO_2 is added to sodium bicarbonate-containing medium to maintain the pH via the following reactions:

$$CO_2(aq) + H_2O = H^+ + HCO_3^- \quad K_1 = 4.2 \times 10^{-7}. \quad (15.2)$$

The pH is then determined from the Henderson-Hasselbalch equation:

$$pH = 6.38 + \log \frac{HCO_3^-}{CO_2(aq)}. \quad (15.3)$$

Since the concentration of dissolved CO_2 determines the culture pH, and since the partial pressure of CO_2 in the gas phase sets the aqueous phase CO_2 level at equilibrium, the gas phase composition of CO_2 determines the culture pH. Three potential problems with this type of control are: (1) high rates of CO_2 production from extremely dense cultures can limit controllability; (2) CO_2 sparging may upset DO control (see section 15.2.3); and (3) the limited buffering capacity of the bicarbonate system may become inadequate at high lactate concentrations or during periods of rapid lactate production.

To elaborate on the last point, as lactate ions (and hence protons) are generated in culture, more bicarbonate ions must be consumed to neutralize the protons and to maintain the pH. As the bicarbonate concentration decreases, the CO_2 partial pressure required to maintain a given pH (i.e., to maintain a constant HCO_3^-/CO_2 ratio) also decreases. Eventually, the level of CO_2 required becomes so low it becomes ineffectual for control purposes, especially at high cell concentrations that produce and maintain a low basal level of CO_2 in the medium. Depending on the medium, this would occur in batch reactor at lactate levels of 15–20 mM (for Roswell Park Memorial Institute [RPMI] media) or 30–40 mM (for DMEM).

For non-CO_2 controlling schemes, liquid base addition is used to maintain the pH. This strategy is preferred if quantification of base addition is desired for on-line estimation of (lactic) acid levels (see section 15.3). It is imperative that the media have some buffering capacity, otherwise the control may oscillate by as much as ± 0.2 pH units. N-[2-hydroxyethyl]piperazine-N'-[2-ethanesulfonic acid] (HEPES) or Tris at 5–10 mM is usually sufficient. Also important is the cation content of the base. Since many cellular functions are affected by sodium and potassium ions (e.g., sodium/potassium ATPase, Darnell et al. 1986), the NaOH and KOH concentrations in the base should reflect that of the medium (usually 20:1; Na:K). Also, the base should be isotonic with the medium (320 mM; Na + K). One drawback to this type of control is the dilution of the culture medium that results from the base addition. In a high cell density fed-batch (glucose) bioreactor in which 70 mM lactate would be allowed to accumulate, about 20% of the liquid volume would be from the base. The lower nutrient and serum levels could reduce bioreactor productivity.

Another disadvantage of this type of control is observed if the liquid base is allowed to drip onto the liquid surface, since cell death may occur near the addition point due to poor mixing (resulting from the low agitation required to avoid shear disruption). This is easily solved by introducing the base deep into the liquid surface, directly under the impeller.

15.2.3 DO Control

The rate of transfer of oxygen from the gas to liquid phase is proportional to the equilibrium driving force; i.e.,

$$N_o = K_La \, (C_G - C_L^*) \qquad (15.4)$$

where

N_o = oxygen transfer rate (mmol O_2/l-hour);
K_La = mass transfer coefficient (mmol O_2/atm-l-hour)
C_G = oxygen concentration in the gas phase (atm); and
C_L^* = the oxygen concentration in the gas phase that would be in equilibrium with the actual concentration of oxygen in the liquid phase (atm).

At a steady state, the rate of transfer from the gas phase equals the rate of cellular consumption:

$$K_La \, (C_G - C_L^*) = q_{o_2}X \qquad (15.5)$$

where

q_{o_2} = specific rate of cellular oxygen utilization, mmol/cell-hour;
X = cell concentration, cells/l.

If we solve for C_L and take the time derivative of each term, we get

$$C_L^* = C_G - \frac{q_{o_2}}{K_La}X \qquad (15.6a)$$

$$\frac{dC_L^*}{dt} = \frac{dC_G}{dt} - \frac{q_{o_2}}{K_La}\frac{dX}{dt} = \frac{dC_G}{dt} - \frac{q_{o_2}\mu}{K_LaX} . \qquad (15.6b)$$

Since animal cell doubling times are very long ($\sim$12–24 hours), and since the gas phase O_2 concentration can be rapidly manipulated via control devices, then over short time intervals, the second term in equation 15.6.2 is much smaller than the first term so that $dC_L^*/dt \simeq dC_G/dt$. That is, the rate of change of the gas phase oxygen concentration determines the rate of change in the DO level. Therefore, C_G should be the parameter that is directly manipulated by the controller. If the oxygen uptake rate (OUR = $q_{o_2}X$) of the bioreactor can be measured continuously (see section 15.3), then the steady-state gas phase concentration ($C_{G,s}$) required to achieve a desired DO level ($C_{L,S}^*$) may be estimated if K_La is known:

$$C_{G,s} = C_{L,S}^* + \frac{OUR}{K_La} . \tag{15.7a}$$

Gas phase concentrations may be adjusted from this value based on the difference (ϵ) between the present DO and the set point and from the standard proportional-integral-derivative (PID) controller equation (Coughanowr and Koppel 1965):

$$C_G = K_c\epsilon + \frac{1}{K_I}\int \epsilon dt + K_D\frac{d\epsilon}{dt} + (C_{L,S}^* + \frac{OUR}{K_La}) . \tag{15.7b}$$

Note that if DO equals the set point, C_G equals its steady-state value, $C_{G,s}$.

A computer or microprocessor would be required to perform this calculation on-line. The computer would send on-off signals to oxygen, air, and nitrogen valves to obtain the desired gas mixture. For example, if a C_G of 42% oxygen (0.42 atm) was required, then in each 10 sec interval, the air solenoid value would be on for 7.3 sec, while the oxygen solenoid value would be on for 2.7 sec. If a C_G below 21% (0.21 atm) was required, nitrogen and air should be mixed. This split level control, i.e., O_2 and air mixtures for $C_G > 0.21$ atm and N_2 and air mixtures for $C_G < 0.21$ atm, minimizes the oscillation in C_L^* that may occur if only mixtures of N_2 and O_2 were used. The flowrate for each gas, when on, should be equal, so that the gas flowrate through the bioreactor would always be constant. This would minimize any mild upsets to the pH caused by uneven CO_2 stripping by the gas mixture (if the pH were controlled by CO_2 addition). The value of C_G should be increased to account for CO_2 addition for pH control. For example, the computer could be instructed to keep track of the cumulative time the CO_2 valve was activated during each 10 min period (y). If the flowrate of CO_2 were equal to that of the other gases, the actual value of C_G should be corrected by multiplying by $10/(10-y)$.

Our experience with this method of control (without the CO_2 correction) is that it is very robust and results in very stable control with minimal oscillation. If a dedicated computer is unavailable and/or if the oxygen uptake rate can not be readily estimated on-line, a standard PID controller with twin relay outputs could be used. The controller signal from the PID algorithm would determine the length of time the relays (solenoid values) should be activated. In this case, the best control would result if three-way solenoids instead of two-way solenoids were used. Air would flow through the bioreactor until the solenoid valve is activated; then oxygen (or nitrogen) would flow. Again the response would be less oscillatory than if only oxygen and nitrogen were used.

It is important to point out that sophisticated control schemes may be compromised by poor hardware design. For example, the response time for control increases as the volume of gas downstream of the point of mixing air, N_2 or O_2 increases. Since this would result in increasingly poor control, this volume should be minimized. Therefore, all gases should be humidified separately and mixed just prior to being introduced into the bioreactor. This section did not discuss the different methods for oxygenating bioreactors (see the chapter on oxygenation by Jurgen Lehman), since the control schemes presented would apply equally well with all oxygenation methods.

15.2.4   Redox Control

The oxidation/reduction potential (ORP) of cell culture media can easily be monitored with a specific probe that operates similar to a pH probe. In fact, probes are now available that provide both measurements simultaneously. The problem, however, is interpreting the data. The ORP is a measurement of the propensity of the media to either donate or accept electrons for oxidative or reductive reactions. However, there are many chemical species in cell culture medium that can participate in such reactions. Consequently, the ORP measurement is, at best, a bulk property dependent on pH and DO levels.

Still, ORP-based control should not be dismissed out of hand. Investigators have long noted the existence of ORP levels that result in optimal growth (Daniels et al. 1970a and 1970b; Griffins 1984; Hwang and Sinskey 1989). They have also shown that mammalian cells reduce their environment as they grow. More recently, we have demonstrated that the growth-promoting activity of serum spontaneously decays with time, due possibly to the oxidation of serum thiols (Glacken et al. 1989a). This rate of decay was substantially decreased by lowering the ORP with the addition of reducing agents, such as cysteine, dithiothreitol (DTT), or glutathione, or by the elimination of oxidative agents such as cystine. For hybridoma fed-batch bioreactors, there was strong evidence to indicate that culture productivity was limited by both ammonium accumulation and the degradation of serum activity. Consequently, it is likely that control of the ORP at a predetermined optimal value via the addition of reducing agents such as DTT may be beneficial.

15.3   PROCESS CONTROL FROM INFERRED MEASUREMENTS

Not all important process variables can be quantitated, and thus controlled, via direct measurements. These include lactate concentration and production rate, cell concentration and growth rate, and the respiration rate. (Recently, a method using acoustic densitometry has been described that can measure cell concentrations of 10^6 cells/ml or more on-line; Kilburn et al. 1989). All of these values, however, have been estimated on-line with the application of simple algorithms and models, and were consequently used to minimize lactate production via manipulation of the glucose feed-rate (Fleischaker 1986; Glacken et al. 1986).

The lactic acid concentration may be estimated from the quantity of base required to maintain a constant pH. Note that in the absence of $NaHCO_3$, very little carbonic acid would exist. Consequently, the moles of base added would equal the total moles of lactate produced to within at least ± 0.2 mM if less than 10 mM HEPES or Tris buffer was used and if the pH control was accurate to within ± 0.01 pH units. The amount of base added may be measured via a load cell (weight measurement) or via a dose monitor that calculates the cumulative on-time of the pump. Many commercial bioreactors now offer the latter as a standard feature.

The OUR (mmol/l-hour) may be measured by the dynamic method (Wang et al. 1979): oxygen flow to the bioreactor is stopped and the decrease in DO with time is recorded. The OUR is then calculated as:

$$\text{OUR} = \frac{C_o - C_f + \int_{t_o}^{t_f} K_L a(C_G - C_L^*) dt}{t_f - t_o}. \qquad (15.8)$$

Note that even if the gas flow was zero, oxygen could still be transferred across the liquid surface/headspace interface or through silicone tubing, especially as C_L^* decreases. Therefore, it is critical for accurate measurements of the OUR that the $K_L a$ be precisely known. Unfortunately, at low cell levels and in small fermentors, $K_L a$ is large compared to the total oxygen demand and, therefore, small errors in the assumed value of $K_L a$ can lead to large errors in calculating the OUR. One way to minimize this problem is to reduce C_G to 75% of the set-point value of C_L^* immediately upon initiation of the OUR measurement. Initially, $C_G - C_L^*$ will be negative, but will become positive as C_L^* (measured on-line with a DO probe) decreases. The OUR measurement is stopped when the integral is zero. In this way, the inaccuracies in the $K_L a$ estimate are rendered unimportant.

Another method to avoid this problem is to perform the OUR measurement off-line in a completely liquid filled chamber containing a DO probe. Since there would be no gas phase, the second term in equation 15.8 would be zero and the OUR could be directly measured. This measurement could be automated for bioreactor operation with the aid of a computer and a few solenoid valves. Note that it would be important to maintain the temperature at 37 °C in the off-line respirometer.

The ATP production rate (APR, mmol/l-hour) may be estimated from measurements of the OUR and the lactic acid production rate (LPR) by the following reasoning. Glutamine is oxidized in the Kreb's cycle of mammalian cells to produce NADH and FADH (see Figure 15–2). One mole of oxygen is required to oxidize 2 mol of either molecule in the electron transport chain: oxidation of 2 mol of NADH produces at most, 6 mol of ATP (assuming $P/O = 3$), while oxidation of 2 mol of FADH produces 4 mol of ATP. Fortunately, FADH is always produced simultaneously with GTP in the Kreb's cycle, so for ATP accounting purposes, the utilization of 1 mol of O_2 would produce 6 mol of ATP, regardless of whether NADH or FADH was oxidized.

The production of lactate from glucose via aerobic glycolysis also results in net ATP synthesis: 1 mol of ATP per 1 mol of lactate excreted. Unfortunately, lactate can also be derived from glutamine. Since the production of lactate from glutamine results in no substrate level phosphorylation (other than the GTP lumped with FADH oxidation), some knowledge of the relative fraction of lactate resulting from glucose instead of glutamine must be known ($F_{L/C}$). The APR may then be calculated as:

$$APR = 6OUR + (F_{L/C}) LPR. \qquad (15.9)$$

Glacken et al. (1986) have shown that, in most cases, $F_{L/C}$ may be assumed to be 1 without significant error. The reasoning is that if a great fraction of the ATP formed resulted from aerobic glycolysis, most of the lactate produced would be from glucose and $F_{L/C}$ would be very close to 1. On the other hand, if almost all the ATP resulted from glutamine oxidation, i.e., $6OUR \gg LPR$, then the actual value of $F_{L/C}$ would be immaterial.

If the assumption is made that the specific ATP productivity, q_{ATP} (mmol ATP/g DCW/hour, where DCW = dry cell weight or mmol ATP/cell-hour) is constant, then the cell concentration may be calculated from the APR by:

$$X = APR/q_{ATP}. \qquad (15.10)$$

This method was used to accurately predict, on-line, the *exponential* phase cell concentration of microcarrier cultures of human fibroblast cells grown with glucose or galactose (Fleischaker 1986; Glacken et al. 1986). The OUR and the LPR were estimated on-line, as described previously. However, since q_{ATP} decreased monotonically from 3.5 mmol ATP/g DCW/hour just after inoculation to the steady-state level of 1.7 mmol ATP/g DCW/hour at the end of the lag phase, the cell concentrations during this period were overestimated. This may have been due to the high maintenance energy requirements caused by trypsinizing the anchorage-dependent cells off culture surfaces prior to inoculation onto microcarriers. The maintenance energy model developed by Pirt (1975) decomposes q_{ATP} into two terms: a growth associated term, μ/Y_{ATP}, where μ is the specific growth rate and Y_{ATP} is the yield of cells from ATP, and a maintenance energy term (m_{ATP}):

$$q_{ATP} = \frac{\mu}{Y_{ATP}} + m_{ATP}. \qquad (15.11)$$

Maintenance energy reflects the ATP requirements of cellular processes that operate at both growth and nongrowth conditions and includes protein turnover, futile cycles, and ion-gradients. Consequently, equations 15.9 and 15.10 may accurately predict cell levels during steady growth, but may be in error during periods of cellular stress, reflected by nonconstant q_{ATP}. This was also seen with hybridomas (Miller et al. 1987), from which q_{ATP} was found to decrease rapidly from 3×10^{-8} to 2×10^{-8} mmol ATP/cell-day at DO levels $\leq 0.5\%$. The authors gave the following three possible explanations: (1) the reduced growth rates (μ) at these low DO levels may have resulted in decreased q_{ATP} (see the first term in equation 15.11; (2) maintenance requirements were lower at DO levels less than 0.5% due to lower auto-oxidation rates (although cell viabilities did not increase as drastically as the decrease in q_{ATP}); or (3) the P/O ratio abruptly changed from 2 to 3 as the DO was decreased below 0.5%. Indeed, for the hybridomas in question, a constant value of q_{ATP} was obtained if it was assumed that P/O = 2 for DO >0.5% and P/O = 3 for DO <0.5%.

Why is it desirable to be able to estimate the OUR, LPR, and the cell concentration on-line without removing a sample for analysis? There are two reasons. (1) Cell samples from heterogeneous systems, such as hollow-fiber bioreactors, can not be directly obtained, so the overall, averaged progress of the culture may only be followed via indirect methods; and (2) to aid in process control. An example of the last item is the glucose control scheme developed to minimize lactic acid production in microcarrier cultures of human fibroblast cells (Fleischaker 1986; Glacken et al. 1986). The scheme involved feedback and feed-forward control loops. In the feed-forward loop, glucose was fed to the culture at a rate determined by the estimated glucose utilization rate (GUR):

$$GUR = 1/2 \ LPR + 1/6 \ OUR \qquad (15.12)$$

where the LPR and OUR were estimated on-line, as previously discussed.

The feedback loop included measurement of the glucose level via an on-line glucose analyzer. The glucose feed rate calculated in equation 15.12 was adjusted depending on whether the measured glucose concentration was higher or lower than the set point. The feed-forward control loop was required since the glucose concentration could only be measured once every four hours, whereas the LPR and OUR could be measured every hour. The rate of lactic acid accumulation was minimized by continually lowering the glucose set point until the desired specific lactate productivity, q_L, was reached. This control scheme was able to reduce the specific lactate productivity from 1.2–0.2 mmol/g DCW-hour.

15.4 FEED-FORWARD CONTROL USING EMPIRICALLY DERIVED MATHEMATICAL RELATIONS

Unfortunately, in mammalian cell bioreactors, the biochemical engineer can not measure all the process states that are important to control. We instead have to rely on empirically derived expressions that relate cellular function with bioreactor operation to guide our control strategy a priori. Two examples of such feed-forward control schemes are presented.

15.4.1 Reduction of Ammonia Production via Glutamine Control

As stated previously, cultured mammalian cells catabolize glutamine to produce ATP. A by-product of this catabolism is the formation of ammonia. It is therefore reasonable to assume that the rate of ammonia production may be related to the rate of cellular glutamine utilization; that is, if the rate of glutamine utilization per cell is q_G, than the rate of ammonia production per cell is $Y_{A/G}q_G$, where $Y_{A/G}$ is the yield of ammonia from glutamine. Since glutamine contains two amino groups per molecule, $Y_{A/G}$ can range from 0–2 mol NH_4/mol glutamine.

If the first-order spontaneous decomposition of glutamine to ammonia and pyrrolidone-carboxylate is included (Tritsch and Moore 1962), then the rate of change of glutamine and ammonia levels in a batch bioreactor may be described by (Glacken et al. 1986):

$$\frac{-dG}{dt} = KG + q_GX \qquad (15.13)$$

$$\frac{dA}{dt} = KG + Y_{A/G}q_GX \qquad (15.14)$$

where

  G = glutamine concentration, mM;
  A = ammonia concentration, mM;
  X = cell level, cells/l; and
  K = first-order glutamine decomposition rate, per hour.

By measuring the rate of glutamine disappearance at various glutamine levels in MDCK microcarrier cultures, q_G was found to be related to the glutamine level by "Michalis-Menton-type" kinetics (Glacken et al. 1986):

$$q_G = \frac{(q_G)_{max}G}{K_{q_G} + G}. \qquad (15.15)$$

It is clear from equations 15.13 through 15.15 that the rate of ammonia production should decrease if the glutamine level is decreased. This implies that ammonia formation may be minimized by controlling glutamine at low concentrations. This concept is shown in Figure 15–3. Glutamine was fed to the cultures every 12 hours. The amount of glutamine delivered to the culture at each feeding was based on the criterion that the concentration should not be allowed to become less than 0.2 mM at any time, and was calculated from equations 15.13 and 15.15 using the measured values of the cell concentration and previous estimates of the glutamine concentration.

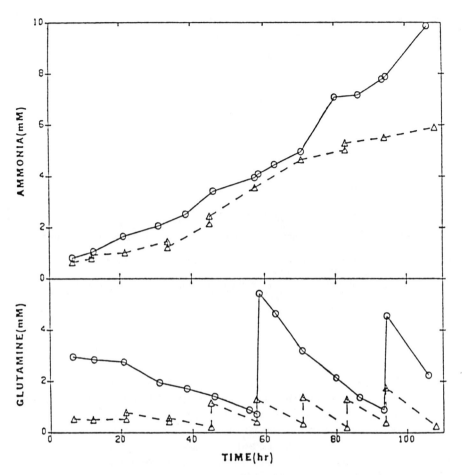

FIGURE 15-3 Cultures of MDCK controlled at low glutamine concentrations (△) via intermittent feeding of glutamine every 12 hours produce less ammonium (○) than uncontrolled cultures. The feeding schedule was formulated using equations 15.13 and 15.15. Reproduced with permission from Glacken et al. 1986.

Table 15-1 demonstrates that this manual feed-forward/feedback type control strategy was successful in reducing the average glutamine concentration from 2.5 mM to 0.8 mM. Consequently, this lower glutamine level reduced ammonia production by approximately 40%. Note that the glutamine concentration at the end of each 12 hour period was kept reasonably constant at approximately 0.2 mM. This reflects the relative accuracy of the feeding strategy determined from equations 15.13 and 15.15.

Obviously, more precise and robust control may be obtained if the glutamine concentration could be measured or estimated on-line and fed back to a controller. However, on-line measurement of glutamine is not

TABLE 15–1   Ammonium and Glutamine Kinetics from the Batch and Low Glutamine Cultures of MDCK Cells Presented in Figure 15–3[1]

	Batch[2]	Low Glutamine Culture[3]
Average glutamine concentration, mmol/l	2.5	0.8
Total glutamine depletion, mmol/l	9.1	6.0
Glutamine utilized by cells, mmol/l	7.9	5.6
Glutamine lost via spontaneous decomposition, mmol/l[4]	1.2	0.4
Total ammonium produced, mmol/l	9.1	5.2
Ammonium excreted by cells, mmol/l	7.8	4.8
Cellular yield of ammonium on glutamine, $Y_{A/G}$	0.99	0.86

[1]The growth curves for both cultures were similar. The final cell concentration attained for both cultures was 1.2×10^6 cells/ml.
[2]Glutamine was added only to avoid depletion. The initial glutamine concentration utilized (4 mM) is standard for DMEM medium.
[3]Glutamine was added every 12 hours to maintain low glutamine concentrations.
[4]Estimated from the expression for the first-order decomposition of glutamine (0.0048) (glutamine)Δt, for each time interval Δt.

straight-forward. Glutamine measuring devices consisting of glutaminase immobilized on a standard ammonia probe have been developed for some time, but have not found even limited use (Arnold and Rechnitz 1980). The main disadvantages of these probes are (1) the limited lifetime of the enzyme (1 day); (2) the signal interference from ammonia in the medium; and (3) the requirement that samples be withdrawn for measurement outside the bioreactor, although this could be automated with pumps and a computer. An HPLC could conceivably be automated to measure bioreactor glutamine and amino acid concentrations. However, the sample would need considerable preparation (cell, particle, and protein removal) before it could be injected into the HPLC.

For these reasons, direct measurement of glutamine for control purposes is not yet feasible. Consequently, indirect estimations are the only alternative. Since ammonia generation results solely from glutamine degradation and catabolism, ammonia measurements can be used to estimate glutamine levels. This may be achieved by mathematically describing ammonia and glutamine levels in a bioreactor with continuous glutamine feed at a variable rate, F:

$$\frac{d(GV)}{dt} = FG_i - KGV - q_GXV \qquad (15.16)$$

$$\frac{d(AV)}{dt} = KGV + q_G\,Y_{A/G}XV \qquad (15.17)$$

where G_i = the glutamine concentration in the feed.

Substituting for $q_G XV$ from equation 15.16 we get:

$$\frac{d(AV)}{dt} = KGV + Y_{A/G}\left(FG_i - \frac{d(GV)}{dt} - KGV\right). \qquad (15.18)$$

Integration and rearrangement gives

$$\Delta(GV) = \frac{(1 - Y_{A/G})\, K \int (GV)dt + Y_{A/G}G_i \int F\, dt - \Delta(AV)}{Y_{A/G}}. \qquad (15.19)$$

The glutamine concentration may then be estimated by simply measuring the ammonia concentration of a small slip stream that is continually removed from the bioreactor via a pump. An ammonia probe with a flow-through cap would be required for the measurement. This equation is very convenient for the following reasons: (1) an estimate of q_G is not required; (2) an estimate of the cell concentration is not required; (3) every term in the equation, except $Y_{A/G}$, is either constant and known to high accuracy (K, G_i) or is a measured value (F, $\Delta(AV)$); and (4) if $Y_{A/G}$ is close to 1, the results are somewhat insensitive to the actual value of $Y_{A/G}$. To illustrate the last point, let $1 - Y_{A/G} = \epsilon$, where ϵ is small, then

$$\Delta(GV) = \left(\frac{\epsilon}{1-\epsilon}\right) K \int (GV)dt + G_i \int F dt - \frac{\Delta(AV)}{1-\epsilon}. \qquad (15.20)$$

If we assume that $Y_{A/G} = 1$, then ϵ represents the error between the actual and assumed values. As ϵ increases (or decreases) the first and last terms would both increase (or decrease), such that small errors in ϵ would tend to cancel. Obviously, as ϵ increases, the accuracy of the prediction would deteriorate. Note that the term $\int (GV)dt$ would not be known directly, but would be calculated using previous estimates of the glutamine concentration.

An example that illustrates the predictive ability of this estimator is shown in Figure 15–4 for a batch culture of ATCC-CRL-1606 hybridomas. Based on 16 independent CRL-1606 cultures from a statistically designed experiment, the average yield was calculated to be 0.71 and was the value used in the estimator (see Table 15 of Glacken et al. 1988). Considering that the glutamine estimations shown in Figure 15–4 were based only on ammonium measurements, the predictive ability of the estimator is reasonably good.

It must be pointed out that the use of this estimator in a glutamine control scheme has not, as yet, been experimentally implemented. One potential problem is the sensitivity of the ammonia probe. This probe operates exactly like a pH probe, with a $\sim$59 mV output for each decade change in the ammonia level. Consequently, the sensitivity of the probe (at 5 mM NH_4^+) is about ± 0.1 mM. Obviously, this may be a problem if it were desired to control the glutamine level at 0.1–0.2 mM in medium containing 5 mM or so ammonium, since the culture could run out of glutamine before the

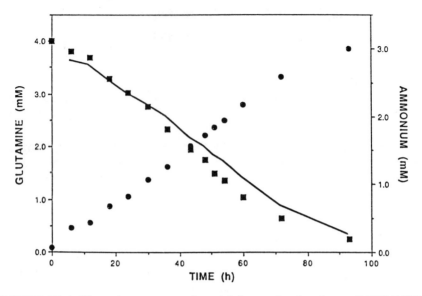

FIGURE 15–4 Glutamine concentrations (■) from a batch culture of ATCC-CRL-1606 hybridomas are compared with glutamine predictions (—) using equation 15.19. Ammonium levels (●) are shown for reference.

ammonia probe registered a detectable change. One way to minimize this problem is to use a combination of feed-forward and feedback control. The steady-state glutamine feed rate could be set equal to the estimated glutamine utilization rate (GLNUR), by using a predetermined value for q_G, as follows:

$$GLNUR = q_G XV = \frac{(q_G APR)V}{q_{ATP}} = \frac{q_G(6\ OUR + F_{L/C} LPR)V}{q_{ATP}}. \quad (15.21)$$

If the actual glutamine utilization rate was similar to the glutamine feed rate determined in equation 15.21, then the feed-forward control loop should allow more time for the ammonia probe to respond before possible glutamine depletion occurred. This steady-state feed rate could then be adjusted faster or slower depending on whether the estimated glutamine level was above or below the set point.

15.4.2 Control of Fed-Batch Bioreactors from Mathematical Simulations

The control schemes described so far have sought to maximize or minimize a specific effect: minimize ammonia accumulation, minimize lactate production, etc. However, for commercial processes, the ultimate goal is to minimize production costs, as represented by equation 15.1. How can this be achieved? First, a mathematical simulation of the performance of a given mammalian cell production process, which includes the concentration profiles of all of the relevant medium constituents, must be available. As recently as a few years ago, a data base did not exist to develop such simulations. Recently, however, we have formulated a strategy for quickly developing mathematical descriptions that relate cellular function to bioreactor performance. The model cell line used to develop this strategy was ATCC-CRL-1606 hybridomas producing antifibronectin. The details of this strategy are presented elsewhere (Glacken et al. 1988) and will only be summarized here.

Critical to any simulation is the accurate mathematical description of the specific growth, death, and product synthesis rates as a function of the relevant bioreactor variables such as the glutamine, ammonia, lactate, and serum levels. The strategy developed to formulate these descriptions is as follows.

1. Reduce the Variable Set with Statistically Designed Experiments that Measure Initial Metabolic Rates. Cell culture medium contains approximately 30 different chemical species in addition to the waste products that the cells excrete into the medium. The concentration of these components, either singularly (linear) or in concert with other variables (co-linear), may affect the culture kinetics. This variable set is clearly too large to attempt to mathematically relate the metabolic rate parameters to each variable. Well-designed factorial experiments can reduce this variable set to more manageable dimensions. This reduced variable set may then be examined fully to develop new mathematical descriptions. Variables may be combined to reduce the factorial designs. For example, amino acids, vitamins, and mineral salt levels relative to the normal levels found in DMEM may be examined together. All experiments should be initiated at low cell levels, since this ensures that the measured metabolic rates are directly attributable to the initial conditions. In this way, the rates intrinsic to the initial environment are "decoupled" from cellular influences. Relations developed from the initial rate data are pseudo-steady-state (PSS) relations, since the rates are measured over brief time intervals and at low cell concentrations ($<$50,000 cells/ml). Steady-state relations may be rigorously developed from chemostat data; however, the initial rate approach is often preferable, since more experiments, and hence more conditions, may be examined in a given time period. For commercial processes, however, it would be wise to compare the PSS relations with actual steady-state data derived from chemostats.

2. Initiate Additional Experiments Using the Reduced Variable Set to Develop Functional Relationships. As an example of step number 1 above, the statistical experimental design of CRL-1606 hybridomas showed that: (1) the growth rate, μ, was not affected by glucose levels greater than 0.1 mM; (2) serum and ammonia levels had the greatest affect on μ, and there was a significant interaction between the two variables; i.e., ammonia in-hibited growth to different extents at various serum levels; (3) μ increased with increasing glutamine levels above 1.0 mM; (4) lactate slightly inhibited growth, but significantly inhibited antibody production (q_p); and (5) all other components in cell culture medium had no effect on μ or q_p at levels above 60% of the concentration found in DMEM. Since the variable set was re-duced considerably, the number of experiments that needed to be performed to develop functional equations were also reduced. For example, the growth rate was examined as a function of the glutamine concentration, indepen-dent of all the other variables [$\mu = f_1(GLN)$], while the effects of serum and ammonia levels were examined simultaneously due to their internal inter-action [$\mu = f_2(serum, NH_4)$]. The total growth rate equation was then for-mulated by superimposing all the individual equations ($\mu = f_1 \times f_2$).

3. Coupling Initial Rate Kinetics to Other Culture States.   The results from initial rate studies would be of little value if the mathematical descriptions could not be applied to other culture states, especially at higher cell levels. Therefore, it is critical to compare the initial rate predictions to actual results at noninitial states. In the case of CRL-1606 hybridomas, three discrepancies were identified: (1) the hybridomas grew faster at higher cell levels, especially in low serum medium (Glacken et al. 1989a); (2) the cells did not respond instantaneously to rapid changes in their environment (i.e., there was a dead time; Glacken et al. 1989b); and (3) the growth rate of low serum, low cell density cultures decreased much more rapidly than predicted by the initial rate equations (Glacken et al. 1989a). These discrepancies were coupled to the initial rate equations by (1) a cell level-dependent Monod constant in the expression relating μ to the serum level (see equation 15.29); (2) meas-uring the dead time and the transient response of cultures and representing the response via a differential equation in μ (see equation 15.30); and (3) relating the rapid decline in the growth rate of low serum cultures to the spontaneous degradation of the growth promoting activity of serum (see section 15.2.4 and equation 15.24).

4. Use Existing Data Bases to Design Experiments and Develop Functional Relationships.   Experimental designs for the development of mathematical descriptions of metabolic rates would be much more efficient if the ap-proximate form for the relations were known a priori. At the initiation of our research, the availability of quantitative relationships in the literature to serve as guides were nonexistent. Based on this work, a general starting point for developing mathematical descriptions of other cell systems may be assumed (Glacken et al. 1988). (a) Nutrients such as glutamine and glu-cose stimulate the growth rate via Monod-type kinetics. The Monod con-stants are probably low ($\sim$0.2 mM). (b) Ammonium and lactate are non-competitive-type inhibitors. The apparent inhibition constants are inversely proportional to the concentration of the inhibitor. (c) The growth rate should also demonstrate Monod-type kinetics with respect to the concentration of serum or specific growth factors. The Monod constant in this case might be a decreasing function of the cell concentration. (d) Each relation in (a) through (c) is independent of the other; that is, the form for the growth rate equation can be developed by simply multiplying together relations (a) through (c); (e) lactate may inhibit excretion of molecules (i.e., antibody and ammonium) nonlinearly. Finally, (f) the death rate may be a decreasing function of the growth rate (see equation 15.23c) (Glacken et al. 1989b).

15.4.2.1 Optimization Once the metabolic rate equations are related to the process variables, differential equations may be written to describe the rate of change for all of the relevant molecular species in a given bioreactor configuration. For example, a fed-batch CRL-1606 process may be described by (see Nomenclature) (Glacken et al. 1989b):

$$\frac{d(XV)}{dt} = \mu XV \tag{15.22}$$

$$\frac{d(X_vV)}{dt} = (\mu_v X_v - K_d X_v)V \tag{15.23a}$$

$$\frac{d(X_dV)}{dt} = K_d X_v V \tag{15.23b}$$

$$K_d = 0.051 \; e^{-101.2 \, \mu} \tag{15.23c}$$

$$\alpha = \frac{X - X_d}{X} = \frac{X_v}{X} \tag{15.23d}$$

$$\frac{d(SV)}{dt} = -0.016 S^{0.7} \tag{15.24}$$

$$\frac{d(GV)}{dt} = FG_i - KGV - q_G \alpha XV \tag{15.25}$$

$$\frac{d(AV)}{dt} = KGV + q_A \alpha XV \tag{15.26}$$

$$\frac{d(PV)}{dt} = q_p \alpha XV \tag{15.27}$$

$$\frac{dV}{dt} = F(t) \qquad 0 \le F \le F_{max} \tag{15.28}$$

$$\mu = \frac{\mu_{max} SG}{[(K_S)_0 X^{-\beta} + S]\left[1 + \dfrac{A^2}{K_A}\right][K_G + G]} \tag{15.29}$$

$$\frac{d\mu}{dt} = \frac{\mu(t-\tau_L) - \mu(t)}{\tau} \tag{15.30}$$

X_0, S_0, G_0, A_0, V_0, and P_0 are all given at $t = t_0$.

For CRL-1606 hybridomas, the values of the constants were determined from independent experiments to be:

$$(K_s)_0 = 26.5 \pm 2\% \ FCS/(cells/l)^{-0.21},$$
$$q_p = 2.0 \pm 0.7 \ pg/cell\text{-}hour, \ \beta = 0.21,$$

$$\mu_{max} = 0.055 \pm 0.003/hour,$$
$$q_A = 5.0 \times 10^{-11} \pm 2 \times 10^{-11} \ mmol/cell\text{-}hour,$$

$$q_G = 7.0 \times 10^{-11} \pm 2 \times 10^{-11} \ mmol/cell\text{-}hour$$

$K_A = 26 \ mM^2$, $K_G = 0.15 \ mM$, $\tau = 0.01 \ hour$, and $\tau_L = 12 \ hour$.

If all the constants have been previously determined, the equations may be solved to either simulate the bioreactor performance or to optimize the process. Obviously, the rate of nutrient flow to the bioreactor is a control parameter that can be changed with time. The question is: How does one manipulate the flow rate with time in a manner that minimizes equation 15.1; or in other words, what is the optimal function for the flowrate, F? This involves the subject of optimal control theory, which will not be discussed here, but is described in detail elsewhere (see Bryson and Ho 1976 for general discussion and Modak et al. 1986 for optimization of biological reactors). Using control theory, the optimal volume profile (which is directly related to the nutrient flow rate) was determined for the CRL-1606 hybridoma fed-batch bioreactor simulated above, and is shown in Figure 15–5. In order to perform the optimization, it was necessary to assume that there was no serum degradation ($dS/dt = 0$) and that the growth rate could respond instantaneously to medium manipulations (see Glacken 1987 for optimization details). Also, the inhibition of the antibody productivity by lactate accumulation was not included. Despite these limitations, experimental implementation of the control policy shown in Figure 15–5 resulted in cell and antibody yields (mg Mab per l per % serum) more than 20 and 10 times that obtained in typical (10% serum) batch cultures and more than three and 1.6 times that obtained from low serum (1.5%) cultures (Table 15–2). Note that the fed-batch culture produced more antibody, even though q_p for the batch cultures (2.6 pg/cell-hour) was much greater than that of the fed-batch culture (1.3 pg/cell-hour). [It should be noted that q_p values were always consistent between cultures grown concurrently, but tended to vary greatly between cultures cultivated at different times (from 1.1–2.7 pg/cell-hour). The reasons for this are not clear. Since the two batch cultures in Table 15–2 were grown concurrently (q_p values of 2.5 and 2.7) while the fed-batch culture was cultivated separately, the difference in q_p between the batch and fed-batch cultures was most likely due to normal variance.]

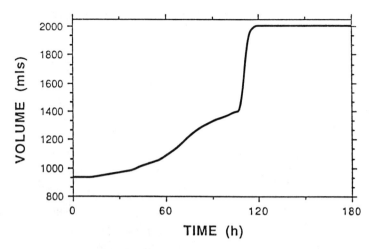

FIGURE 15-5 The optimal volume profile determined from control theory for a fed-batch culture of ATCC-CRL-1606 hybridomas. Reproduced with permission from Glacken et al. (1989).

TABLE 15-2 Comparison of an Optimized Fed-Batch and a Typical Batch Culture of ATCC-CRL-1606 Hybridomas

Bioreactor Type	X_{final} 10^9 cells/l	P_{final} mg/l	t_{final} hours	$(q_P)_{avg}$ (pg per hour per cell)	$Y_{X/S}$ 10^9 cells/l % FCS	$Y_{P/S}$ mg/l % FCS
Fedbatch	3.1	202	177	1.3	2.6	168
Batch, 10% FCS	1.6	155	141	2.5	0.16	15.5
Batch, 1.5% FCS	1.3	160	150	2.7	0.87	107

The last two columns represent the yield of cells and antibody on serum, respectively.

It should be noted, however that since invalid assumptions were made, the cell and antibody levels predicted by the optimization routine were considerably higher than the actual results. If all of these assumptions are relaxed, (i.e., include serum degradation, time lags, and lactate inhibition of q_p), then the predicted cell, antibody, glutamine, and ammonia profiles are seen to be reasonably close to the actual profiles (Figures 15-6 and 15-7). This is encouraging, since if accurate simulations are available, production costs of commercial processes can be minimized if optimization routines can be developed that include all the dynamic characteristics of mammalian cell bioreactors (especially serum degradation and time lags).

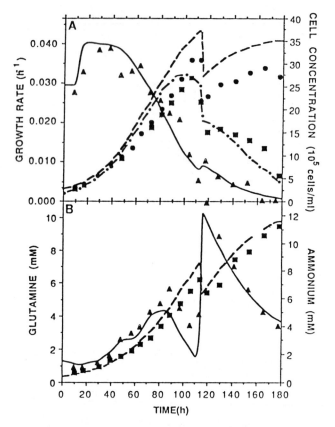

FIGURE 15–6 Results from the experimental fed-batch culture are compared with the model predictions. The model included both serum degradation and lags in the growth response. Growth rates were calculated for every combination of three contiguous measurements of the total cell concentration. (A) Growth rate, actual (▲), predicted (—); viable cell level, actual (■), predicted (—■—); total cell level, actual (●), predicted (----), (B) Ammonium, actual (■), predicted (----); glutamine, actual (▲), predicted (—). Reproduced with permission from Glacken et al. (1989).

15.5 SPECULATIVE STRATEGIES FOR MANIPULATING CELLULAR METABOLISM TO IMPROVE BIOREACTOR PERFORMANCE

In the preceding examples, empirical descriptions of bioreactor dynamics were used to develop control strategies. Although these relations have been utilized with some success, their potential for optimizing bioreactor performance is limited. This is because these descriptions do not make use of any knowledge concerning the regulatory properties of intracellular energy pathways. These pathways are directly responsible for nutrient utilization

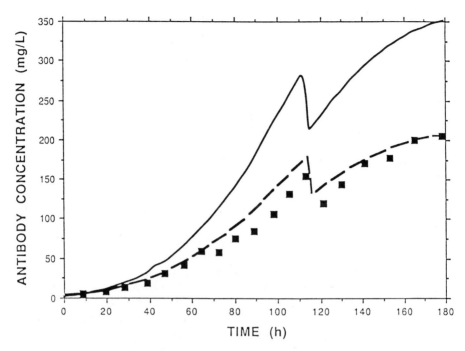

FIGURE 15-7 The actual (■) and predicted (—) antifibronectin monoclonal antibody concentration from the fed-batch culture are compared. Also shown is the predicted monoclonal antibody profile assuming lactate inhibition of the specific antibody productivity (----). Reproduced with permission from Glacken et al. (1989).

and waste product excretion and must be the targets of any rational design to manipulate cellular function.

But the question now is: How does one use knowledge concerning the regulation of cellular energy pathways to improve bioreactor performance? The first step is to formulate metabolic objectives that are to be either maximized or minimized. For example, suppose we wish to minimize the specific rate of ammonia production; that is, the amount of ammonia excreted per cell per hour. This may be expressed as minimizing the moles of ammonia excreted per mole of ATP synthesized. From the glutamine catabolic pathways shown in Figure 15-2, it may be seen that this ratio may be minimized by maximizing the relative activity of alanine or aspartate aminotransferase with respect to glutamate dehydrogenase (GDH). The reason is as follows. If all the glutamate from glutamine is processed via GDH, the terminal products may be either lactate, pyruvate, or acetyl-CoA and the ATP yields would be 9, 12, and 15 mol of ATP per mole of glutamine catabolized, respectively. Higher ATP yields may not normally be obtained, since acetyl-CoA has been shown not to be oxidized via the tricarboxylic acid cycle (TCA) cycle of many cultured cells (Kuchka et al. 1981; Lanks

1987; Reitzer et al. 1979; Zielke et al. 1978), although Lanks and Li (1988) have recently observed extensive excretion of citrate into the medium. Assuming that glutamine is catabolized via glutaminase (instead of an aminotransferase), the ratio of ammonia produced per ATP synthesized for GDH-processed glutamate is either 0.22, 0.17, or 0.13, depending on the form of the terminal product (e.g., lactate, pyruvate, or acetyl-CoA). On the other hand, if glutamate is processed via aspartate or alanine aminotransferase, alanine and/or aspartate are the terminal products, and 9 mol of ATP are formed per mole of glutamine catabolized. However, the ratio of ammonia produced per ATP synthesized is significantly lower (i.e., 0.11). Note that the α-ketoglutarate formed by the transaminase must be stoiciometrically converted to either pyruvate or oxaloacetate for the reaction to continue, unless, of course, sufficient glucose-derived pyruvate or oxaloacetate is available.

This analysis implies that ammonia formation in bioreactors might be reduced by adding specific allosteric inhibitors of GDH to the medium, such as fumarate and pyridoxal phosphate (Smith et al. 1976). Fumarate can enter the cell (Nakano et al. 1982), whereas pyridoxal is a component of standard cell culture media. Although pyridoxal phosphate probably cannot enter the cell, increased concentrations of pyridoxal in the medium could possibly inhibit GDH since pyridoxal must be phosphorylated intracellularly. It should be noted that pyridoxal phosphate is a co-factor for transaminase and, as such, might act as a reciprocal regulator of GDH and the transaminases. Since leucine activates GDH (Smith et al. 1976), it may be worthwhile to control the concentration of this essential amino acid to as low a level as possible.

Other strategies for reducing ammonium production may be formulated if we assume a constant yield of cell mass is produced per mole of ATP utilized. In this case, we can surmise that as more of the cell's energy is produced from glucose via aerobic glycolysis, less glutamine would need to be catabolized and, consequently, less ammonia would be produced. Indeed, this has been observed experimentally by Miller et al. (1989) with glucose-limited hybridoma chemostats. A large increase in the glucose feed concentration resulted in a 30% lower steady-state specific ammonium production rate. Since the specific rate of glutamine utilization also decreased, and since there was no significant change in the yield of ammonium from glutamine, we may conclude that the lower ammonium production rates were due to a sparing effect from glucose and did not indicate a drastic change in the relative rates of carbon flux through the various glutaminolytic pathways. Maintenance of bioreactor glucose concentrations at high levels via on-line control (see section 15.3) may therefore improve productivity by reducing ammonium production.

One might suppose that if the rate of aerobic glycolysis can be increased sufficiently, this sparing effect may be taken to an extreme and no glutamine need be catabolized at all. Indeed, studies comparing the energy metabolism of wild-type Chinese hamster ovary (CHO) cells to respiration-deficient mutants showed that the total energy requirements of the mutants may be met solely by aerobic glycolysis (DeFrancesco et al. 1975; Donnelly and Scheffler 1976). Although ammonia production rates were not measured, the mutants utilized oxygen and glutamine at rates more than 10 times slower than the wild-type cells. Since the glutaminolytic pathway was essentially inoperative, these mutants strictly required the addition of two glutamine-derived intermediates: aspartate and asparagine. These data suggest that ammonia generation in bioreactors may be drastically reduced by controlling the glutamine concentration at extremely low levels (<0.1 mM) in the presence of aspartate and asparagine. The low glutamine levels should effectively shut down glutaminolysis, but the addition of these nonessential amino acids plus an increased rate of aerobic glycolysis might enable the cells to continue to grow rapidly. This strategy has not been tested and it must be pointed out that inhibiting respiration by restricting the glutamine supply may not mimic the metabolic effects caused from defects in the respiratory system. For example, one possible difference between the two cases may be the internal NADH/NAD ratio; the mutant cells would probably have a larger ratio than would the glutamine-limited wild-type cells. Consequently, the regulatory effects on various metabolic pathways may differ to the extent that, under glutamine limitation, the rate of aerobic glycolysis might not rise to a level sufficient to meet the cell's energy demands.

An additional potential benefit of this strategy would be the reduction of the oxygen utilization rate. As stated previously, since the rate of oxygen transfer in any given bioreactor is limited, lower oxygen utilization rates can result in higher theoretical maximum cell levels, provided all nutrients may be supplied and all waste products may be adequately removed. Indeed, Donnelly and Scheffler (1976) demonstrated that CHO cells could grow normally in the presence of a respiration inhibitor (rotenone) provided asparagine was provided. It may be anticipated that similar results could be obtained by inhibiting respiration via glutamine limitation. Our laboratory has shown that rotenone-inhibited lymphocytes can grow at near normal rates provided pyruvate and at least 1% serum is provided. These cells produced almost no ammonium and consumed little glutamine (Lee and Glacken 1989).

One disadvantage to inhibiting glutaminolysis would be the extremely high lactate production rate that would result. The rotenone-inhibited cultures, mentioned previously, produced 40% more lactate than uninhibited cells (Lee and Glacken 1989). Unfortunately, high lactate levels can also inhibit growth and product synthesis (Glacken et al. 1988). If glucose could be oxidized at reasonable rates in the Kreb's cycle, less lactate would be produced and the cell's energy requirements could still be met without glutamine. Unfortunately, only a very small fraction of glucose (<5%) is typically oxidized by cultured mammalian cells (Kuchka et al. 1981; Lanks

1987; Reitzer et al. 1979; Zielke et al. 1978). Various hypothesis have been offered to explain this observation. One hypotheses is that the shuttle systems that transfer reducing equivalents from the cytosol to the mitochondria, such as the malate-aspartate shuttle, operate inefficiently for proliferating cells in culture (Boxer and Devlin 1961; Kovacevic 1972). If the NADH generated by glycolysis in the cytosol can not be transported to the mitochrondria, this surplus hydrogen may only be eliminated by reducing pyruvate to lactate via lactate dehydrogenase.

Support for this hypothesis has been found in our laboratory from the observation that oxaloacetate addition drastically reduces glutamine utilization and ammonia production from cultured lymphocytes (Gayton and Glacken 1989). Oxaloacetate is a limiting substrate for aspartate aminotransferase, a critical enzyme in the malate-asparate shuttle. Oxaloacetate stimulation of aminotransferase activity would reactivate NADH transport into the mitochondria from the cytosol and may possibly permit the oxidation of pyruvate in the TCA cycle. Increased ATP synthesis from pyruvate oxidation would imply that less glutamine would need to be catabolized and less ammonia would consequently be produced. This intriguing hypothesis requires further verification.

Another hypothesis may be formulated from the observation that tumor cells have low superoxide dismutase and catalase levels (Oberlay et al. 1980). These low levels are responsible for high intracellular superoxide concentrations (O_2^-), which is known to inhibit both pyruvate and isocitrate dehydrogenase, but not succinate dehydrogenase (Hornsby and Gill 1981; Hornsby 1982). Consequently, glucose oxidation is inhibited, but glutamine catabolism can proceed normally. Additionally, high concentrations of superoxide radicals cause peroxidation of lipid membranes. This would imply that the acetyl-CoA formed from glucose would be required for lipid replenishment and would consequently not be available for oxidation in the Kreb's cycle.

Based on these hypotheses, strategies to reduce lactate accumulation during severe glutamine limitation may be formulated, and may include the following:

1. Provision of adequate levels of shuttle intermediates such as asparate, oxaloacetate, malate, and citrate. It has been shown that malate and citrate reduce lactate excretion and increase pyruvate oxidation in SV-40 transformed 3T3 cells (Nakano et al. 1982).
2. Minimizing lipid peroxidation and peroxide levels by maintaining a low ORP via the controlled addition of reducing agents such as glutathione, DTT, tocopherols, and ubiquinones, and/or the addition of selenium, a co-factor for glutathione peroxidase: (see section 15.2.4).
3. Controlling oxygen levels at low concentrations, e.g., $\sim 1\%$ of air saturation.
4. The addition of lipid precursors, such as fatty acids and ethanolamine.

TABLE 15–3 How the Ammonia Concentration Can Be Kept Constant at Varying Ammonium Levels by Adjusting the pH

Ammonium (NH_4^+)	Ammonia (NH_3)	pH
1 mM	9 μM	7.2
2 mM	9 μM	6.9
4 mM	9 μM	6.6

Finally, another strategy to improve bioreactor performance would be to manipulate the cell's metabolism in a manner that minimizes the toxicity of the ammonium produced. Ammonium is known to increase the pH of intracellular organelles (Suput 1984), since un-ionized ammonia (NH_3) can diffuse through lipid membranes, whereas ionized ammonium (NH_4^+) may not (Benjamin et al. 1978). Once ammonia crosses the membrane, it can rapidly pick up a proton on the other side to form ammonium. This rise in pH may disrupt the operation of enzymes in organelles, especially those that operate at an acidic pH, such as lysosomes. This implies that ammonia and not ammonium may actually be the inhibitory agent. If this is the case, a bioreactor control strategy that lowers the pH of the medium as ammonium accumulates could be beneficial. This strategy would keep the ammonia concentration constant even though the ammonium level would be increasing (see Table 15–3). Obviously, the inhibitory effect of low pH levels must be more than offset by the decreased NH_3 concentrations for this strategy to have any merit.

15.6 NOMENCLATURE

A	= Ammonium concentration, mM
APR	= Volumetric ATP production rate, mmol/l-hour
C_G	= Oxygen concentration in the gas phase, atm
$C_{G,S}$	= Steady-state gas phase oxygen concentration, atm
C_L^*	= Oxygen concentration in the gas phase which would be in equilibrium with the actual concentration of oxygen in the liquid phase, atm
$C_{L,S}^*$	= Set point for dissolved oxygen level, atm
C_O, C_f	= Initial and final DO levels during an OUR measurement, atm
F	= Flow rate of medium into a fed-batch bioreactor, l/hour
$F_{L/C}$	= Fraction of lactate derived from the carbohydrate source
G	= Glutamine concentration, mM

G_i	= Concentration of glutamine in the medium feed to a fed-batch process, mM
GUR	= Volumetric glucose utilization rate, mmol/l-hour
K	= First order glutamine decomposition rate, per hour
K_d	= Specific death rate, per hour
K_C, K_D, K_I	= Controller constants
$k_L a$	= Mass transfer coefficient, mmol/atm-l-hour
LPR	= Volumetric lactic acid production rate, mmol/l-hour
m_{ATP}	= Rate of cellular ATP utilization required for maintenance processes, mmol/cell-hour
N_o	= Oxygen transfer rate, mmol/l-hour
OUR	= Volumetric oxygen uptake rate, mmol/l-hour
q_A	= Specific rate of cellular ammonium production, mmol/cell-hour
q_{ATP}	= Specific rate of cellular ATP production, mmol/cell-hour
q_G	= Specific rate of cellular glutamine utilization, mmol/cell-hour
q_L	= Specific rate of cellular lactate production mmol/cell-hour
q_{O_2}	= Specific rate of cellular oxygen utilization, mmol/cell-hour
q_p	= Specific rate of cellular product (antibody) production, mmol/cell-hour
S	= Serum concentration, % fetal calf serum (FCS)
P	= Product concentration, g/l, pg/l, mol/l
t	= Time, hour
V	= Volume of liquid in the bioreactor, l
X	= Total cell concentration, cells/l
X_d	= Dead cell concentration, cells/l
X_V	= Viable cell concentration, cells/l
$Y_{A/G}$	= Yield of ammonia derived from glutamine catabolism
Y_{ATP}	= Yield of cell mass from ATP, g cells/mmol ATP
α	= Fraction of total cells that are viable
μ	= Apparent specific growth rate, per hour

REFERENCES

Adamson, S.R., and Schmidli, B. (1986) *Can. J. Chem. Eng.* 64, 531–539.

Arnold, M.A., and Rechnitz, G.A. (1980) *Anal. Chem.* 52, 1170–1174.

Benjamin, A.M., Kamoto, K.O., and Quastel, J.H. (1978) *J. Neurochem.* 30, 131–143.

Boxer, G.E., and Devlin, T.M. (1961) *Science* 134, 1495–1501.

Bree, M.A., Dhurjati, P., Geoghegan, R.F., and Robnett, B. (1988) *Biotechnol. Bioeng.* 32, 1067–1072.

Bryson, A.E., and Ho, Y.C. (1976) *Applied Optimal Control* 2nd ed., Blaisdell, Waltham, MA.

Commoy-Chevalier, M.J., Robert-Galliot, B., and Chany, C. (1978) *J. Gen. Virol.* 41, 541–547.

Coughanowr, D.R., and Koppel, L.B. (1965) *Process Systems Analysis and Control*, McGraw-Hill, New York.

DeFrancesco, L., Werntz, D., and Scheffler, I.E. (1975) *J. Cell. Physiol.* 85, 293–306.

Daniels, W.F., Garcia, L.H., and Rosensteel, J.F. (1970a) *Biotechnol. Bioeng.* 12, 409–417.

Daniels, W.F., Garcia, L.H., and Rosensteel, J.F. (1970b) *Biotechnol. Bioeng.* 12, 419–428.

Donnelly, M., and Scheffler, I.E. (1976) *J. Cell Physiol.* 89, 39–52.

Eaton, M.D., and Scala, A.R. (1961) *Virology* 13, 300–307.

Fleischaker, R.J. (1986) in *Mammalian Cell Technology*, (Thilly, W.G., ed.), pp. 199–211, Butterworths, Boston.

Fleischaker, R.J., and Sinskey, A.J. (1981) *Eur. J. Appl. Microbiol. Biotechnol.* 12, 193–197.

Furusawa, E., and Cutting, W. (1962) *Proc. Soc. Exp. Biol. Med.* 111, 71–75.

Gayton, M., and Glacken, M.W. (1989) *Analysis of Alanine and Aspartate Aminotransferase Activity of JURKAT Cells from Steady State Chemostats*, AIChE Annual Meeting, San Francisco, November 8.

Glacken, M.W. (1987) Doctoral Dissertation, MIT, Cambridge, MA.

Glacken, M.W., Adema, E., and Sinskey, A.J. (1988) *Biotechnol. Bioeng.* 32, 491–506.

Glacken, M.W., Adema, E., and Sinskey, A.J. (1989a) *Biotechnol. Bioeng.* 33, 440–450.

Glacken, M.W., Fleischaker, R.J., and Sinskey, A.J. (1983) *Trends Biotechnol.* 1, 102–108.

Glacken, M.W., Fleischaker, R.J., and Sinskey, A.J. (1986) *Biotechnol. Bioeng.* 28, 1376–1389.

Glacken, M.W., Huang, C., and Sinskey, A.J. (1989b) *J. Biotechnol.* 10, 39–66.

Griffiths, B. (1984) *Develop. Biol. Stand.* 55, 113–116.

Hornsby, P.J. (1982) *J. Cell Physiol.* 112, 207–216.

Hornsby, P.J., and Gill, G.N. (1981) *J. Cell Physiol.* 109, 111–120.

Hwang, C., and Sinskey, A.J. (1989) The Role of Medium Oxidation-Reduction Potential in Monitoring Growth of Cultured Mammalian Cells. Presentation at Cell Culture Engineering II, Engineering Foundation Conference, Santa Barbara, CA, December 3–8.

Ito, M., and McLimans, W.F. (1981) *Cell Biol. Int. Rep.* 5, 661–666.

Jensen, E.M., and Liu, O.C. (1961) *Proc. Soc. Exp. Biol. Med.* 107, 834–838.

Kilburn, D.G., Fitzpatrick, P., Blake-Coleman, B.C., Clarke, D.J., and Griffiths, J.B. (1989) *Biotechnol. Bioeng.* 33, 1379–1384.

Kovacevic, Z. (1972) *Eur. J. Biochem.* 25, 372–378.

Kuchka, M., Markus, H.B., and Mellman, W.J. (1981) *Biochem. Med.* 26, 356–364.

Lanks, K.W. (1987) *J. Biol. Chem.* 262, 10093–10097.

Lee, H.-C., and Glacken, M.W. (1989) *Growth and Metabolic Characterization of Mammalian Cells in the Presence of Respiration Inhibitors*, AIChE Annual Meeting, San Francisco, November 8.

Miller, W.M., Blanch, H., and Wilke, C. (1988) *Biotechnol. Bioeng.* 32, 947–965.

Miller, W.M., Wilke, C., and Blanch, H. (1987) *J. Cell Physiol.* 132, 524–532.

Miller, W.M., Wilke, C.R., and Blanch, H.W. (1989) *Biotechnol. Bioeng.* 33, 477–486.

Modak, J.M., Lim, H.C., and Tayeb, Y.J. (1986) *Biotechnol. Bioeng.* 28, 1396–1407.

Nakano, E.T., Cianpi, N.A., and Young, D.V. (1982) *Arch. Biochem. Biophys.* 215, 556–563.

Oberley, L.W., Oberley, T.B., and Buettner, G.R. (1980) *Med. Hypotheses* 6, 249–268.

Paul, J. (1975) *Cell and Tissue Culture*, Longman Group Limited, New York.

Pirt, S.J. (1975) *Principles of Microbe and Cell Cultivation*, Blackwell Scientific Publications, Cambridge, MA.

Reitzer, L.J., Wice, B.M., and Kennell, D. (1979) *J. Biol. Chem.* 254, 2669–2676.

Ryan, W.L., and Cardin, C. (1966) *Proc. Soc. Exp. Biol. Med.* 123, 27–30.

Smith, E.L., Austin, B.M., Blumenthal, K.M., and Nye, J.F. (1976) *The Enzymes*, Vol. 11 (Boyer, P.B., ed.), pp. 294–366, Academic Press, New York.

Suput, D. (1984) *Biochim. Biophys. Acta* 771, 1–8.

Tritsch, G.L., and Moore, G.E. (1962) *Exp. Cell. Res.* 28, 360–364.

Van't Riet, K. (1983) *Trends Biotechnol.* 1, 113–119.

Wang, D.I.C., Cooney, C.L., Demain, A.L., et al. (1979) *Fermentation and Enzyme Technology*, John Wiley and Sons, New York.

Zielke, H.R., Ozand, P.T., Tildon, J.T., Sevdalian, D.A., and Cornblath, M. (1978) *J. Cell. Physiol.* 95, 41–48.

Instrumentation of Animal Cell Culture Reactors

Winfried Scheirer
Otto-W. Merten

Animal cell culture technology has shown a rapid development within the last few years. As the economy of the process becomes more and more important, there is a clear movement toward more efficient fermentation systems and methods. The advantages of high cell density systems have been reported recently (Grdina and Jarvis 1984; Takazawa et al. 1988; Velez et al. 1987). One common problem arises with such high-efficiency-systems; the control of these cultures, which becomes more critical and difficult as the cell density rises. There are relatively high volumetric nutrient consumption rates and very steep substrate gradients within the system compared to conventional reactors that reach cell densities of 1–3 millions/ml as a maximum.

In high density cultures, reaching cell counts to 100 millions of cells/ml (Runstadler and Cernek 1988), dissolved oxygen, for example, will be depleted within 20 sec after starting from the air-saturation level. In addition to the principal problem of maintaining an appropriate nutritional supply, there is the obvious difficulty of finding sensor systems that are both fast and accurate and the appropriate control set-up for critical parameters.

However, in addition to maintaining the minimal culture conditions, it is obviously necessary to properly keep the culture within the parameters required for optimal productivity. This includes, in addition to the physical and nutritional parameters, some physiological controls due to frequent occurrence of correlation between one or more metabolic parameters and optimal productivity.

Some of these can be measured directly, such as lactate dehydrogenase (LDH) or glutamic-oxaloacetic transaminase (GOT) concentrations in the medium, both enzymes, which are released from dying cells. They can, for example, be used for monitoring and controlling cell viability. Other parameters must be calculated from different measurements, such as production and consumption rates, growth rate, and specific intracellular concentrations.

In addition to the sensor/amplifier systems needed for the direct parameters, a computerized calculation step for the indirect parameters is also necessary. The situation is particularly complicated by the fact that these systems are yet not readily available from the suppliers.

16.1 THEORETICAL APPROACHES

16.1.1 Standard Controls and Advantageous Systems

Many measurement and control loops are necessary for each fermentor, others are advantageous for distinct processes. We will list those that are—according to our experience or opinion—important for processes in which animal or human cells are cultivated. Our focus is high-efficiency culture, which is usually associated with the need for high cell densities.

16.1.1.1 Temperature Temperature control is quite simple in almost all cell culture systems because of low endogenous heat. The only critical point is the local overtemperature, because of the low Reynolds-numbers, which are involved in cell culture fermentations. In a free suspension fermentation, there is a possibility of temperature limitation from a heating water circuit.

16.1.1.2 pH The measurement of pH is also easy by standard systems because of the slow rate of change in cell culture systems. pH control can be performed by CO_2/air titration. In most cases of well-balanced fermentation media, this can be achieved through surface aeration only. For some applications, an additional-base titration may be necessary. Care must be taken for membrane fouling of the electrode, which may be caused by protein-precipitation during runs of long duration.

16.1.1.3 Agitation The maximum value of stirring speed must be carefully controlled because of the mechanical fragility of the cells. This value has to be determined for each cell line in each system.

16.1.1.4 pO$_2$ The oxygen tension is one of the most sensitive control problems because of the narrow range of useful working concentrations, the high consumption rates, which cause steep gradients in slowly agitated systems, and the need for a quick response time. Oxygen tension is, in practice, maintained close to the minimum level, which is in the range of 15 mm Hg, to obtain maximum transfer rates and to avoid high local concentrations, which are toxic. Oxygen consumption rates are in the range of 1.5–15 pg oxygen/cell-hour (De Bruyne 1988), which corresponds to a volumetric demand of 150–1,500 mg/l-hour at 10^8 cells/ml. This may result in complete oxygen depletion within seconds or, taking the low liquid bulk speed and the smooth mixing characteristics into account, within a few millimeters distance from the oxygenator. Therefore, a well designed oxygenation system should be applied that would include measurement and control of oxygen at different critical points of the reactor.

16.1.1.5 Filling Level The control of the filling level is another point that becomes important when continuous systems are applied. Since most retention systems have a critical liquid level, which influences separations and the perfusion rate and sometimes allows only a few millimeters of variation, this may become a serious problem.

16.1.1.6 Cell Density Cell density is important as a direct control parameter and is also the basis for all calculations of specific growth, uptake and production rates, as well as specific contents. Since cells tend to grow on glass windows, most commercial systems exhibit inadequate drift in operations of any duration. Therefore, most researchers are still using external sample counting. However, this provides very few measuring points and creates problems of inaccuracy due to rapid aging of the sample.

16.1.1.7 Cell Viability Cell viability is obviously a parameter that has to be monitored as accurately as possible. The viability value can provide the proper course of the fermentation, the correct time to harvest, or the overall nutritional state of a continuous culture.

16.1.1.8 Liquid Mass Flow (Perfusion Rates) In highly efficient continuous systems, the feeding rate must be kept as close as possible to optimal medium utilization. This means that cells can be starved quite easily especially animal cells, which are quite sensitive. Therefore, the feeding rate must be maintained as accurately as possible. With peristaltic pumps, for example, there are potential problems with huge flow variation caused by different influences such as temperature changes, tube age, changes in differential pressure, etc.

16.1.1.9 Vessel Pressure Vessel pressure is necessary for maintaining control when employing very high density cultures in which oxygen transfer is insufficient without pressurization. There is a great deal of appropriate equipment for measurement and control available. Because of differences between systems designed for use with or without pressurization, this parameter must be defined early in the system set-up.

16.1.1.10 Biochemical Parameters Generally, the medium has to provide the cells with the necessary nutrients and take all released metabolites. The basic medium contains amino acids, organic acids, one or several sources of carbon (glucose and glutamine in most cases), vitamins, inorganic ions, and frequently serum as a source of growth factors, hormones, and other factors, which might support or inhibit cellular proliferation, depending on the cell lines used.

The use of cultivation systems with cell densities up to 5×10^6 cells/ml generally pose no problem with respect to the medium used (e.g., DMEM, IMDM, etc.). However, switching from a low to a high cell density culture system (up to 10^8 cells/ml), necessitates the control of one or more growth or production limiting substances. The above-mentioned medium compositions are sufficient for low density culturing systems in most cases, but may be inadequately composed for high cell density culturing systems. With the development of new culture system configurations that provide the possibility of high cell densities, it is evident that cells in a high cell density environment have additional nutritional demands and that their metabolism and physiology are changed.

Some examples with respect to medium composition are medium components, such as a carbon source (mainly glucose) and some amino acids such as glutamine, which are used in high concentrations and can cause some problems. Too high a glucose concentration causes the production of high amounts of lactic acid mainly during the growth phase (Velez et al. 1986; Luan et al. 1987a; unpublished results). This cause, in addition to the production of CO_2 by the metabolism of glucose and glutamine, a strong acidification of the medium and an unnecessary waste of the energy source with respect to the production of lactic acid. Cells cultivated in a medium with a concentration of glutamine that is too high [often main energy source (Reitzer et al. 1979)] produce high amounts of NH_4^+, which is toxic for cultured cells (Butler and Spier 1984; Butler 1985 and 1986). A good feedback control system for these compounds would provide a means of adding other carbon sources (glucose and glutamine) to the culture according to the demand of the cells and waste products would thus be minimized. Hu et al. (1987) presented a feeding system for a forward regulation model system for hybridomas by keeping the concentrations of glucose and glutamine lower than usual. The reduction of lactic acid generation, which with CO_2 is the main cause of acidification of the supernatant in bioreactors, is possible by replacing glucose with other sugars if the cell line of choice can be adapted to the new conditions. Replacing glucose with other sugars, such as galactose, maltose, or fructose, would provide in many cases the same influence on the cell growth as glucose, but would be accompanied by a lower production of lactate (Eagle et al. 1958; Butler 1986). The reduction of the sugar concentration in general is only possible when the glutamine concentration is elevated in parallel, which often causes elevated NH_4^+ production.

The presence of amino acids is also necessary but there is the question as to which amino acid must be present and at what concentration. Until recently, much work was done in this field with the result that different cell lines have different demands and produce different amino acids. In addition, these demands depend on the cultivation system used and the physiological state of the cells. Roberts et al. (1976) presented some results for the MOPC-31C mouse plasmacytoma cell line in a stationary batch culture. The cells consumed glutamine, isoleucine, methionine, valine, and some tyrosine and phenylalanine. Aspartic acid, glutamic acid, glycine, proline, and serine were produced. The other amino acids were not influenced. Polastri et al. (1984) published results from virus production using Vero cells on microcarriers in a spinner culture system. During cell growth, glutamine, histidine, arginine, tryptophane, and methionine were consumed, as well as glutamine during the virus production. Alanine and serine were produced during cell growth and virus production. The concentration of the remaining amino acids were influenced slightly.

Generally, hybridomas utilize glutamine and produce alanine (Seaver et al. 1984; unpublished results). For instance, one mouse-mouse hybridoma consumed leucine, serine (totally), isoleucine, methionine, arginine (totally), glutamine (totally), phenylalanine, tyrosine, and some valine in a static batch culture. In parallel, it produced alanine, asparagine, glycine (slightly), and glutamic acid (during the lag and the beginning of the log phase), other amino acids were not affected. In comparison with these results, a human-mouse-human hybridoma consumed only glutamine and arginine and produced glutamic acid, alanine, and proline in a static batch culture. The concentration of the other amino acids decreased slightly during and increased at the end of the log phase and at the beginning of the stationary phase (unpublished results).

Some examples concerning the consumption and production of amino acids by different cell lines and/or clones of recombinated BHK21 and mouse Ltk cells, respectively, in comparison with glucose consumption and lactate production in a perfusion process were given by Wagner et al. (1988), which showed the complicated biochemical network of synthesis, conversion, and transport phenomena.

In general, all biochemical parameters that might have an inhibitory or promoting effect on growth and/or production of the cells in a certain bioprocess have to be considered as important. In all cases, there is the pO_2 (16.1.1.4), the pH (16.1.1.2.), the cell density (16.1.1.6.), the cell viability (16.1.1.7.), carbon source(s), amino acids, which are essential for the cell line used, waste products, and, of course, the product itself.

The examples mentioned above imply that the development of a control system is very important for process control and optimization. But it is also evident that controls have to be established for each process and that changing a process system also requires changing the control system, because of each cell's physiological requirements. Before these control systems can be made available, the application of sensing devices has to provide enough data on the different culturing systems in order to allow the establishment of useful mathematical models.

16.1.1.11 Calculated Parameters The use of calculated parameters can be very important for obtaining more information about a process and for controlling a bioprocess. Of course, these parameters have to be captured by measuring devices and sequently processed and calculated by microprocessors. They can later be used for regulating relevant process parameters (see sections 16.1.1.1 to 16.1.1.10) and for optimizing the bioprocess.

An old and largely used parameter is the respiratory quotient (RQ = CO_2-production rate/O_2-uptake rate), which can be used as a valuable indicator of the total viable biomass in a bioreactor system. The uptake of oxygen and the production of CO_2 can be measured by off-gas analysis and dynamic assessment. For details, see Fleischacker et al. (1981).

In principle, the consumption of nutrients or the production of metabolites can also be used as an estimation of the viable biomass (expressed, e.g., as μmol glucose/l $\times$ d if glucose consumption is taken as the parameter). In this case, the glucose concentration has to be determined and, in the case of continuous cultures, the dilution rate and the glucose concentration of the medium have to be taken into account. However, it should be mentioned that the consumption of nutrients or the production of metabolites depends on the physiological state of the cells. Therefore, these indirect methods should be used only when other, more direct methods are not applicable.

Knowing the viable biomass, specific consumption or production rates can be calculated and expressed as, e.g., μmol glucose/10^9 cells $\times$ d. For the determination of viable and total cell density, respectively, see sections 16.2.1.6 and 16.2.1.7. For biochemical parameters see section 16.2.1.10.

Two very important calculated values are the specific growth rate and the viability, which are calculated as:

$$\text{Specific growth rate} = \frac{\ln \times 2 - \ln \times 1}{t2 - t1} \text{ (per hour)} \qquad (16.1)$$

and

$$\text{Viability} = \frac{\text{viable cell count}}{\text{total cell count}} \times 100 \text{ (\%)} \qquad (16.2)$$

where $\times 2$ is the number of living cells/ml at time t2 in hours, $\times 1$ is the number of living cells per milliliter at time t1 in hours, and t1 and t2 are the times in hours at two different sampling points. Note that in continuous culture systems, the dilution rate has to be taken into account for the determination of the specific growth rate.

Luan et al. (1987b) introduced the viability index ($V(t)$), which can be used for quantifying the cell viability in batch cultures:

$$V(t) = \int_0^t X_v dt, \qquad (16.3)$$

where X_v is the viable cell concentration in 10^5 cells/ml and t is the time in days.

The determination of viable and total cell number is described in sections 16.2.1.6 and 16.2.1.7, respectively.

The specific productivity is a very important calculated parameter and is expressed as, e.g., pg mAb $-$ Ig/cell $\times$ h:

$$\text{Specific productivity} = \frac{(c2 - c1)(\ln \times 2 - \ln \times 1)}{(t2 - t1)(\times 2 - \times 1)} \qquad (16.4)$$

$$\text{or simpler:} \qquad \frac{2 (c2 - c1)}{(t2 - t1)(\times 2 + \times 1)} \qquad (16.5)$$

where $\times 2$ is the number of living cells per milliliter at time 2, $\times 1$ is the number of living cells at time 1, t2 is the time in hours at sampling point 2, and t1 is the time in hours at sampling point 1, c2 is the antibody concentration of supernatant at time 2 in picograms per milliliter, and c1 is the antibody concentration of the supernatant at time 1 in picograms per milliliter. The simpler formula (equation 16.5) should be used in those cases when $\times 2 = \times 1$, because the specific productivity automatically becomes 0 when equation 16.4 is used for the calculation of the specific productivity.

The values for the living cell concentration and the product concentration have to be determined at two different sampling points, as detailed in sections 16.2.1.8. and 16.2.1.10, respectively.

Using a continuous and/or a perfusion system, the dilution or perfusion rate (D) is an important parameter, which is calculated as:

$$D = \frac{f}{V} \text{ (per hour)} \tag{16.6}$$

where f is the rate of addition of new medium in liters per hour and V is the reactor volume in liters. f is determined by mass flow meters, which is described in sections 16.1.1.9 and 16.2.1.9.

Another group of calculated substances are the specific intracellular concentrations of distinct substances. For example, one can use the specific ATP content as a parameter for the growth rate (see also section 16.2.1.6) and the specific DNA content as a parameter for the S-phase distribution, which again gives some information on the growth of the cell population (see also section 16.2.1.6 and Klöppinger et al. 1989).

16.1.2 Optional Controls
There are some additional control possibilities that—according to our opinion or experience—are not as important as the points above.

16.1.2.1 pCO$_2$ This parameter can be measured very well, but the high CO$_2$ content of the carbonate-buffered media varies strongly in dependence on aeration, pH changes, and feeding rates. Therefore, we do not see a correlation with cell metabolism.

16.1.2.2 Redox Potential Even though there is no problem with measurement, there is strong interaction of redox potential with oxygen sparging. When using a very low oxygen tension, this parameter may be useful for control of aeration. However, we do not see any application of a microaerophilic operation range in animal cell culture fermentations. Other applications of redox-measurements are possible (see section 16.2.1.7).

16.1.2.3 Osmolarity Osmolarity becomes increasingly critical with high-efficiency fermentation systems because of its rise during the course of fermentation. Easily determined by an off-line device, one should investigate its importance and monitor osmolarity carefully on a case-by-case basis (Öyaas et al. 1989).

16.2 STATE OF THE ART

16.2.1.1 Temperature Commercial control loops are well-standardized systems and are readily supplied by all fermentor manufacturers. Usually, there is a working sensor of a Pt-100 type and a safety overtemperature switch of another type. Many manufacturers offer an additional temperature control of the heating water circuit to avoid local overtemperatures, particularly during the period of heating up.

16.2.1.2   pH   Commercial, combined glass electrodes in combination with pressurized housings are a widely used, standard set up. The operation time can last up to several months until replacement is necessary. However, additional safety measures can be gained by the use of housings to allow the change of probes under sterile conditions. (Ingold). A better cell-culture-specific solution for measurement may be necessary for controlling pH in carbonate buffered media. Recently, Geahel et al. (1989) proposed the use spectrocolorimetric measurements of the pH-dependent color change of the pH indicator, phenol red, which is a very rapid (i.e., instantaneous) and precise method (accuracy of 0.01 pH units). This method can be used in the pH range from 6.4 to 8 and does not depend on the cellular density. Another method to overcome stability problems of pH electrodes was demonstrated by Junker et al. (1988). They added pH sensitive fluorophors to yeast cultures and monitored pH development during these cultures by measuring the fluorescence intensities at 405 nm, which is proportional to the nondissociated fluorophor, and at 460 nm, which is proportional to the dissociated form of the fluorophor; in their case: 8-hydroxy,1,3,6, pyrene trisulfonic acid trisodium salt. The log of this ratio is proportional to the pH of the solution. However, this system was only used for yeast cultures; during higher cell densities there was a constant sensing error of 0.2 pH units. Also, the fluorophors may be toxic to animal cells. Both propositions may hamper the possible growth of cells on the optical windows and may therefore change the signals considerably.

The traditional addition of acid and base may result in a change of osmolarity or Na/K ratio. A useful and gentle alternative, which is adequate in virtually all cases where well balanced media are used, is the control by CO_2 and air, respectively. The gasses are introduced by sparging or headspace aeration, using similar control loops as that used for oxygenation. Frequently, strong acidification is caused by excessive, glucose-dependent lactate generation (see also section 16.1.1.10).

16.2.1.3   Agitation   The use of slow-motion stirring systems needs a higher gear ratio for microbial fermentors to ensure a powerful and uniform movement within the operation range, usually 20–200 rpm. Since the operation speed usually approaches the limit for damaging the cells, this is a particularly important measure. For this reason, an accurate control and recording of the stirring speed is advisable for avoiding speed oscillations. A widely used setup is a variable speed drive that is controlled by an electronic setpoint controller in combination with a tachogenerator mounted to the drive.

The agitation by a fixed-speed drive is a very useful and cheap alternative but requires knowledge of the optimal speed for the particular process as a prerequisite.

16.2.1.4 pO$_2$ Polarometric and amperometric electrodes (Clark-type) show a similar feasibility and operation time as do pH electrodes, which is in the range of months or weeks (W. Scheirer, unpublished results). They also can be used in combination with housings, which allow a sterile exchange. However, there is some trouble with exact recalibration of ready-mounted electrodes after sterilization and during operation. However, even if recalibration cannot be done exactly, it can be done with sufficient accuracy for practical work by sampling and off-line comparison. These electrodes have a 90% response in the range of 1 min and requires sophisticated electronic controllers and gassing systems to keep the oscillation of the actual oxygen tension in an acceptable range.

Similar to monitoring the pH, Junker et al. (1988) proposed the use of fluorophors (pyrene butyric acid) for measuring pO$_2$ in the bioreactor. The same problems may arise here as they did for the fluorophor-based pH measurements (see section 16.2.1.2). In addition, their actual method of operation also depends, to some extent, on the optimal character of the culture suspension. Actually, the effect of the growing biomass on the response is not characterized properly and is therefore a source for error. The advantage is the faster response versus galvanic dissolved oxygen electrodes.

For high density cultures, the use of multiple high efficiency oxygenation systems may be necessary. Some of these have been described in the literature (Lehmann et al. 1988; Katinger 1988). In any case, even with small reactors, some more electrodes should be mounted for recording the oxygen tension in different reactor zones.

16.2.1.5 Liquid Level   Measuring the liquid level by electronic means has been difficult because of the sensitivity of the probes of conductive- and capacitive-type against foaming, leveling, and other fermentor installations. An alternative is the use of electronic balances, which are included in the fermentor base. These are very accurate and reliable, but in smaller scale fermentors there is a marked influence from forces introduced by piping and by changes in the fitting installations. Three alternative principles recently arose to alleviate this problem.

1. The first is a ultrasonic device, called X'SONAR. It is mounted to the vessel bottom and gives the liquid height by sounding the liquid surface (Moore).

2. The second is the measurement of differential pressure between the headspace and the bottom of the fermentor by means of very sensitive piezoelectric sensors (Hottinger-Baldwin Messtechnik, Vienna, Austria).
3. A third device, based on electrical admittance, has been reported by Anderson et al. (1985).

All three principles are quite new, but future experience will prove the usefulness of those devices.

16.2.1.6 Cell Density During the fermentation process, the determination of the viable and total cell count is one of the most important measurements; it is done discretely by counting the trypan-blue stained and unstained cell suspensions in a haemocytometer. This method cannot easily be automated. Therefore, other methods must be utilized: i.e., physical and chemical. By comparing both possibilities, the use of microscopic and chemical methods have disadvantages, such as the need for taking samples, the addition of reagents, off-line rather than real time character, and the discontinuous mode of determination. In comparison, physical methods show advantages in that they operate in real time, in situ, and they are non-destructive.

Two chemical methods will be briefly mentioned here. First, the determination of ATP by bioluminescence assays; second, the determination of double-stranded DNA by mithramycin staining and measurement of the fluorescence. Details are shown in Table 16–1. It should be noted that the NADH content of the cells can be measured by luminescence measurements (for details see Girotti et al. 1984), but problems arising from sampling, etc., are similar to those for measurement of ATP.

In summary, the determination of the cellular ATP content is possible to obtain automatically, but the problem of the influence of the physiological state remains (unpublished results). Despite this drawback, the determination of the ATP content and of the NAD/NADH relation in the cells (by using fluorescence probes, see below) provides information about the physiological state. The optimal production of a product is linked to a certain stage of the cell physiology and to the cell growth (e.g., monoclonal antibody production is often dependent of the growth stage of the culture). Using continuous cultivation (low and high cell density systems), information about the physiological state of the culture is necessary for optimal control. Unfortunately, on-line determination of ATP has been used only in the case of yeast fermentations (Siro et al. 1982), not for control of animal cell fermentations.

The determination of the cellular DNA content is an exact method because of the stable content of DNA in cells. The major problem in automation is the necessity for cell separation, due to the influence of the medium, color, and sonication. However, these problems should be solved

TABLE 16–1   Details of the Measurements of ATP and DNA for the Chemical Determination of Biomass

	Determination of ATP	Determination of DNA
Principle	ATP^1 + luciferin + O_2 − oxyluciferin + AMP + PP_i + CO_2 + *light*	*DNA* + mithramycin − interaction of DNA with mithramycin − *fluorescence*
Separation		Necessary, if phenol red containing media are used
Cells	Not necessary	
Medium		
Extraction	Necessary	Necessary
Addition of reagents	1. Extraction buffer	1. Mithramycin-solution
	2. Luciferin/luciferase	
	3. Internal standard: option	
Range	1×10^5–2×10^6 cells/ml	1×10^5–1.5×10^6 cells/ml
Problems	Variation of cellular ATP content with the change of the physiological conditions	Centrifugation is necessary if phenol red-containing media are used, sonication for cell disruption is necessary
Application	Only in well-known standard processes possible because of the influence of the physiological state	Possible after solution of the problems
References	Chapman et al. 1971 Siro et al. 1982; automation	Hill and Whatley 1975 Himmler et al. 1985

[1]The italicized analytes (ATP, DNA) are measured by the underlined parameters (light, fluorescence).

by the use of phenol red-free media and the use of DNA extracting reagents. The advantage of the systems is the determination of the viable cell count. As shown in sections 16.1.1.12 and 16.2.1.11 in greater detail, the use of the consumption rate of nutrients such as glucose or glutamine and/or the production of metabolites such as lactate or alanine, for instance, can be used to estimate the viable cell count as well as the other chemical methods. However, these estimations are again dependent on the physiological state of the cells.

Many different physical methods have been described for the determination of the viable or total microbial biomass, however, only a few systems were also used for animal cell fermentation. Therefore, only the latter detection systems will be discussed here.

Many papers have been published on the use of fluorescence sensors for the determination of intracellular NADH, mainly in microbial fermentations (Beyeler et al. 1981; Meyer and Beyeler 1984; Scheper et al. 1984; Luong and Carrier 1986). Although the NADH content correlates well with the cell density during the lag and exponential growth phase of a batch culture, the intracellular NADH content rises during the stationary phase, indicating changes in the physiology of the microorganisms (Beyeler et al. 1981; Luong and Carrier 1986). This was found to be valid for animal cells too (Leist et al. 1986). The advantage is that the NADH content of the cells reacts immediately to changes in the medium, which was shown by Meyer and Beyeler (1984) in a continuous culture of yeast. This fast reaction of the cells, which is superior to the classical control by volumetric oxygen uptake rate (OUR) and RQ, allows faster and, therefore, more effective control and regulation of the fermentation process (Meyer and Beyeler 1984). Leist et al. (1986) used an on-line fluorometer for the control of a fermentation of Bowes 4 melanoma cells. The response correlated with the cell number, but was not directly caused by the cells. The authors suggested that the application of a fluorometer probe might be useful, but that the results would have to be carefully analyzed.

Other physical methods that have been used for animal cell fermentations are nephelometry, infrared nephelometry, turbidimetry, electronic counting based on the Coulter Counter, and acoustic resonance densitometry. The principles are shown in Table 16–2. The photometric methods and acoustic resonance densitometry give only a determination of the total cell biomass; the Coulter Counter-based method gives the viable cell number and can be used for determining the cell concentrations of single cell suspensions. However, the latter has the problem of the cell diameter being a variable that depends on the physiological state of the culture (unpublished results). In addition the Coulter-Counter based method can be used for determining the cell density of microcarrier cultures with the advantage that the cell density and the presence of mono- or multilayers of cells on the microcarrier can be detected (Miller et al. 1986).

The main disadvantages of all photometric systems is the fouling of the optical surface. Certain improvements have been envisaged by the use of fibre-optic techniques and/or laser optics (Jeannesson et al. 1983). In addition, environmental light and air bubbles have an influence on the output signal.

The main advantages of the acoustic resonance densitometry are the high response stability of the system, the independence from cell size, flow rate, and viscosity of the fluid. The use of cross-flow devices provides on-line filtration, which is necessary for compensating the influences of the supernatant on the response of the device. The main disadvantages are the possible filtration problems, the relative insensitivity (usable starting with 10^6 cells/ml), the dependence of the signal on pH, medium conditions in

TABLE 16–2 Comparison of the Different Photometric Methods for the Determination of Biomass (Cell Count)

Principle	Advantages
Nephelometry: Straight-forward measurements of light scattered by the cells in a suspension give signals, which are directly proportional to the cell count	Background automatically 0, linear correlation between the signal and the cell count; detection limit: 1.6×10^5 cells/ml (*E. coli*)
Infrared Nephelometry: Light: 900 nm	Linear correlation between the signal and the cell count; detection limit: 10^5 cells/ml (hybridomas)
Turbidimetry: Measurement of the transmitted light or the optical density	Rapid and easy measurements
Electronic Counting Device: The electronic counting technique monitors the effect on cells on an electric field as the cells traverse the field. The cells are suspended in the growth medium, which is electronically conductive and have to flow through a small aperture across which an electric field is applied via a constant current source. When the relatively nonconducting cells pass through the field, the electrical resistance within the aperture increases, giving rise to a transiently increased voltage drop across the aperture. Under certain conditions this magnitude is proportional to the size of the cells. The pulses per time correspond with the cell number per milliliter.	Differentiation between living and dead cells, possible counting of adherent cells directly on microcarriers. It is possible to measure particles in the range of 0.4–800 μm (Coulter Electronics Inc.)
Acoustic Resonance Densitometry: The amplifier, electromagnets, and sample test cell constitute a closed oscillatory circuit whose frequency of oscillation depends on the mass of the flow cell; any change in the density of the contents of the test cell causes a change in the resonant frequency of the system.	Non-invasive, as a flow cell, independent from cell size, viscosity and flow rate of the fluid, tremendous stability, the use of cross flow devices reduce clogging problems caused by the filtrated cells.

Disadvantages	On-line	References
No discrimination between living and dead cells; temperature control necessary; long use in fermentors is impossible because cells can grow on the optical surface	Yes	Koch 1961 Mallette 1969 Harris and Kell 1985
Like normal nephelometry; additionally: the influence of the environmental light has to be constant	Yes	Merten et al. 1987
Gas bubbles may disturb the measurements; the cells can grow on the optical surface; no discrimination between living and dead cells; detection limit: 2.4×10^6 cells/ml (*E. coli*)	Yes	Mallette 1969
The sample signal has to be much greater than the background signal. Unusually large pulses have to be minimized. The cell concentrations increase the occurrence of coincident counts; low cell concentrations decrease the signal-to-noise ratio. The path of the cells through the field should be uniform. The magnitude of the applied field must be below a critical value; above this value dielectric breakdown occurs. The on-line application of this system might be disturbed by the time-dependent changes in the medium conductivity. Air bubbles, cell debris, and cell aggregates may have some influence on the signal. For suspension cells, the cellular diameter depends on the physiological state of the cells.	Yes	Matsushita et al. 1982 Harris and Kell 1985 Miller et al. 1986 Thebline et al. 1987
A tangential filtration unit is necessary for using this system in order to compensate for influences of the medium by making differential measurements of cell suspension and culture supernatant. Influences from air bubbles, differences in pH, temperature, and other medium conditions are relatively insensitive. Starting point: 10^6 cells/ml.	Yes	Blake-Coleman et al. 1984 Kilburn et al. 1989

general, temperature, and air bubbles (D.G. Kilburn, personal communication).

The electrical counting may be hampered by foreign particles, which does not pose large problems because all cell culture media are filtered. Cell debris, cell aggregates, and air bubbles, however, have some influence on the results.

A photometric method (infrared nephelometry) was used by Merten et al. (1987) for the control of a repeated batch culture of a hybridoma cell line, which showed good correlation between sensor output and total cell count when viability did not change too much. Thebline et al. (1987) successfully used the Coulter-Counter based in situ cell number estimation device for monitoring the cell density of HTC, 3T3 NIH, and HeLa 229 microcarrier batch cultures.

Kilburn et al. (1989) successfully used the acoustic resonance densitometry for monitoring the cellular density of continuous cultures of a hybridoma cell line and of U937 in an on-line mode.

The principles, advantages, and disadvantages of these methods are shown in Table 16–2. Other physical methods, such as dielectric monitoring (Blake-Coleman et al. 1984), and electrochemically based systems (Matsunaga et al. 1979 and 1980) have not yet been used for animal cell fermentations, but they could be adapted for this application.

16.2.1.7 Cell Viability   Cell viability is directly related to cell density and should not be regarded as a separate parameter. Because it is difficult to monitor viable and total cell count together, in many cases, it is more interesting and sufficient if the viable cell count only is monitored. All methods can be used to estimate the viable cell count (see also section 16.2.1.6). In addition, parameters that are dependent on the physiological state of the cells, such as ATP content, NADH content, consumption of nutrients, or production of metabolites, can equally be used, particularly in cases when the right time of harvest, the overall nutritional state of a continuous culture, or the right time for infecting a culture with viruses for virus-production, etc., are important. In this instance, the use of redox probes should be mentioned, which can be used for following the physiological state of batch cultures (Griffiths 1984).

16.2.1.8 Liquid Mass Flow   Getting a uniform flow in the correct range for pilot scale cell culture fermentors requires a setup that was not obtainable as a complete unit up to this point. The devices available are either not stable and accurate enough or cannot be sterilized in situ. We tried to use two new assemblies with good success, both of which are made up by combination of the following custom-made devices.

1. One system consists of a peristaltic pump equipped with MapreneR-tubing and remote control, which is accomplished by a set-point operated proportional controller. The controller receives an input signal that corresponds to the actual flow from a thermoelectric flowmeter (Fluid Components Inc., San Marcos, CA, USA) mounted within the medium inlet pipe. The real flow can therefore be monitored and controlled with a precision that is approximately 2% independent from any environmental changes. Additional recording of flow and alarm settings are easily performed. This system is very useful and reliable, but has some problems when operating under pressure because of the very limited pressure range applicable for pump tubings.
2. The second system is a hydraulically operated membrane metering pump that can be sterilized in situ (Burdosa 6305 Buseck, FRG) in combination with a pulse driving unit. The use of this driving unit allows the operator to reduce the flow rate to the correct range with an acceptable pulsation (about one pulse per minute), and can be controlled remotely. This system is stable up to 6 bars of differential pressure, but has no simultaneous monitoring of actual flow. To add this element, the measuring device described immediately above (number 1) can be used.
3. A third possibility is the use of a combination of an electronic balance and pumps/valves, which are interfaced with a computer that controls the exact weight of feeding medium in appropriate portions.

16.2.1.9 Vessel Pressure   There are many pressure sensors based on various principles. They can be obtained as standard equipment from fermentor manufacturers and other suppliers.

16.2.1.10 Biochemical Parameters   Traditional methods of monitoring bioprocesses rely either on removing samples from the bioreactor with subsequent analytical chemistry or indirect monitoring by sampling the gas phase. However, with the development of biosensor technology, more direct methods are available for monitoring a bioprocess. Following the definition of Gronow et al. (1985):

> A biosensor is an analytical tool or system consisting of an immobilized biological material (e.g.: an enzyme, antibody, whole cell, organelle, or combinations thereof) in intimate contact with a suitable transducer device which will convert the biochemical signal into a quantifiable electrical signal.

This signal can be electrically amplified, stored, and subsequently displayed. The principle construction of a biosensor is shown in Fig. 16–1. The transducer element with the biocatalyst usually contains the chemistry that provides the selectivity of the device. The biochemical reaction system is

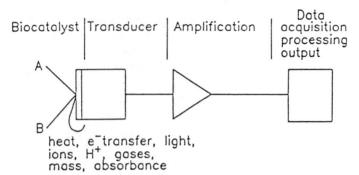

FIGURE 16-1 Schematic of a generalized biosensor. The biocatalyst converts substrate A into product B with a concurrent change in a physicochemical parameter, which is converted into an electrical signal by the transducer, amplified, and suitably processed and outputted. Reproduced with permission from Lowe (1984).

chosen to generate a readily detectable species from the specific analyte of interest. Depending on the generated species, different types of transducers can be employed. Table 16–3 lists typical transducers that have been exploited. Generally, they can be categorized into the following groups: potentiometric, amperometric, optical, calorimetric, conductimetric, and gravimetric. For more details concerning the transducer constructions, etc., the reader is referred to Guilbault (1984), Clarke et al. (1984), and Merten (1988). Finally it should be mentioned that in some cases the enzyme does not have to be coupled directly to the transducer when flow injection analysis (FIA) systems are employed (e.g., enzyme reactor with subsequent detector, see Fig. 16–4D).

Unfortunately, the application of biosensors for the control of bioreactors is difficult due to the following drawbacks. Generally, the biosensors are not steam sterilizable or autoclavable; their working lifetime is limited, which depends mainly on the biocatalyst used, the immobilization method, and the quantity and the purity of the immobilized biocatalyst (Guilbault 1984), and can be influenced by autoinactivation, as is known for glucose oxidase (Bourdillon et al. 1982 and 1985) and inactivation by H_2O_2, which is a product of oxidases (Bourdillon et al. 1985).

Because of the limited lifetime of the sensors and fouling, clogging, and poisoning of the membranes, which is used for immobilizing the biocatalyst or for separating the biosensor from the bioreactor (i.e., sterile barrier), drifting phenomena can arise, which have to be controlled and corrected and, in some cases, may rapidly disturb a biosensor application. Therefore, the possibility for recalibration should be available. The response characteristics should be fast in cases where a rapid change for a parameter is expected. In addition, the sensor should work in real time or nearly real time in order to increase the possibility of using the output directly or regulating important bioreactor parameters. Finally, the sensor has to pro-

TABLE 16-3   Biosensor Transducers, Operation Modes, and Applications

Measurement Mode	Transducer System	Species Detected	Typical Applications
Potentiometric	Ion-selective electrode (ISE)	H^+, K^+, Na^+, NH_4^+, Ca^{2+}, Li^+, J^-, CN^-	Ions in biological media, enzyme electrodes, enzyme immunosensors
	Gas-sensing electrodes	NH_3, CO_2	Gases, enzymes, organelles, cell or tissue electrodes for substrates and inhibitors, enzyme immunoelectrodes
	Field effect transistor (FET)	H^+, H_2, NH_3	Ions, gases, enzyme substrates, and immunological analytes
Amperometric	Enzyme electrodes	O_2, H_2O_2, J_2, NADH, mediators	Enzyme substrates and immunological systems
Optical	Photodiode (optoelectronic, fiber optic, and wave guide devices in conjunction with a light-emitting diode)	Light absorption, fluorescence	pH, enzyme substrates, immunological analytes
	Photomultiplier (in conjunction with fiber optic)	Light emission, bio- and chemi-luminescence	Enzyme substrates
Calorimetric	Thermistor	Heat of reaction	Enzymes, organelles, whole cell or tissue sensors for substrates, products and inhibitors, gases, pollutants, antibiotics, vitamins, etc., immunological analytes
Conductance	Conductimeter	Increase of solution conductance	Enzyme substrates
Mass change	Piezoelectric crystals	Mass absorbed	Volatile gases, vapors, immunological analytes

Reproduced with permission from Merten (1988).

vide the correct range of sensitivity in order to monitor the whole concentration range of the analyte of interest.

Unfortunately, these drawbacks and requirements prevent the direct use of biosensors in a fermentor. Therefore, sampling and filtration systems (especially designed as stick-in probes) have to be employed in order to use the biosensor without steam sterilization by maintaining the sterile barrier of the bioreactor. These devices also provide the possibility to recalibrate the sensor, to change the sensor when it fails, and to adapt the working range if necessary. In addition, these sampling devices increase the lifetime of the biosensor because of sample filtration for providing a sterile barrier in the bioreactor, and because of the sample dilution for adapting the detection range of the sensor to the concentration range of the analyte (very often a 10- to 1,000-fold dilution is necessary). These two features increase the sensor's lifetime because fouling and clogging of the sensor by cell debris and cells is avoided by filtration, and poisoning and fouling of the sensor is reduced by diluting the substances that cause these negative effects. For more details, the reader is referred to Clarke et al. (1985).

The following sampling and filtration devices have been used for microbial fermentations and most can be applied to animal cell fermentations. (In the case of high density culture systems where the cells are immobilized and separated from the supernatant, these sampling devices only have to provide the sterile barrier. Problems with respect to fouling and clogging caused by cell debris and cells are not present. However, real on-line detection of biochemical parameters in animal cell fermentations has not yet been done or published.) In general, two different types of filtration systems can be distinguished: the invasive (Fig 16–2A) and the external systems (Fig. 16–2B) (Table 16–4). The invasive system is characterized by the use of a bioreactor-internal filtration or dialysis system, which may be static or agitated. Although directly agitated (Madenius et al. 1984) or rotated (Tolbert et al. 1981; Tolbert and Feder 1983) systems are superior to indirectly agitated or static systems, where a high liquid velocity surrounds the device (Zabriskie and Humphrey 1978; Cleland and Enfors 1984a and 1984b) and the devices may or may not be backflushed periodically (Schmidt et al. 1984), both systems can become fouled or clogged sooner or later. Therefore, invasive systems are useful for only a limited time because they are not changeable. Their advantages are that the cell separation is done in situ, wherein the cells do not have to be taken out of the system as is done in external systems. Also, the cellular physiology does not change, which can be a problem in external systems when the cells are cycled through a sampling loop. The following systems have already been used for microbial systems: dialysis systems and/or units (Zabriskie and Humphrey 1978; Mandenius et al. 1984a); dialysis systems with incorporated biosensors (Cleland and Enfors 1984a and 1984b; Brooks et al. 1987/88), which is shown as schematic in Fig. 16–3; flat membrane devices equipped with a stirrer (Schmidt et al. 1984); and radial flow devices (see Figure 16–2A), which

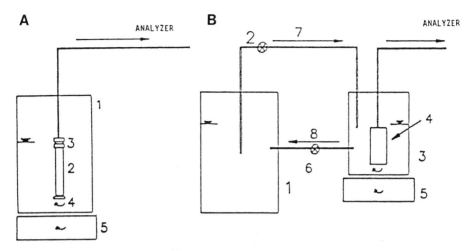

FIGURE 16–2 Possible filtration systems for automatic cell-free sampling. (A) Invasive system: (1) fermentor, (2) rotating porcelain filter (1 μm), (3) rotating seal, (4) magnetic bar, (5) magnetic stirrer. (B) External system: (1) fermentor, (2, 6) pump, (3) filtration unit, (4) rotating filter (1 μm), (5) magnetic stirrer, (7) cell suspension, (8) cell suspension minus cell-free sample.

TABLE 16–4 Advantages and Disadvantages of Invasive versus External Filtration or Sampling Systems

Invasive Systems	*External Systems*
Advantages	
Cell separation in situ	Changeable
No changes in the cell physiology	
Disadvantages	
Limited use, when fouling or clogging	Cells are recirculated
Not changeable	Cell physiology may be changed
	Pumps may be disastrous
Systems Available	
Dialysis systems/units	Hollow fiber systems
Dialysis systems with incorporated bio-sensor	Amicon filtration cell
	German Acroflux cell
Radial flow ceramic filter devices	Millipore Pellicon
	Radiale flow ceramic filter modules
	Rotating filter devices

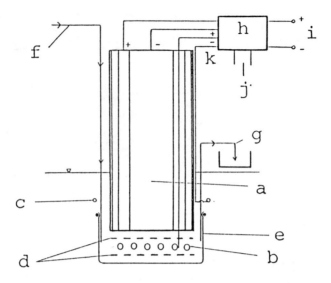

FIGURE 16-3 Main parts of an externally buffered enzyme electrode: (a) oxygen electrode, (b) Pt gauze with immobilized enzymes, (c) Pt coil (cathode), (d) nylon nets, (e) dialysis membrane, (f) in-going buffer stream, (g) buffer effluent, (h) PID controller, (i) reference potential, (j) connection to recorder, (k) electrolysis current. Reproduced with permission from Cleland and Enfors (1984a).

have been used in homogeneous perfusion systems for retaining the biomass within the bioreactor. Due to the disadvantages of invasive systems, external systems were mainly used. The cell suspension or the supernatant is cycled via an external filtration/sampling device and a certain quantity is taken out of the system and transferred to the analyzer or sensor. Of course, the sampling device can be placed in the outflow stream of a continuous or perfusion culture system, where the cycling of supernatant or cell suspension can be avoided. The main disadvantage is that the cells have to be recirculated, which can change the cellular physiology and may be disastrous for the cells when the wrong pumps are used. This can be avoided if the sampling device is placed in the outflow stream. The advantage is that the filtration system is changeable in a sterile manner when it is clogged or fouled. The following systems have already been used: nonagitated, such as hollow fiber systems (McLaughlin et al. 1985), filtration cells (e.g., Amicon, which is not steam sterilizable) (Chotani and Constantinides 1982; Kroner and Kula 1984), Gelman Acroflux filtration cells (not steam sterilizable) (Dinwoodie and Mehnert 1985), and agitation, such as radial flow ceramic filtration (see Figure 16–2B) (Tobert et al. 1981; Tobert and Feder 1983) and rotating filter devices (Rebsamen et al. 1987). The first is used in high cell density perfusion systems for changing the medium while the last example is used as a filtration system for cell suspensions. A special construction was published by Ghoul et al. (1986), in which the filter was changed after each use. In addition, the flow line between the bioreactor and the filtration unit was steam sterilized after each sampling. A minor disadvantage was the loss of cells, which might be a problem in small-scale laboratory bioreactors.

Connections of biosensors to the fermentor and applications have been used. Because the biosensor is not steam sterilizable, has to be recalibrated, has to be changed if it fails, and the concentration range of the analyte must be adapted to the working range of the sensor, the following five constructions (Fig. 16–4A through 16–4E) are recommended. Cell flow line systems are used for monitoring microbial fermentations and can be used for animal cell fermentations with little difficulty. All systems, with the exception of the dialysis system (see Figure 16–4A), should be equipped with a filtration device because of the reasons mentioned above. The dialysis system with subsequent sensors or with an integrated biosensor (Cleland and Enfors 1984a, 1984b; Brooks et al. 1987/88) have the following features. The analyte diffuses through the dialysis membrane. Depending on the surface and the permeability of this membrane, the flow rate of the dialysis buffer and the volumetric ratio of dialysis buffer to the bioreactor volume, the analyte is diluted in the flow stream to a higher or lower degree. Such constructions were used by Clarke et al. (1982 and 1984), Mandenius et al. (1984b), and Zabriskie and Humphrey (1978).

A somewhat different diffusion system was proposed by Cleland and Enfors (1984a and 1984b). Here the biosensor was directly inserted into the diffusion system. The whole device has a similar conception to the Ingold pCO_2 electrode (see Figure 16–3). The detection range of the sensor can be changed depending on the dialysis buffer flow. Figure 16–3 shows a sensor that was used as a glucose probe, based on the use of glucose oxidase. In order to avoid dependence on oxygen, O_2 is reproduced by using catalase and electrolysis of water. This stick-in probe is steam sterilizable and the enzyme is added after the sterilization. The dialysis membrane functions as a sterile barrier. A description of a similar device was published by Brooks et al. (1987/88); however, the sensor was based on a glucose-oxidase-ferrocene electrode, which was almost independent of oxygen.

The second, very general system (see Figure 16–4B) is based on FIA. Using an injection valve, a certain quantity of the supernatant is injected into the buffer flow of the sensor. By changing the injection volume, the concentration range of the analyte can be adapted to the detection range of the sensor. Two somewhat similar systems are shown in Figures 16–4D and 16–4E, using an enzyme reactor with a subsequent detector (e.g., Holst et al. 1988; Nopper and Wichmann 1988) or a separation system (HPLC, gas chromatography [GC]) and a subsequent detection system (I. Rousseau, personal communication; McLaughlin et al. 1985; Mathers et al. 1986). Like the FIA system, the detection range can be adapted by changing the injection volume.

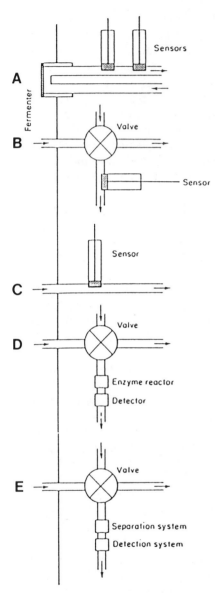

FIGURE 16–4 Current possible connections of sensing devices to the fermentor. All connections, based on flow systems, are shown without filtration devices. (A) In situ dialysis system where the sensors are fixed in the downstream. (B) Semicontinuous sampling by flow injection analysis principles (Ruzicka and Hansen 1980). (C) Continuous flow-line sampling. (D) Flow system with the use of an enzyme reactor and a detector (e.g. Appelqvist et al. 1985), with only a filtration system applicable. (E) Flow system equipped with a separation system (HPLC, GC) and detector, with only a filtration system applicable.

The simplest construction is shown in Fig. 16–4C. The sensor is placed in the out-flow of the bioreactor. No possibilities exist for adapting the working range of this device.

Some applications are already used for bioreactor monitoring. Published applications of the on-line monitoring and control of microbial fermentations are shown in Table 16–5. With respect to important parameters for animal cell culture, only glucose, sucrose, lactate, isoleucine, (hydroxy bu-tyric acid), NH_4^+, and phosphate are shown. Of course, there are many other parameters such as ethanol concentration (Mattiasson et al. 1981; Mandenius et al. 1987) in the broth or the off-gas analysis, which are significant in microbial fermentations. With the exception of two publications, one by Nopper and Wichmann (1988), which showed the on-line monitoring of glucose, lactate, and isoleucine with a utilization time of the sensor of up to 100 days, and the other by Ghoul et al. (1986), who used an on-line chemical method that has no limitations in lifetime, all other biosensors have been used for short-term fermentations, which lasted only a few hours (Mattiasson et al. 1981; Cleland and Enfors 1984a; Mizutani et al. 1987; Holst et al. 1988) and a few days (Mandenius et al. 1987; Mizutani et al. 1987; Brooks et al. 1987/88). In general, batch or fed batch fermentations are used for microbial cultures. Because of the relative short generation time of microbes (as short as 20 min), fermentations last a relatively short time. However, in the case of animal cell cultures, culture duration can be extended to several months when continuous or perfusion culture systems are used. For these types of cultures, biosensors with high stability and high lifetime have to be available. Therefore, biosensors have to be constructed so that they can be recalibrated and changed (see above). In addition, they should be used in order to increase their lifetime. Certainly, it is easier and better for the working lifetime of the sensor to use FIA systems where a sample is injected in an on-line mode at intervals (discrete mode) (e.g., Nopper and Wichmann 1988) than it is to use sensors in a continuous on-line mode, which is performed without interruption. This was done by Mattiasson et al. (1981) and Mandenius et al. (1987). The concentration range of the published applications does not pose problems because this can be changed by changing the injection volume, the parameters of the dialysis system or the dilution system. Parker et al. (1986) published a commercial on-line filtration and sampling system that is based on FIA. The Yellow Spring Instruments glucoseanalyzer is used as one channel for the glucose detection, for example. As a second channel, a colorimetric/diffusion procedure for monitoring phosphate or NH_4^+ can be used. The glucose electrode remains viable for about four weeks and can be changed in four minutes. As for most of the applications shown in Table 16–4, this apparatus can be recalibrated automatically.

Sensors of the determination of other medium compounds, such as galactose, lactate, pyruvate, NH_4^+, and amino acids will not be discussed in detail, since the detection principles are often similar to those mentioned above. For more information the publications of Guilbault (1982 and 1984), Aston and Turner (1984), Karube and Suzuki (1984), and Merten et al. (1986) are recommended. Nevertheless, some applications are shown in detail in Table 16–6, which seem to be the most suitable for application in animal cell cultivation with respect to lifetime, influences from other compounds, or detection range.

TABLE 16-5 On-Line Biosensor Applications Used for Monitoring and Controlling Microbial Fermentations

Application	Organism	Analyte	Biosensor	Filtration/ Sampling System	Utilization Time	Response Time	Concentration Range	Comments	References
Regulation of sugar concentration for ethanol production	Saccharomyces cerevisiae	Sucrose	Thermistor-invertase	On-line in a continuous mode, cells were immobilized—no filtration necessary	5 hours	—	2–100 mmol	Long-time use possible?	Mattiasson et al. 1981
					35–40 hours	—			Mandenius et al. 1987
Regulation of sugar concentration for ethanol production	Saccharomyces cerevisiae: fed batch	Glucose	Oxygen electrode, glucose oxidase	On-line in a discrete mode, external filtration, injection valve	10 hours	?	2–20 g/l	Drift: 1 g/l: 1 mV/h, 10 g/l: 3.4 mV/h. Fluctuations: 0.3 g/l: 0.08–0.54 g/l, 10 g/l: 10%	Mizutani et al. 1987
	Micrococcus ruteus fed batch	Glucose			60 hours	?	0–4 g/l		
Regulation of sugar concentration for galactosidase production	E. coli	Glucose	As for Mizutani et al. 1987		?	?	0–15 g/l		Kobayashi et al. 1987
Monitoring	?	Isoleucine[1]	Enzyme reactor with subsequent fluorometer	FIA with external filtration unit, injection valve, enzyme reactor, and detector	100 days	25[2] min	1–2 decades (linear, ± 3%)	Interferences: leu, val	Nopper and Wichmann 1988
		Glucose[3]			100 days	25[2] min	1–2 decades (linear, ± 3%)		
		Lactate[4]			100 days	25[2] min	1–2 decades (linear, ± 3%)		
		Hydroxy butyric acid[5]			100 days	25[2] min	1–2 decades (linear, ± 3%)	Interferences: hydroxy acids	

Monitoring, controlling	Baker's yeast (batch, continuous SCP, production of methanol)	Glucose NH$_4$, up to 16 parameters	Microprocessor controlled autoanalyzer		—	?	?		Pécs et al. 1987
Monitoring, controlling	?	Glucose	Enzyme reactor?	FIA	?	?	2 decades		Garn et al. 1987
Monitoring	S. cerevisiae, E. coli	Glucose	Enzyme reactor, oxygen electrode	External dialysis system, commercial glucose analyzer	450 min	6 min	0–5 g/l (± 7%)	Clogging of dialysis membrane	Holst et al. 1988
Monitoring	E. coli	Glucose	Glucose-oxidase-ferrocene electrode	In situ autoclavable electrode; immobilized enzyme is applied after autoclavation; dialysis flow cell	Some days	1–2.5 min (95% over a 6 day period)	0–30 mM	Sensibility drift, change in baseline, correctable by predictive software	Brooks et al. 1987/88
Monitoring	E. coli	Glucose	Oxygen-electrode glucose-electrode	In situ autoclavable electrode; immobilized enzyme is applied after autoclavation; dialysis flow cell	5 h	5–10 min	up to 30 g/l	5% drift within 24 h	Cleland and Enfors 1984a
Monitoring, controlling	Yeast	Glucose	Autoanalyzer, equipped with an external filtration device based on neocuproin method		No limitation	7 min	0–5 g/l		Ghoul et al. 1986

[1]Leucine dehydrogenase.
[2]Time: includes filtration, calibration, several repeated injections, etc.; one sample alone takes 3 min.
[3]Glucose dehydrogenase.
[4]Lactate dehydrogenase.
[5]α-Hydroxyisocaproate dehydrogenase.

TABLE 16-6 Determination Methods for Different Medium Compounds

Analyte, Enzyme	Principle	Determined Parameter	Dependency of pH	Dependency of °C	Dependency of dO_2	Influences of Other Medium Compounds	Connection to the Fermentor (Figure)	Drift	Range mM/l (Linear range)	Lifetime	Reference
Galactose Galactose oxidase, EC1.1.3.9	Mediator-electrode (enzyme reactor)	Increasing H_2O_2, reduction of mediator[a]	7.0	Ambient	Yes	pO_2, H_2O_2, di-hydroxyacetone of the sample	FIA system (1 ml/min) (16-4D)	0.3%/day	0.002-60	>1 m[b]	Lundbäck and Olsson 1985
Lactate Lactate oxidase[i], EC1.1.3.2	O_2 electrode[i]	Decreasing O_2	5.5-7.5	?	Yes	pO_2 of the sample	Dialysis or flow system (16-3, 16-4A,B,C)	?	0.02-0.2	>1 m[b]	Mascini et al. 1985
Lactate oxidase, EC1.13.12.4	O_2-electrode[i]	Decreasing O_2	6.0	+10%/+3°C	Buffer contains enough O_2	pO_2 of the sample	FIA system (1 ml/min) (16-4B)	No	0.02-0.2	>1 m[b]	Mascini et al. 1984
Three enzymes²	Thermistor (enzyme reactor)	Increasing temperature	7.0	Tempered	O_2-saturated buffer	Solvation heat prevented by reference system or having no differences in sample and buffer matrix	FIA system (1 ml/min) (16-4D)	?	0.025-1 (WONA[d]) 10^{-5}- 10^{-1} (WND[e])	?	Scheller et al. 1985
Lactate dehydrogenase, EC 1.1.1.27	Mediator electrode (enzyme reactor)	Increasing NADH	?	?	—	?	FIA system (0.5 ml/min) buffer contains NAD (16-4D)	?	10^{-2}-1	300 days	Schelter-Graf et al. 1984
Lactate dehydrogenase, EC 1.1.1.27	Photometer (340 nm), soluble enzyme is added	Increasing NADH	?	?	—	Sample color	FIA system (2 ml/min)	—	0.2-2	—	Rydevik et al. 1982
Lactate dehydrogenase, EC 1.1.1.27	Pt-electrode[a]	Increasing $Fe(CN)_6^{4-}$	7.4	+3.5%/+°C	No	?	Dialysis or flow system (16-3, 16-4A,B,C)	?	0-1.5	6 w[f]	Racine et al. 1975

	Detector	Effect	pH	Temperated	O_2-saturated buffer	Interferences / notes	System	Stability	Range	Lifetime	Reference
Pyruvate Three enzymes[c]	Thermistor (enzyme reactor)	Increasing temperature	7.0	?		Solvation heat prevented by reference system or having no differences in sample and buffer matrix	FIA system (1 ml/min) buffer contains NAD (16–4D)	?	10^{-5}–10^{-1}	?	Scheller et al. 1985
Pyruvate oxidase, EC 1.2.3.3 *NH_b, NH_4^+*	O_2-electrode[i]	Decreasing O_2	7.05–7.35	?	Yes	pO_2 of the sample	Dialysis or flow system buffer, g	10 days stable	0.06–0.8	?	Mizutani et al. 1980
—	NH_3-gas electrode	pH has to be >10: $NH_4^+ + OH^- > NH_3 + H_2O$	—	—	No	Influences of the NH_4^+ of the medium	Dialysis or flow system (16–3, 16–4A,B,C)	Stable	0.005–10	?	com[h]
(Nonactin based)	Ion-selective electrode	NH_4^+	—	—	—	K^+: 15%, Na^+: 0.13%, influences of the NH_4^+ of the medium	Flow system (16–4A,B,C)	Yes	1–100	Some w[f]	Guilbault and Nagy 1973; Fogt et al. 1985
Alanine Alanine dehydrogenase, EC 1.4.1.1	Mediator electrode (enzyme reactor)	Increasing NADH	?	?	No	?	FIA system (0.5 ml/min) buffer contains NAD (16–4D)	?	0.1–1	120 days	Schelter-Graf et al. 1984
Asparagine Asparaginase, EC 3.5.1.1	NH_4^+ electrode[i]	Increasing NH_4^+	7.5–8.7 optimum	?	—	As for NH_4^+ electrode, NH_4^+ of the medium, compensated by reference system	Dialysis or flow system (16–3, 16–4A,B,C)	?	?	2–4 w[f]	Guilbault and Hrabankova 1971

(continued)

TABLE 16-6 (continued)

Analyte, Enzyme	Principle	Determined Parameter	Dependency of			Influences of Other Medium Compounds	Connection to the Fermentor (Figure)	Drift	Range mM/l (Linear range)	Lifetime	Reference
			pH	°C	dO_2						
Glutamate Glutamate dehydrogenase, EC 1.4.1.3	Mediator electrode (enzyme reactor)	Increasing NADH	?	?	—	?	FIA system (0.5 ml/min) buffer contains NAD (16-4D)	?	0.1-10	146 days	Schelter-Graf et al. 1984
Glutamine Glutaminase, EC 3.5.1.2. + glutamate oxidase	O_2 electrode	Decreasing	Up to 6.5	?	Air wash between each sample	pO_2 of the sample, glutamate, which has to be compensated by a reference electrode	Dialysis or flow system, air flushing, $-O_2$ equilibration of electrode	at 0.5 mM, stable for 300 d	0.2-2	?	Romette and Cooney 1987
Leucine Leucine dehydrogenase, EC 1.4.1.9.	Mediator electrode (enzyme reactor)	Increasing NADH	?	?	—	?	FIA system (0.5 ml/min) buffer contains NAD (16-4D)	?	0.1-1	>150 days	Schelter-Graf et al. 1984
Lysine Lysine-α-oxidase, EC?	O_2 electrode[i]	Decreasing O_2	8.7	25 °C	Air wash between each test	Arginine, phenylalanine, L-ornithine	Dialysis or flow system (16-3, 16-4A,B,C)	?	0.2-4	3,000 times	Romette et al. 1983
Lysine Lysine decarboxylase EC 4.1.1.18.	CO_2 electrode[i]	Increasing CO_2	6.0	?	—	CO_2 of the medium, compensated by reference system	Dialysis or flow system (16-3, 16-4A,B,C)	No	0.05-100	60 days	White and Guibault 1978

434

Methionine										
Methionine-lyase, EC 4.4.1.11	NH_3-gas electrode[i]	Increasing NH_3	8.7	?	—	NH_4^+ of the medium has to be compensated by reference system	Dialysis or flow system (16-3, 16-4A,B,C), buffer[j]	Stable	0.01–10	3 m[b]	Fung et al. 1979
Tyrosine										
Tyrosine-carboxylase, EC 4.1.1.25	CO_2 electrode[i]	Increasing CO_2	4.8	23 °C, 8 min for 2 mM; 37 °C, 5 min for 2 mM	CO_2 of the medium has to be compensated by reference systems	Dialysis or flow system (16-3, 16-4A,B,C)	-1.4×10^{-2} kPa/h	0.04–2.6	90 days	Havas and Guilbault 1982

Reproduced with permission from Merten (1988).

[a] Hexacyanoferrate (III) as mediator.

[b] m, Month(s).

[c] Enzymes: lactate oxidase (EC 1.1.3.2.), catalase (EC 1.11.1.6.), lactate dehydrogenase (EC 1.1.1.27.), all are immobilized on an enzyme reactor. (EC: Enzyme Commission number.)

[d] WONA, without NADH in flow buffer.

[e] WNA, with NADH in flow buffer.

[f] w, Week(s).

[g] Buffer contains: phosphate (0.5 mM), thiamine pyrophosphate (0.06 mM).

[h] com, Commercial.

[i] Enzyme is immobilized onto the electrode.

[j] Buffer contains: pyrophosphate (0.05 M), p-5'-p (10^{-4} M).

435

Detection ranges can be adapted easily by changing the injection volume, the parameter of dialysis, or dilution system.

If a group of substances, such as amino acids, peptides, etc., have to be monitored by one analysis, HPLC or GC or mass spetrometer systems are applicable. Thus, McLaughlin et al. (1985) used on-line GC for monitoring and controlling butanol, butyrate, acetate, acetone, ethanol, and acetoin in an acetone-butanol fermentation. Dinwoodie and Mehnert (1985) used on-line HPLC for monitoring glucose, glycerol, acetate, and ethanol in a yeast-ethanol fermentation.

Actually, on-line amino acid analysis, on-line analysis of organic acids, etc., using HPLC should be possible using the new Waters Millipore on-line HPLC system.

The parameter of greatest interest in technical cell culture is product concentration. Because the product is generally a protein, immunological methods should be employed. Almost all systems, for example, thermistor enzyme-linked immunosorbent assay (TELISA), enzyme sensors working with an oxygen-electrode, potentiometric ionophore modulation immunoassay (PIMIA), microgravimetric immunoassay (MGIA), and optical systems, require the addition of reagents (for exceptions see North 1985 and Merten 1988). They can be constructed as competitive, direct, or two-site sensors. The competitive and the direct types have the advantage of requiring only the addition of reagents, maximally once; the two-site type, twice. All types require additional washing steps and elution steps for preparation or reuse. Therefore, all systems are based on the flow injection principle. The immobilized antibodies must have a certain affinity for the analyte. This affinity must be high enough to achieve a sufficient sensitivity, but it has to be in such a range that the elution of the analyte is possible without damaging the immobilized antibody. This is a very important consideration for the lifetime of the systems mentioned. Until recently, only one automatic procedure for the detection of a fermentation product (human proinsulin produced by *E. coli*) has been published, which can be adapted to an on-line system very easily. Birnbaum et al. (1986) used a TELISA for monitoring the proinsulin concentration. This system could be used for several days, even up to three weeks, the response time was 7 min, the cycle time was 13 min, and the detection range was between 0.1 and 50 μg/ml. After each test, this affinity system had to be prepared for reuse by eluting the analyte and the reactive and by reequilibrating the column. A schematic in Figure 16–5 shows the principle of this detection system. No other system has been described in the literature in connection with an on-line application for monitoring and controlling fermentation processes. However, all flow-line based immunosensors should be adaptable to the on-line application for monitoring and controlling animal cell fermentors. Except for the energy transfer immunoassay (Lim et al. 1980), which is a FIA-based homogeneous system, all others are heterogeneous assays that require a solid phase for the antigen-antibody reaction, and their lifetime and applicability depends

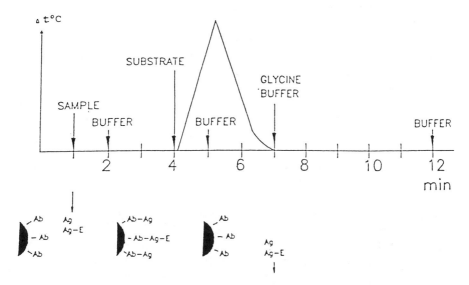

FIGURE 16–5 Schematic presentation of a reaction cycle in the TELISA procedure. The arrows indicate changes in the perfusing medium (flow rate 0.8 ml/min). The cycle starts with potassium phosphate buffer pH 7.0 (0.2 M). At this time the thermistor column contains only immobilized antibodies. At the arrow "sample" a mixture of antigen and catalase-labeled antigen is introduced. The system is then washed with potassium phosphate buffer for 2 min. Now the sites on the antibodies of the column are occupied by antigen as well as by catalase-labeled antigen (i.e., competitive assay). The amount of catalase bound is measured by registering the heat produced during a 1 min pulse of 1 mM H_2O_2. After the heat pulse is registered, the system is washed with glycine/HCl (0.2 M, pH 2.2) to split the complex. After 5 min of washing, phosphate buffer is introduced and the system is ready for another assay. Reproduced with permission from Mattiasson et al. (1977).

mainly on the effectiveness of the elution step after the detection/monitoring step (see above). For more information about the principles of possible immunological test procedures that might be used in an on-line mode, the reader is referred to Heineman and Halsall (1985), North (1985), and Merten et al. (1986).

16.3 FUTURE DEVELOPMENTS

The development of biosensors for the on-line monitoring of biochemical parameters and control of animal cell cultures is just beginning. Most of the research has been done for the development of disposable devices for medical applications. However, in contrast to these types of devices (often disposable, with a short lifetime, fast response, sterility and only for in vivo applications, the fermentation industry also needs reliable and sterilizable devices with a sufficient lifetime (weeks to months) that can be recalibrated and changeable if necessary. The last requirement is not a problem because commercial changeable pH electrodes are already available from Ingold, Urdorf, Switzerland.

The continuation of research in the microbial field and in molecular biology may provide new enzymes that are more thermostable, perhaps even steam sterilizable (see, e.g., Comer et al. 1979), and/or recombined enzymes with features particularly designed for biosensor application.

Another approach, shown by Ho and Rechnitz (1987) was the development of artificial enzymes that may provide the possibility of tailoring the necessary characteristics of artificial enzyme for a certain application. Their system (artificial oxalacetate decarboxylase-based pCO_2 biosensor) showed some advantages by comparing it with the natural oxalacetate carboxylase based-pCO_2 sensor. It was stable for at least six months versus less than one week, had no co-factor requirement against Mn^{2+} and Mg^{2+} requirement, and a broad pH profile with optimum at pH 4.5 (compared to pH 7 to 8 where the pCO_2 electrode is of marginal utility).

New developments such as the on-line Fourier transformed infrared (FTIR) spectrometry will provide a new and more sophisticated method of the on-line monitoring and control of several groups of substances in one analysis step.

The increase of animal cell culture and the importance of the animal cell fermentation for the production of drugs will increase the necessity of on-line monitoring and control of all relevant parameters in order to achieve reliable models, to optimize a pilot process, and to control a production plant, not only for optimal production but also perhaps for regulatory purposes and as good manufacturing practice (GMP) documentation.

16.4 CONCLUSIONS

Mammalian cell fermentation has been developed very rapidly within the last few years, at least with respect to the cultivation vessel (Handa-Corrigan 1988). The instrumentation, however, has not improved that much. But fermentation with good economy means high efficiency by optimal medium utilization and high plant output. In practice, this means high volumetric yields from high density cultures, which are kept well within the optimal production conditions by using standard and process-specific parameters. An appropriate process control is therefore a prerequisite, which should not be underestimated in its importance.

The relevance of distinct physiological parameters, which can be used as guide parameters must be investigated on a case-to-case basis. In any case, the application of a computerized process control system in addition to the sensor devices is an unavoidable prerequisite for a distinct degree of automatization.

Even if a system is not obtainable for ready use, one can obtain proper components and can prepare a process specific setup that meets the challenges resulting from the potency of the most modern reactors.

Using all the possibilities can show a surprising rise in production and economy, which allows at least some of the animal cell culture products to compete with established fine chemicals. Monoclonal antibodies, for example, can be produced much cheaper today as their polyclonal counterpart. Even compared to the economy of biotechnology products in microorganisms, there are cell culture products that show, at least when calculated after purification, the same production costs. Considering many other advantages, such as proper glycosylation and folding, etc., the production of many products in cell culture should have similar or better feasibility than do microorganisms.

REFERENCES

Anderson, K.W., Grulke, E., and Gerhardt, P. (1985) *Biotechnol. Bioeng.* 27, 917–920.

Appelqvist, R., Marko-Varga, G., Gordon, L., Torstensson, A., and Johansson, G. (1985) *Anal. Chim. Acta* 169, 237–247.

Aston, W.J., and Turner, A.P.F. (1984) *Biotechnol. Genet. Eng. Rev.* 1, 89–120.

Beyeler, W., Einsele, A., and Fiechter, A. (1981) *Eur. J. Appl. Microbiol. Biotechnol.* 13, 10–14.

Birnbaum, S., Bülow, L., Hardy, K., Danielsson, B., and Mosbach, K. (1986) *Anal. Biochem.* 158, 12–19.

Blake-Coleman, B.C., Calder, M.R., Carr, R.J.G., Moody, S.C., and Clarke, D.J. (1984) *Trends Anal. Chem.* 3, 229–235.

Bourdillon, C., Hervagault, C., and Thomas, D. (1985) *Biotechnol. Bioeng.* 27, 1619–1622.

Bourdillon, C., Thomas, V., and Thomas, D. (1982) *Enzyme Microbiol. Technol.* 4, 175–180.

Brooks, S.L., Ashby, R.E., Turner, A.P.F., Calder, M.R., and Clarke, D.J. (1987/88) *Biosensors* 3, 45–56.

Butler, M. (1985) *Develop. Biol. Stand.* 60, 269–280.

Butler, M. (1986) Presented at the Society of Chemical Industry, *Symposium on Large-Scale Production of Monoclonal Antibodies*, London, December 9, 1986.

Butler, M., and Spier, R.E. (1984) *J. Biotechnol.* 1, 187–196.

Chapman, A.G., Fall, L., and Atkinson, D.E. (1971) *J. Bacteriol.* 108, 1072–1086.

Chotani, G., and Constantinides, A. (1982) *Biotechnol. Bioeng.* 24, 2743–2745.

Clarke, D.J., Blake-Coleman, B.C., Calder, M.R., Carr, R.J.D., and Moody, S.C. (1984) *J. Biotechnol.* 1, 135–158.

Clarke, D.J., Calder, M.R., Carr, R.J.G., et al. (1985) *Biosensors* 1, 213–320.

Clarke, D.J., Kell, D.B., Morris, J.G., and Burns, A. (1982) *Ion Select. Elec. Rev.* 4, 75–133.

Cleland, N., and Enfors, S.-O. (1984a) *Anal. Chem.* 56, 1880–1884.

Cleland, N., and Enfors, S.-O. (1984b) *Anal. Chim. Acta* 163, 281–285.

Comer, M.J., Burton, C.J., and Atkinson, A. (1979) *J. Appl. Biochem.* 1, 259–270.

De Bruyne, N.A. (1988) in *Animal Cell Biotechnology* Vol. 3 (Spier, R.E., and Griffiths, J.B., eds.), pp. 142–174, Academic Press, London.

Dinwoodie, R.C., and Mehnert, D.W. (1985) *Biotechnol. Bioeng.* 27, 1060–1062.

Eagle, H., Barban, S., Levy, M., and Schulze, H.O. (1958) *J. Biol. Chem.* 233, 551–558.

Fleischacker, R.J., Weaver, J.C., and Sinskey, A.J. (1981) *Adv. Appl. Microbiol.* 27, 137–167.

Fogt, E.J., Cahalan, P.T., Jeyne, A., and Schwinghammer, M.A. (1985) *Anal. Chem.* 57, 1155–1157.

Fung, K.W., Kuan, S.S., Sung, H.Y., and Guilbault, G.G. (1979) *Anal. Chem.* 51, 2319–2324.

Garn, M., Thommen, C., and Gisin, M. (1987) in *Proc. 4th European Congress on Biotechnology* Vol. 3 (Neijssel, O.M., et al., eds.), p. 92, Elsevier Science Publishers B.V., Amsterdam.

Geahel, I., Dordet, Y., Duval, D., Dufau, A.F., and Hache, J. (1989) in *Advances in Animal Cell Biology and Technology for Bioprocesses* (Spier, R.E., et al., eds.), pp. 134–135, Butterworths, UK.

Ghoul, M., Ronat, E., and Engasser, J.-M. (1986) *Biotechnol. Bioeng.* 28, 119–121.

Girotti, S., Roda, A., Ghini, S., et al. (1984) *Anal. Lett.* 17(B1), 1–12.

Grdina, T.A., and Jarvis, A.P. (1984) *Biotech '84 USA*, pp. A235–A246, Online Publications, Pinner, UK.

Griffiths, B. (1984) *Develop. Biol. Stand.* 55, 113–116.

Gronow, M., Kingdon, C.F.M., and Anderton, D.J. (1985) in *Molecular Biology and Biotechnology* (Walker, J.M., Gingold, E.B., eds.), pp. 295–324, The Royal Society of Chemistry, Burlington House, London.

Guilbault, G.G. (1982) *Ion. Select. Elec. Rev.* 4, 187–231.

Guilbault, G.G. (1984) *Analytical Uses of Immobilized Enzymes*, Marcel Dekker, New York.

Guilbault, G.G., and Hrbankova, E. (1971) *Anal. Chim. Acta* 56, 285–290.

Guilbault, G.G., and Nagy, G. (1973) *Anal. Chem.* 45, 417–419.

Handa-Corrigan, A. (1988) *Bio/Technology* 6, 784–786.

Harris, C.M., and Kell, D.B. (1985) *Biosensors* 1, 357–360.

Havas, J., and Guilbault, G.G. (1982) *Anal. Chem.* 54, 1991–1997.

Heineman, W.R., and Halsall, H.B. (1985) *Anal. Chem.* 57, 1321A–1331A.

Hill, B.T., and Whatley, S. (1975) *FEBS Lett.* 56, 20–23.

Himmler, G., Palfi, G., Rüker, F., Katinger, H., and Scheirer, W. (1985) *Develop. Biol. Stand.* 60, 291–296.

Ho, M.Y.K., and Rechnitz, G.A. (1987) *Anal. Chem.* 59, 536–537.

Holst, O., Håkanson, H., Miyabayashi, A., and Mattiasson, B. (1988) *Appl. Microbiol. Biotechnol.* 28, 32–36.

Hu, W.S., Dodge, T.C., Frame, K.K., and Himes, V.B. (1987) *Develop. Biol. Stand.* 66, 279–290.

Jeannesson, P., Manfait, M., and Jardillier, P.M. (1983) *Anal. Biochem.* 129, 305–309.

Junker, B.H., Wang, D.I.C., and Hatton, T.A. (1988) *Biotechnol. Bioeng.* 32, 55–63.

Karube, I., and Suzuki, S. (1984) *Ion. Select. Elec. Rev.* 6, 15–58.

Katinger, H.W.D. (1988) Presented at the Engineering Foundation Conference on Cell Culture Engineering, Palm Coast, FL.

Kilburn, D.G., Fitzgerald, P., Blake-Coleman, B., Clarke, D., and Griffiths, J.B. (1989) *Biotechnol. Bioeng.* 33, 1379–1384.

Klöppinger, M., Fertig, G., Fraune, E., and Miltenburger, H.G. (1989) in *Advances in Animal Cell Biology and Technology for Bioprocesses* (Spier, R.E., et al., eds.), pp. 125–128, Butterworths, UK.

Kobayashi, T., Iijima, S., Shimizu, K., and Matsubara, M. (1987) in *Proc. 4th Eur. Congr. Biotechnol.* Vol. 3 (Neijssel, O.M., et al., eds.), pp. 114–117, Elsevier Science Publishers B.V., Amsterdam.

Koch, A.L. (1961) *Biochim. Biophys. Acta* 51, 429–471.

Kroner, K.H., and Kula, M.-R. (1984) *Anal. Chim. Acta* 163, 3–15.

Lehmann, J., Vorlop, J., and Büntemeyer, H. (1988) in *Animal Cell Biotechnology* Vol. 3 (Spier, R.E., and Griffiths, J.B., eds.), pp. 222–237, Academic Press, London.

Leist, C., Meyer, H.-P., and Fiechter, A. (1986) *J. Biotechnol.* 4, 235–246.

Lim, C.S., Miller, J.N., and Bridges, J.W. (1980) *Anal. Chim. Acta* 114, 183–189.

Lowe, C.R. (1984) *Trends Biotechnol.* 2, 59–65.

Luan, Y.T., Mutharasan, R., and Magee, W.E. (1987a) *Biotechnol. Lett.* 9, 751–756.

Luan, Y.T., Mutharasan, R., and Magee, W.E. (1987b) *Biotechnol. Lett.* 9, 535–538.

Lundbäck, H., and Olsson, B. (1985) *Anal. Lett.* 18, 871–889.

Luong, J.H.T., and Carrier, D.J. (1986) *Appl. Microbiol. Biotechnol.* 24, 65–70.

Mallette, M.F. (1969) in *Methods in Microbiology* Vol. 6B (Norris, J.R., and Ribbons, D.W., eds.), pp. 319–566, Academic Press, London.

Mandenius, C.F., Danielsson, B., and Mattiasson, B. (1984a) *Anal. Chim, Acta* 163, 135–141.

Mandenius, C.F., Danielsson, B., Winquist, F., Mattiasson, B., and Mosbach, K. (1984b) *Appl. Biochem. Biotechnol.* 7, 141–146.

Mandenius, C.F., Mattiasson, B., Axelsson, J.P., and Hagander, P. (1987) *Biotechnol. Bioeng.* 29, 941–949.

Mascini, M., Fortunati, S., Moscone, D., et al. (1985) *Clin. Chem.* 31, 451–453.

Mascini, M., Moscone, D., and Palleschi, G. (1984) *Anal. Chim. Acta* 157, 45–51.

Mathers, J.J., Dinwoodie, R.C., Talarovich, M., and Mehnert, D.W. (1986) *Biotechnol. Lett.* 8, 311–314.

Matsushita, T., Brendzel, A.M., Shotola, M.A., and Groh, K.R. (1982) *Biophys. J.* 39, 41–47.

Matsunaga, T., Karube, I., and Suzuki, S. (1979) *Appl. Environ. Microbiol.* 37, 117–121.

Matsunaga, T., Karube, I., and Suzuki, S. (1980) *Eur. J. Appl. Microbiol. Biotechnol.* 10, 125–132.

Mattiasson, B., Borrebaeck, C., Sanfridson, B., and Mosbach, K. (1977) *Biochim. Biophys. Acta* 483, 221–227.

Mattiasson, B., Danielsson, B., Mandenius, C.F., and Winquist, F. (1981) *Ann. N.Y. Acad. Sci.* 369, 295–305.

McLaughlin, J.K., Meyer, C.L., and Papoutsakis, E.T. (1985) *Biotechnol. Bioeng.* 27, 1246–1257.

Merten, O.-W. (1988) in *Animal Cell Biotechnology* Vol. 3 (Spier, R.E., and Griffiths, J.B., eds.), pp. 75–140, Academic Press, London.

Merten, O.-W., Palfi, G.E., Stäheli, J., and Steiner, J. (1987) *Develop. Biol. Stand.* 66, 357–360.

Merten, O.-W., Palfi, G.E., and Steiner, J. (1986) in *Advances in Biotechnological Processes* Vol. 6 (Mizrahi, A., ed.), pp. 111–178, Alan R. Liss, New York.

Meyer, C., and Beyeler, W. (1984) *Biotechnol. Bioeng.* 26, 916–925.

Miller, S.J.O., Henrotte, M., and Miller, A.O.A. (1986) *Biotechnol. Bioeng.* 28, 1466–1473.

Mizutani, F., Tsuda, K., Karube, I., Suzuki, S., and Matsumoto, K. (1980) *Anal. Chim. Acta* 118, 65–71.

Mizutani, S., Iijima, S., Morikawa, M., et al. (1987) *J. Ferment. Technol.* 65, 325–331.

Nopper, B., and Wichmann, R. (1988) Presented at the 8th Int. Biotechnol. Symp., Paris, July 17-22, 1988.

North, J.R. (1985) *Trends Biotechnol.* 3, 180–186.

Öyaas, K., Berg, T.M., Bakke, O., and Levine, D.W. (1989) in *Advances in Animal Cell Biology and Technology for Bioprocesses* (Spier, R.E., et al., eds.), pp. 212–220, Butterworths, UK.

Parker, C.P., Gardell, M.G., and Di Biasio, D. (1986) *Int. Biotechnol. Lab.* June, 33–40.

Pécs, M., Szigeti, L., Nyests, L., and Hollo, J. (1987) in *Proc 4th Eur. Congr. Biotechnol.* Vol. 3 (Neijessel, O.M., et al., eds.), pp. 103, Elsevier Science Publishers B.V., Amsterdam.

Polastri, G.D., Friesen, H.J., and Mauler, R. (1984) *Develop. Biol. Stand.* 55, 53–56.

Racine, P., Engelhardt, R., Higeline, J.C., and Mindt, M. (1975) *Med. Instrum.* 9, 11–14.

Rebsamen, E., Goldinger, W., Scheirer, W., Merten, O.-W., and Palfi, G.E. (1987) *Develop. Biol. Stand.* 66, 1273–1277.

Reitzer, L.J., Wice, B.M., and Kennell, D. (1979) *J. Biol. Chem.* 254, 2669–2676.

Roberts, R.S., Hsu, H.W., Lin, K.D., and Yang, T.J. (1976) *J. Cell. Sci.* 21, 609–615.

Romette, J.L., and Cooney, C.L. (1987) *Anal. Lett.* 20, 1069–1081.

Romette, J.L., Yang, J.S., Kusakabe, H., and Thomas, D. (1983) *Biotechnol. Bioeng.* 25, 2557–2566.

Runstadler, P.W., and Cernek, S.R. (1988) in *Animal Cell Biotechnology* Vol. 3 (Spier, R.E., and Griffiths, J.B., eds.), pp. 306–320, Academic Press, London.

Ruzicka, J., and Hansen, E.H. (1980) *Anal. Chim. Acta* 114, 19–44.

Rydevik, U., Nord, L., and Ingman, F. (1982) *Int. J. Sports Med.* 3, 47–49.

Scheller, F., Siegbahn, F., Danielsson, B., and Mosbach, K. (1985) *Anal. Chem.* 57, 1740–1743.

Schelter-Graf, A., Schmidt, H.-L., and Huck, H. (1984) *Anal. Chim. Acta* 163, 299–303.

Scheper, T., Gebauer, A., Sauerbrei, A., Niehoff, A., and Schügerl, K. (1984) *Anal. Chim. Acta* 163, 111–118.

Schmidt, W.J., Meyer, H.-D., Schügerl, K., Kuhlmann, W., and Bellgardt, K.-H. (1984) *Anal. Chim. Acta* 163, 101–109.

Seaver, S.S., Rudolph, J.L., and Gabriels, Jr., J.E. (1984) *BioTechniques* 2, 254–260.

Siro, M.-R., Romar, H., and Lövgren, T. (1982) *Eur. J. Appl Microbiol. Biotechnol.* 15, 258–264.

Takazawa, Y., Tokashiki, M., Hamamoto, K., and Murakami, H. (1988) *Cytotechnology* 1, 171–178.

Thebline, D., Harfield, J., Hanotte, O., Dubois, D., and Miller, A.O.A. (1987) in *Modern Approaches to Animal Cell Technology* (Spier, R.E., and Griffiths, J.B., eds.), pp. 504–512, Butterworths, London.

Tolbert, W.R., and Feder, J. (1983) in *Annual Reports on Fermentation Processes* Vol. 6 (Tsao, G.T., ed.), pp. 35–74, Academic Press, New York.

Tolbert, W.R., Feder, J., and Kimes, R.C. (1981) *In Vitro* 17, 885–890.

Velez, D., Reuveny, S., Miller, L., and Macmillan, J.D. (1986) *J. Immunol. Meth.* 86, 45–52.

Velez, D., Reuveny, S., Miller, L., and Macmillan, J.D. (1987) *J. Immunol. Meth.* 102, 275–278.

Wagner, R., Ryll, T., Krafft, H., and Lehmann, H. (1988) *Cytotechnology* 1, 145–150.

White, W.C., and Guilbault, G.G. (1978) *Anal. Chem.* 50, 1481–1486.

Zabriskie, D.W., and Humphrey, A.E. (1978) *Biotechnol. Bioeng.* 20, 1295–1301.

Large-Scale Process Purification of Clinical Product from Animal Cell Cultures

Christopher P. Prior

To date, mammalian cells provide the only viable production source for many complex proteins, offering exciting potential as therapeutic and diagnostic products (i.e., monoclonal IgG and IgM, growth hormones, erythropoietin, blood clotting factors, plasminogen activators, etc.). Essentially, mammalian cells have the ability to synthesize proteins with the proper configuration, correct disulfide bonding, and arrays of sugar side chains, which result in the desired activity of the naturally occurring protein. Therefore, many proteins derived from mammalian cells are more likely to be efficacious and are less likely to be immunogenic in humans than if they are produced by bacterial or yeast fermentation.

It is critical that once the desired protein is secreted structurally intact from the cell that it does, in fact, reach the vial in a "native" unadulterated state. This requires the implementation of controlled pharmaceutical manufacturing processes capable of producing safe material with high lot-to-lot consistency. The final product must possess the following properties: (1) freedom from denaturation that can occur as a result of harsh downstream processing conditions; (2) purity specifications consistent with U.S. Food and Drug Administration (FDA) guidelines for extraneous protein, DNA, and endotoxin levels; (3) both structural and activity stability in the final container sufficient for clinical application and subsequent marketing activities; (4) freedom from high molecular weight aggregates that could reduce activity and induce harmful immunogenic side effects; (5) the ability to pass the sterility test as described in the 21 CFR 610.12; and (6) freedom from mycoplasma and viral contamination.

In pursuit of these requirements, the optimal purification strategies conducive to pharmaceutical manufacturing standards will be discussed. However, treatment of this subject must also include a discussion on the effect of cell culture conditions on downstream processing. The opportunity for degradative events in the bioreactor should be minimized to improve downstream recovery. In many ways, both cell culture and purification operations should be considered as one carefully coordinated process. On this point, data provided in this chapter suggest that perfusion culture can offer several distinct advantages over batch culture for optimizing both production levels and the structural integrity of the desired biopharmaceutical. (For a description of the various perfusion and batch operated cell culture reactors, see reviews in Spier and Griffiths (1988), and Feder and Tolbert (1985).

First, in a state of continuous perfusion, cell-secreted products that are vulnerable to biodegradation and aggregation at 37°C can be removed expeditiously from the bioreactor to a 4°C environment in readiness for subsequent purification. The rate of product removal from this degradative cellular environment is in terms of hours for perfusion systems as compared to many days and weeks for semibatch and batch systems, respectively. This continuous harvesting process minimizes the opportunity for proteolytic degradation and aggregation processes, as will be discussed in this chapter for the production of monoclonal IgM and human rtPA. The data presented are based on production experience using a perfusion chemostat. Details of these reactors have been described elsewhere (Feder and Tolbert 1983; Tolbert et al. 1988). Data on the effect of prolonged batch culture on the structure and biological activity of an IgM and IgG will also be presented (Macmillan et al. 1987).

Second, in perfusion culture, cells are constantly provided with fresh nutrients and depleted of toxic by-products, resulting in cell densities as high as 10-fold above those normally obtained in batch cultures. Thus, a 100 l perfusion reactor at maximum density can contain roughly the same number of cells as a 1,000 l batch tank.

Third, at such high cell densities and in a state of perfusion, the serum content is decreased dramatically or eliminated using serum-free conditions. Typical objectives are to reduce growth rates and to maintain a steady-state culture as close as possible, as measured by glucose/lactate, oxygen uptake, productivity, and other key operating parameters. Only sufficient nutrients are provided to sustain viability and product expression. This improves the economics of the system, and for purpose of downstream processing it is extremely important to minimize the level of serum and thus reduce the major source of contaminant proteins such as albumin, transferrin, serum-derived immunoglobulins, proteases, protease inhibitors, etc.

In summary, the effects of cell culture technology on production of one- and two-chain human rtPA, as well as monoclonal IgG and IgM, will be reviewed in terms of downstream recovery, activity, and structural integrity. The human rtPA described in this chapter is derived from a proprietary Invitron recombinant Chinese hamster ovary (CHO) cell line. Finally, the strategy for large-scale purification of biopharmaceuticals under FDA guidelines as defined by current good manufacturing practices (cGMP) will also be reviewed.

17.1 POTENTIAL PRODUCT CONTAMINANTS DERIVED FROM ANIMAL CELL CULTURE PROCESSES

As seen in Table 17–1, conditioned media consist of various cell culture components, namely, serum proteins and the powdered media required for cell growth, together with components contributed from the cells themselves such as secreted and cellular proteins, nucleic acid, and lipids. By far, the most difficult contaminants to remove are the serum proteins, of which albumin accounts for roughly 50%. Since it is at best difficult to separate serum immunoglobulins from a given monoclonal antibody, fetal bovine serum prescreened for low IgG levels is generally used in production. Transferrin is also a common contaminant of immunoglobulin preparations since it has an isoelectric point in the same range as many monoclonal antibodies and often migrates similarly during ion exchange chromatography.

The protease inhibitors constitute 10% of plasma proteins and are the third largest group of functional proteins in human plasma after albumin and the immunoglobulins. In the conversion process from plasma to serum, substantial quantities of alpha$_1$-antiplasmin and alpha$_1$-proteinase inhibitor remain, which can inactivate and reduce production levels of tPA under serum-containing cell culture conditions as described by Prior et al. (1988).

Next to serum contaminants, the major concerns are the quality of water and powdered media. Water should be of "water for injection" quality and stored or circulated at high temperatures to prevent microbial contamination. A primary concern for both water and media is the contribution of pyrogens (i.e., endotoxins), which should be kept to an absolute minimum in a pharmaceutical process. Typical endotoxins in the form of lipopolysaccharides can cause a multitude of adverse effects, such as inflammatory or pyrogenic responses if injected into humans. It is important, therefore, irrespective of initial pyrogen concentrations, that purification processes reduce pyrogen levels at each unit operation. As will be discussed later, column sanitization procedures should be evaluated and ultimately validated to ensure that a process purification procedure in itself does not contribute endotoxin.

TABLE 17-1 Cell Culture Components

Component	Potential Contaminants
Water	Organic compounds
	Trace elements
	Pyrogens
Powdered media	Various nutrients
	Phenol red
	Pyrogens
Serum or serum-free culture media	FBS
	(Protein 30–50 mg/ml)
	Albumin (50–60%)
	Immunoglobulins (10% of whole serum, <0.1% in fetal bovine serum)
	Protease inhibitors (10%)
	Transferrin (2–5%)
	Lipoprotein (1–2%)
	Peptide hormones (<0.1%)
	Miscellaneous proteins (20–35%)
	Proteases
	Growth factors
	Lipids
	Cholesterol
	Triglycerides
	Phospholipids
	Steroids
	Vitamins
	Sugars
	Trace elements
	Pyrogens
Hybridoma cells	Secreted proteins (50–200 μg/ml)
	Cellular proteins/debris
	Nucleic acids
	Adventitious Agents (e.g., viruses)
	Endogenous Agents (e.g., retroviruses)

A further concern listed in Table 17-1 is for cell-derived proteins and debris as well as DNA. Host cell DNA sequences can pose a theoretical biohazard. This is a subject that has been extensively reviewed elsewhere (Petricciani 1985) and requires precautions that apply to production of proteins from mammalian cells. Purification processes and specific DNA removal steps should be demonstrated to achieve a reduction of DNA in final product at or below the level of sensitivity of most available DNA assays (<1–5 pg/mg product). Typically, the efficiency of DNA removal is demonstrated by spiking radiolabelled DNA into the purification process and measuring the reduction in counts in the product stream. Also, under optimal culturing conditions, cell debris and DNA should be kept to a minimum, which can be achieved in perfusion technology where cells at high density (i.e., $>2 \times 10^7$/ml) are maintained more intact relative to batch operated reactors. Often in batch cultures, cell lysis occurs as waste products accumulate and nutrients become depleted. Under these circumstances, cell viability decreases (see Figure 17–7; III Batch Case) and growth medium becomes enriched in cytoplasmic and nuclear contaminants. Therefore, it may be preferable to use a perfusion reactor that can be designed to act as a filter in itself and minimize cell-derived contaminants by preventing its contents from entering the product vessel. As an additional precaution, conditioned media are generally filtered prior to or on line with each unit operation. Any extraneous proteins secreted by the cells are present in low concentrations relative to either the serum proteins or to the desired molecule, and therefore, pose less of a contamination threat to the final product.

Last but not least, a growing concern relates to possible contamination from adventitious or endogenous agents produced in the process of culturing animal cells. For example, hybridoma cell lines can carry murine retroviruses that, in principle, are not a direct biohazard to humans, but do constitute a definite undesirable contaminant. Typical precautions involve screening cell lines prior to large-scale culturing conditions in a quarantine stage of operation for retrovirus activity (i.e., reverse transcriptase activity) and viral infectivity assays. An electron microscopic analysis of cell sections and production culture fluid supernatants for visual inspection of potential viral particles together with a determination of viral concentration is also common practice. Purification processes themselves should be evaluated for their ability to achieve greater than a 4–6 log reduction in virus particles beyond the concentration found in conditioned medium. Removal studies can be performed at less than production scale in a dedicated facility and involves spiking each process step with a known viral concentration, at least 10^6 viral particles/ml being desirable and measuring the reduction in the product stream. The overall log removal is simply the cumulative reduction achieved at each of the individual purification steps. For example, the log removal for a four-step purification scheme is presented in Table 17–2. Depending on the type of virus found and intended use of the product, the FDA may recommend implementing specific viral inactivation steps. Typical steps can involve the use of low pH (Stegmann et al. 1987), heat and pH (Lancz and Sample 1985), heat and gamma irradiation (Hussain et al. 1980; Bassin et al. 1978), heat alone (Ng and Dobkin 1985; Ingerslev et al. 1987; Hilfenhaus and Weidmann 1986), dry heat and lyophilization (Wickerhauser and Williams 1987), granulocyte defensins (Daher et al. 1986), detergents (Mitra and Wong 1986), beta-propiolactone/ultraviolet (UV) (Burnouf et al. 1987), and more recently, chaotropic solutions (i.e., urea and salt, guanidine hydrochloride, unpublished results). Viral inactivation procedures can be implemented on purified product prior to formulation or on conditioned medium as well as after the initial concentration step, thus

TABLE 17-2 Inactivation/Removal of Retrovirus Through Purification Process

Column Chromatography Step	Total Virus Particles	Remaining Total Virus Particles (PFU) in Product Fraction	Log Reduction
1	3.1×10^7	2.5×10^3	4.1
2	9.3×10^6	$<2.4 \times 10^3$	> 3.6
3	1.6×10^7	$<1.6 \times 10^3$	> 4.0
4	1.6×10^7	$<1.3 \times 10^3$	> 4.1
		Cumulative Log Reduction	>15.8

Product: IgG_1-murine hybridoma cell line
Purification process: 1/100th scale

Demonstration of a four-step purification process to achieve a >15.8 log removal of virus particles. The model virus used was ecotropic murine retrovirus. The highest viral titer possible in conditioned medium is 10^6–10^7 particles/ml. Therefore, the purification process achieved a greater than 8–9 log reduction beyond the theoretical highest viral titer contained in cell-conditioned medium.

minimizing risk of cross-contamination during downstream processing. Certainly, such inactivation measures must be demonstrated not to be detrimental to the final product and sufficient biological activity (i.e., antigen binding, enzyme assays), and biophysical data (i.e., circular dichroism, HPLC) should be carefully evaluated to ensure that this is the case.

In summary, the removal of potential contaminants seen in Table 17–1 presents an enormous challenge to any downstream process. Preferably, most of these potential contaminants are avoided at the level of cell culture by employing, for example, serum-free medium low in endotoxin and protein levels, and avoiding scaleup with viral-contaminated cells whenever possible. On a final note, the choice of growth medium and general cell culture conditions can determine the level of protein purity, which can influence the safety and efficacy of a product in humans.

17.2 DISCUSSION OF A cGMP ION EXCHANGE PROCESS PURIFICATION SCHEME FOR A MONOCLONAL IgG

To produce multigram quantities of antibodies from a process stream in which they are present in milligram per liter concentrations requires the processing of 100 l to more than 1,000 l of conditioned medium. A typical downstream process flow diagram capable of handling such large volumes (from step 1 to step 4) is presented in Figure 17–1. Essentially, step 1 involves initial volume reduction by ultrafiltration; steps 2 and 3 employ ion exchange procedures followed by a final buffer exchange operation in step 4. In this example, the charge properties of the product were sufficiently different from those of the contaminants that a high degree of purity was

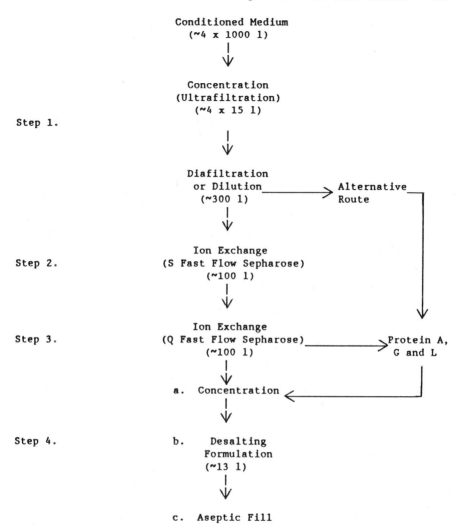

FIGURE 17–1 Process purification flow diagram. A typical purification scheme for the production of clinical product is described (see text).

achieved by two ion exchange steps. If trace contaminants (e.g., protein, DNA) co-purify following ion exchange, then a further purification principle (e.g., phenyl sepharose) should be employed. Variability in expression level and contamination profiles during a cell culture process can compromise the efficiency of a given purification scheme. Therefore, the trend in the industry is to develop a robust process that involves three different column principles to consistently produce pure product irrespective of variability in the starting material. When a robust purification scheme is developed, precise and accurate documentation of all steps is required by cGMPs, as recommended by the FDA.

The process outlined in Figure 17–1 must be economical, produce material with high lot-to-lot consistency, and use equipment and gel media conducive to a pharmaceutical operation. Therefore, treatment of this subject is organized into two parts. The first describes the process chemistry and purification performance and the second part provides information on cGMP operations.

17.2.1 Process Description and Purification Performance

A method of concentration is required as the first step in the process. With sufficient membrane surface area (i.e., >50 square feet), 1,000 l of conditioned media can undergo a 50- to 100-fold concentration in approximately four hours using either hollow fiber, plate and frame, or spiral cartridge systems. All three systems exploit tangential flow of conditioned media across the surface of the membrane to reduce fouling and maintain acceptable flow rates. From a user point of view, such equipment can be seen in a cGMP environment in Figure 17–2. This plate and frame system is operated with sufficient membrane surface area to maintain high flux rates under low inlet and outlet operating pressures. In this way, ultrafiltration proceeds efficiently while minimizing shear forces that can result in surface denaturation of the desired antibody. Hydrophilic ultrafiltration membranes are preferred to hydrophobic surface chemistries to minimize protein-surface interactions and maintain maximum conformational integrity. Biophysical measurements have shown that antibodies may undergo irreversible conformational changes when exposed to hydrophobic membrane filter surfaces as described by Truskey et al. (1987). Such denaturation can result in product instability in the final container in terms of formation of aggregates and reduced biological activity. Such events could pose a potential immunogenic threat to humans.

The concentrate from each ultrafiltration operation is either stored frozen and pooled at some later date with other concentrates to form a larger lot size, or processed individually if a smaller lot size is desired. In the case of perfusion reactors where conditioned medium is received on a continual basis, individual concentrates can be pooled provided the material is derived from a single, cell culture scaleup from the master cell bank. In this example, 4,000 l of starting volume containing approximately 200 g of product were concentrated by ultrafiltration in four separate operations and the concentrates held frozen at $-70°C$. Following a slow thawing process at 4°C, the concentrates were pooled, diluted to the desired salt conditions, adjusted to the required pH, and processed as one homogenous lot through to final purified material using ion exchange chromatography (for equipment see Figures 17–3A and 17–3B). Back dilution is fast and is performed without concern for volume due to the high flow rate properties of the cross-linked

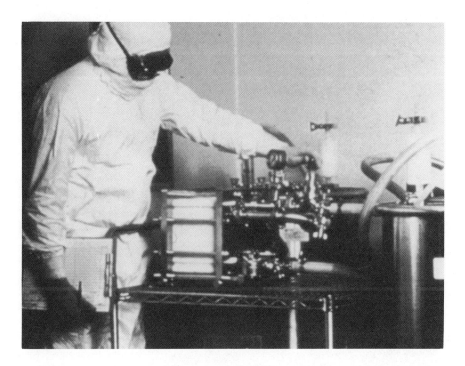

FIGURE 17-2 Plate and frame tangential flow ultrafiltration equipment (step 1). Purification Specialist operating the ultrafiltration unit to concentrate product under cGMPs (see text for detailed picture description).

ion exchange resins used in subsequent steps 2 and 3. Specifically, ion exchange is utilized to take advantage of the fact that the majority of serum proteins are strongly acidic with isoelectric points below 5.5, whereas most immunoglobulins are only weakly acidic with isoelectric points above 6. Therefore, by optimizing both pH and ionic strength at the dilution step, conditions were established to selectively bind the desired antibody to the cation exchange resin (step 2) while most of the serum proteins were not adsorbed. Also, pH conditions were refined during purification development to minimize co-purification of serum-derived immunoglobulins. Protein A HPLC can be used as a rapid in process assay for demonstrating removal of bovine immunoglobulin from the product as described by Moellering and Prior (1990). At extremes of pH, serum proteins and IgG can be bound directly to strong anion exchangers or strong cation exchangers to circumvent the ultrafiltration and back dilution steps. Depending on the charge properties of the particular monoclonal, this can work well, but often the extreme pH conditions required result in poor resolution and possible denaturation of the product. More commonly, conditioned medium must first

A

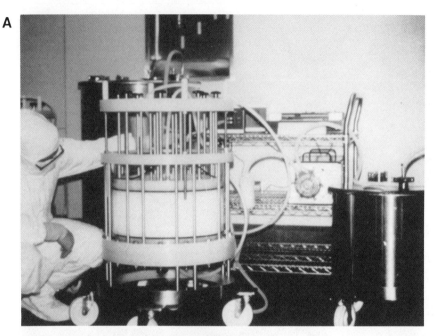

B

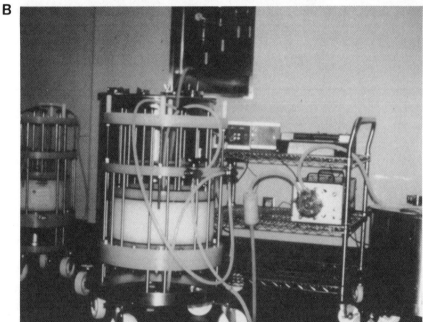

FIGURE 17–3 Large-scale chromatography equipment (steps 2 through 4). Purification specialist performing an ion exchange purification step (200 g scale) under cGMPs. Resins are always protected by in-line sterile filters (see text for detailed picture description).

be concentrated and diafiltered or diluted to reduce the ionic strength prior to ion exchange chromatography.

Figure 17–4 shows a UV absorbance chromatographic profile of the step 2 column (seen in Figures 17–3A and 17–3B) run with product contained in the elution peak and the bulk of contaminants fractionated in the flow-through. Approximately 35 l of S Fast Flow Sepharose (Pharmacia) was packed under pressure in a 440 Industrial column (1520 cm^2 cross-sectional area; Amicon) at a flow rate of 5,400 cm/hour (540 l/hour) and run at 3,000 cm/hour (300 l/hour). Typical binding capacities of the column under most operating conditions are approximately 15 g/l. Therefore, a column of such dimensions is capable of processing 200–500 g of product as one lot.

A 30- to 60-fold purification was achieved using cation exchange chromatography (step 2); however, the purity was insufficient for pharmaceutical purposes. Therefore, a further purification was obtained by optimizing pH and ionic conditions under which the antibody was contained in the flow-through while the remaining contaminants, including DNA, were absorbed to a Q Fast Flow Sepharose column (step 3). A UV absorbance chromatographic profile of step 3 can be seen in Figure 17–5; however, this time product is contained in the flowthrough, and trace protein contaminants

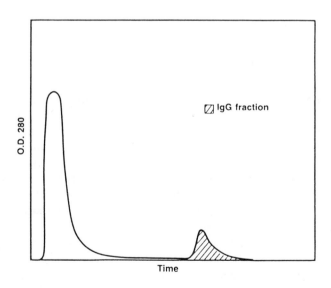

FIGURE 17–4 S fast flow chromatography cation exchange UV absorbance profile (step 2). Approximately 60 l of concentrated medium (representing four concentrates thawed and pooled) were diluted fivefold and applied to the column at a flow rate of 300 l/hour. Under defined pH and salt conditions, approximately 200 g of antibody bound and the bulk of serum proteins and nonproteinaceous material fractionated in the flowthrough. Followed by a five column volume washing procedure, the product was eluted (striped peak) at a purity of 90–95%.

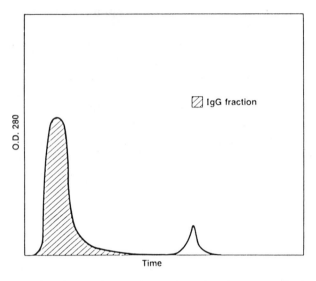

FIGURE 17–5 Q fast flow anion exchange UV absorbance profile (step 3). Monoclonal IgG (approximately 180 g) eluted from the S fast flow column described in Figure 17–4 was applied to the Q Sepharose column. The IgG fraction was eluted in the flowthrough (striped peak) at a purity of >98% and minor contaminants were immobilized. Approximately 160 g of IgG were subjected to desalting procedures and 153.5 g were available for aseptic fill after bulk material had been retained for quality assurance testing.

and DNA are fractioned in the elution peak. Both Q and S Fast Flow columns were of similar dimensions and operated at comparable flow rates.

The column flowthrough (step 3), containing approximately 160 g of product, >98% pure, was concentrated by salt precipitation and followed by centrifugation to obtain a compact pellet. The pellet was subsequently resolubilized at high protein concentration and applied to a desalting column (G25 Sephadex, Pharmacia). This finishing step 4 served to remove final traces of processing chemicals and to exchange the protein into the desired formulation buffer. In other cases, if further purification is required at step 3, additional purification principles may be implemented, typically involving hydrophobic chromatography (e.g., phenyl sepharose) followed by gel filtration. If aggregates are present, high resolution size exclusion chromatography is certainly used (Sephacryl HR, Pharmacia) in place of desalting procedures as the finishing step. However, it is our operating experience that by avoiding harsh pH conditions, aggregate formation is reduced considerably, as determined by HPLC analysis. Therefore, experience dictates that desalting procedures often suffice as a finishing step if aggregates are not detected. Also, desalting procedures are more convenient for processing lot sizes at the multihundred gram to kilogram scale.

Figure 17–6 shows a Coommassie-stained (Laemmli 1970) sodium do-decyl sulfate polyacrylamide gel (SDS-PAGE), depicting the complete cGMP process from step 1 to step 4, as described in the figure legend. As seen in lane 8, the final product contains just two bands under reducing conditions, representing the heavy and light chains of immunoglobulin.

Table 17–3 summarizes the process purification performance of the approximate 200 g scale clinical run from step 1 through step 4. Approximately 153 g of final purified bulk was available for aseptic fill, representing a final recovery of >75%. The levels of pyrogens and other residual contaminants, such as DNA, permitted in final bulk material varies from antibody to antibody and depends on the final dosing regimen. Generally, pyrogen levels can be obtained in the 0.03–0.1 Endotoxin units (EU)/mg antibody range with DNA contamination levels of <1 pg/mg. Contamination of the desired mouse monoclonal antibody preparation with immunoglobulins derived from the growth medium, usually from a bovine source, should be kept to an absolute minimum. By carefully defining the pH and salt process conditions to be highly selective for the desired mouse product, typical background bovine immunoglobulin levels of below 1% of

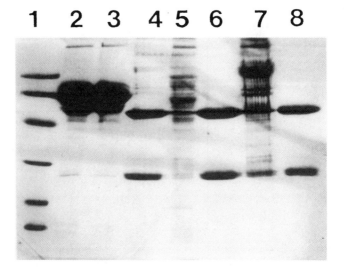

FIGURE 17–6 Process purification performance of a monoclonal antibody. Ten percent SDS-PAGE (reducing) shows the purification performance at each step of the procedure. Lane 1 contains the low molecular weight standards. Lanes 2 through 5 show the purification performance of the first column (step 2); load (lane 2), flow through (lane 3), elution (lane 4), and 1 M NaCl wash (lane 5). Lanes 4, 6, and 7 show the purification performances of the second column (step 3); load (lane 4), flow through (lane 6), and 1 M wash (lane 7). Lane 8 is the final material (step 4) following buffer exchange.

TABLE 17-3   cGMP Process Purification Performance

	Volume (L)	Protein[1] (g/l)	Total Protein (g)	IgG[2] (g/l)	Total IgG (g)	Percent Recovery	Total DNA[3] (ng)	Total Endotoxin[4] (EU)
Starting material	4191.6	2.70	11,317.3	0.048	201.2	100	7.5×10^7	8.0×10^6
Ion exchange steps	112.6	1.47	165.5	1.42	159.9	79.5	3.1×10^2	6.8×10^3
Final purification	12.9	12.0	154.8	11.9	153.5	76.3	<140.0	$<4.6 \times 10^3$

Level of detection, 10 pg <10 pg/11 mg $= <1.0$ pg/mg[3]
1.0 EU = 100 pg/ml   0.35 EU/ml $= <0.03$ EU/mg[4]

[1] Protein was determined using a bicinchoninic acid (BCA) assay from Pierce Chemical.
[2] IgG was determined using ELISA, see Abrams et al. (1984).
[3] DNA was determined using a nonradiometric hybridization assay, see Rigby et al. (1977).
[4] Endotoxin was determined using Limulus Amebocyte Lysate from Haemachem.

total IgG are achieved, as determined by radial immunodiffusion (RID test kits are available from ICN ImmunoBiologicals). Often, it is not possible to achieve bovine immunoglobulin contamination levels such as those obtained with ion exchange when using Protein A as the main purification step due to co-purification of bovine and murine immunoglobulins (see section 17.3).

For a comprehensive description of recommended product purity requirements see *Points to Consider in the Manufacture of Monoclonal Antibody Products for Human Use* (1987), prepared by The Center for Biologics Evaluation and Research. A verbatim duplicate reproduction of this document is conveniently made available from Microbiological Associates, Inc., Rockville, Maryland.

17.2.2 Equipment Design Features and Choice of Purification Gel Media Appropriate for cGMP Processing

Suitable ultrafiltration equipment for cGMP processing of conditioned medium and as it was used in step 1 is shown in Figure 17–2. Certain membrane types and all stainless steel components are routinely cleaned and autoclaved between use. Membranes that cannot withstand autoclave conditions are sanitized using the manufacturer's recommended procedures involving the use of sodium hydroxide, hydrogen peroxide, and/or sodium hypochlorite. Certainly, adequate washing procedures should be validated to ensure removal of sanitization reagents prior to exposing product to membrane and equipment surfaces.

Suitable chromatography equipment of cGMP processing and as it is used in steps 2 and 3 is shown in Figures 17–3A and 17–3B. Note that all buffers and conditioned media are applied to columns through an in-line 0.2 μm hydrophilic membrane to prevent resin fouling and to maintain the highest possible level of sanitary conditions. The peristaltic pumps used here are suitable for cGMP applications in that the fluid contacts only the tubing inside walls, and the possibility of contaminating the pump itself with process fluid is avoided. Concern for tubing rupture and fluid leakage as a consequence of tube wall fatigue is minimized since only new tubing is used for each run. For a discussion on the comparison of various filtration devices and pump designs appropriate for processing under cGMPs see the review article by Knight (1988). Purification personnel operating the equipment shown in Figures 17–2 and 17–3A work fully gowned to maintain the lowest possible bioburden in the processing area. All equipment is on wheels for ease of mobility during use and assembly and/or disassembly. Each piece is thoroughly cleaned following standard operating procedures (SOPs) at the end of each run. All equipment is subsequently returned to storage and the area is cleaned following SOPs before wheeling in new equipment and commencing a different process. These operating practices, together with strict labeling regimens, maintain absolute product segregation and avoid the risk of cross-contamination. Such precautions are essential in manufacturing plants that handle multiple products.

From a process point of view, rigid, chemically cross-linked resins are utilized to facilitate high column flow rates to maximize production throughput and reduce process times. In addition, chemical stability is an essential requirement for cGMPs so that resins can withstand harsh chemical disinfection treatments between purification runs. This procedure involves washing the resin with 0.5 M NaOH to prevent buildup of biological residues, including lipids and proteins. Ion exchange and hydrophobic functional groups attached to cross-linked sepharose resins (Pharmacia, Inc.) are materials that satisfy these requirements.

In a study conducted by Pharmacia, Inc. (see Table 17–4) NaOH was shown to be an effective disinfectant for gram-negative bacteria (i.e., *E. coli* and *Pseudomonas aeruginosa*), gram-positive bacteria (i.e., *Staphylococcus aureus*), and yeast (i.e., *Candida albicans*) at concentrations as low as 0.1 M, while 0.5 M was required to destroy fungal contamination (i.e., *Aspergillis niger*). A challenge of 10^5–10^6 microorganisms/ml was effectively destroyed in all cases. The destruction of *Bacillus subtilis* spores in NaOH at different temperatures has been described by Whitehouse and Clegg (1963).

Martin et al. (1985) have shown that 0.1 M NaOH at room temperature inactivates HTLV-III/LAV in 10 min. (This paper is well referenced and provides a good starting point for the design of an effective virus inactivation scheme). In addition, work conducted at Pharmacia, Inc. (1986) demonstrates that NaOH concentrations of 0.5–1.0 M can effectively inactivate an endotoxin challenge of 100 ng/ml within several hours. Further information concerning endotoxin inactivation can be obtained from Parenteral Drug Association (1985), where chapters 1 and 8 cover this topic. The overall sanitizing properties of NaOH are a function of time, temperature, and concentration. Therefore, these parameters must be considered when de-

TABLE 17–4   Microorganism Destruction by Sodium Hydroxide

Gel Media: S and Q Sepharose® Fast Flow
Challenge: 10^5–10^6 organisms/ml

Microorganism	Conditions
Escherichia coli	0.01 M NaOH (4°C) within 1 hour
Pseudomonas aeruginosa	
Staphylococcus aureus	0.1 M NaOH (4°C) within 1 hour
Candida albicans	0.1 M (4°C) 1–2 hours
Aspergillus niger	0.5 M NaOH (4°C) in 1 hour

Reproduced with permission from Pharmacia, Inc. (1986).

signing sanitary procedures and validation protocols for the cleaning of various gel media.

The validation of sanitary procedures in production operations requires extensive testing and documentation to establish optimal concentrations of cleaning, sterilizing, and decontamination reagents. For further information see the article by Fry (1985) entitled *Process Validation: The FDA Viewpoint.*

The lifetime of gel media following multiple purification/sanitization cycles is a major consideration in the reliability and economics of a process. The Sepharose fast flow, Sephacryl, and Sepharose gel media products have been used for many hundreds of cycles for the fractionation of blood products (information supplied by Pharmacia, Inc.) and for the production of recombinant proteins and monoclonal antibodies for clinical application (information supplied by Invitron Corp.) A cycle is defined as the processing of product on the gel media followed by treatment with sodium hydroxide. In most cases, the upper lifetime limit at which point the resins fail to work has not actually been determined.

For processing under cGMP conditions, it is also important to select resins supported by a manufacturers drug master file (DMF) as well as extensive quality assurance testing results to ensure lot-to-lot consistency, which is essential for reproducible purification efficiency. Most, if not all, of the above-mentioned gel media products are supported by this documentation. A DMF contains detailed information regarding manufacture of the product and its stability characteristics (i.e., leaching properties) together with a description of quality assurance testing procedures for measuring such parameters as flow rate, bead structure, particle size distribution, capacity of functional group, microbiological status, etc. Such data should be included in the certificate of analysis that accompanies each lot. The DMF is filed with the FDA, which in turn can be referenced in an investigational new drug (IND) submission, which provides information on the purification conditions and biochemical characterization data of a particular protein. FDA approval of an IND submission is a mandatory requirement before any protein can be injected into humans. When not in use, resins are typically stored in ethanol to prevent growth of microorganisms and to maintain biostatic conditions.

An additional important aspect of process purification operations under cGMPs is that resins are devoted to one product only. The relatively low cost of ion exchange resins allows them to be discarded when a specific production run is terminated. This policy can be cost-prohibitive with considerably more expensive types of purification media (i.e., Protein A, see next section). If ultrafiltration is required, then those filtration membranes used will only be devoted to that product. All other surfaces in contact with product such as filters and tubing are discarded after use. Electro-polished 316LC stainless steel containers used for buffers are thoroughly cleaned and steam sterilized. These practices and strict labeling procedures are, again, an integral part of any cGMP program designed to ensure product segregation and to minimize risk of cross contamination.

In summary, the design of sanitary downstream processes implemented under cGMPs provides the foundation for achieving controlled and reproducible production of "safe," biologically active material with high lot-to-lot consistency.

17.3 USE OF PROTEIN A AFFINITY CHROMATOGRAPHY FOR MONOCLONAL IgG PURIFICATION

As seen in Figure 17–1, an alternative approach to antibody purification involves the use of Protein A derived from *Staphylococcus aureus* and recognizes the constant heavy chain or Fc portion of most mammalian IgGs with high affinity. Protein A offers a higher-fold purification in a single step than a single ion exchange column is able to perform. Protein A facilitates more of a generic process, requiring less process development time than ion exchange procedures, which have to be specifically designed for each product. For further information on the use of Protein A for IgG purification see the articles by Manil et al. (1986) and Lee and Seaver (1987). However, while the merits of ion exchange for pharmaceutical manufacturing have been discussed, the use of Protein A warrants some concern with regard to the following: (1) significantly higher co-purification of serum-derived immunoglobulins compared to ion exchange (see Table 17–5 and legend); (2) leaching of potentially antigenic breakdown products; (3) contamination with staphylococcal enterotoxins that are known to be mitogenic, as described by Schrezenmeier and Fleischer (1987); (4) inability to sanitize with harsh reagents (i.e., sodium hydroxide); (5) extreme pH conditions for binding and elution, resulting in the potential for formation of aggregates in final product; (6) inability to bind certain IgG subclasses (i.e., human IgG_3, rat IgG_{2A}, and IgG_{2B}, etc.) and IgMs; and finally, (7) substantially greater expense over ion exchange chromatography for use in a limited number of cycles (see Table 17–5 for a comparative cost evaluation for up to 10 uses).

Some of the above concerns may be offset by the recent introduction of a recombinant Protein A produced in *E. coli*, such as that prepared by Repligen (1988). The product is supported by a DMF that documents low protein A leakage levels and freedom from staphylococcal enterotoxin. Also, the recent introduction of Protein G (from a *Streptococcus* strain G), initially by Perstorp Biolytica, provides an affinity ligand with the capability to bind a greater range of IgG subclasses than does Protein A. Unlike Protein A, Protein G can bind the Fc fragment of all human IgG subclasses and virtually all animal IgG subclass equivalents. Perhaps better still, an interesting affinity ligand reported by Bjorck et al. (1987), is a protein that is capable of binding all human IgGs, including IgM, IgA, IgD, IgE, and also rat, mouse, goat, and rabbit IgG. In contrast to Proteins A and G, the so-called Protein

TABLE 17-5   A Comparison of the Purification Efficiencies and Costs of Ion-Exchange Versus Protein A Chromatographic Methods for the Purification of Mouse Immunoglobulins

Comparative Analysis of IgG Purification from 5% FCS

Process Step	Ion Exchange			Protein A		
	Mouse IgG (ELISA)	Bovine IgG (RID)	Bovine IgG to Mouse IgG	Mouse IgG (ELISA)	Bovine IgG (RID)	Bovine IgG to Mouse IgG
Load	103.4 mg	4.86 mg	4.7%	144.6 mg	6.8 mg	4.7%
Flow through	10.4 mg	4.26 mg	41.0%	40.0 mg	2.0 mg	10.0%
Elution	91.1 mg	0.46 mg	0.5%	100.0 mg	4.5 mg	4.5%
Flow through	N/D	N/D	–	–	–	–
Elution	64.1 mg	0.16 mg	0.25%	–	–	–
% Immunoglobulin recovery	62%	3.3%		69%	66%	
Cost per gram for 10 cycles (U.S. $)	$53			$217		

N/D = Nondetectable.

The data show the comparative recovery of serum derived bovine immunoglobulin contaminants using a two-step ion exchange procedure versus Protein A chromatography. The desired mouse antibody is selectively measured by ELISA using an antimouse specific polyclonal (see Abrams 1984). The bovine immunoglobulin is measured by RID using an antibovine specific polyclonal (ICN ImmunoBiologicals, Lisle, IL.)

L shows no affinity for the Fc region and binds instead to Ig kappa and lambda light chains. This means that Protein L could be used as a general binding substance of all Ig classes from different animal species. However, the availability of both Proteins G and L is limited and cost would prohibit their use for large-scale purification at the current time. Certainly, the same concerns with regard to cleaning and leaching rates associated with the use of any protein-based purification column in a cGMP environment would still apply.

For purifying limited quantities of monoclonal IgG for research use, Protein A does offer certain distinct advantages with regard to: (1) high-fold affinity purification in one step and shorter purification development time compared to ion exchange (as previously mentioned); (2) potential for multiple uses in nonfouling conditions; (3) efficient removal of free, light chain contamination; and, (4) fractionation of IgG subclasses as described by Ey et al. (1978). In addition, Protein A offers applications in immunochemical techniques, as described by Langone (1982), and as a method of the rapid quantitation of intact antibody in conditioned medium, as described by Hammen (1988) and Duffy et al. (1989). Often Protein A HPLC techniques offer greater reliability as a physical technique for estimating crude product concentration in conditioned medium where conventional immunological assays [i.e., enzyme-linked immunosorbent assay (ELISA) techniques, described by Abrams et al. 1984] may not distinguish desirable intact antibody from undesirable free light chain.

An example of this is illustrated in Table 17–6. In this case, the cells produced a large amount of free light chain in addition to the whole antibody. The ELISA that was used did not differentiate between the two and quantitated the free light as intact antibody. This translated to a recovery of only 40%, which has an impact on cost-effectiveness, process development, and scheduling. By using the Protein A method, the actual recovery was 75%, which is obviously much more economically acceptable.

Not only can ELISA react with free light, it is entirely possible that degraded antibodies will also react with the assay. As long as the epitope that the immunochemical recognizes is intact, the ELISA will quantitate it as whole antibody. With regard to efficiency and the need to process many samples, Protein A HPLC provides a reliable answer in approximately 30 min compared to hours and days for ELISA techniques.

In summary, Protein A offers potential as an analytical tool and for rapid purification of research material. For large-scale purification of clinical material, it may be desirable to use Protein A where ion exchange steps fail to achieve the desired purity, especially in cases of low product concentration (i.e., $<1–3$ μg/ml) in high serum-containing medium (i.e., $>2\%$ fetal calf serum, approximately 1,000 μg/ml). In such a case, long-term production would be uneconomical irrespective of purification strategy. However, for more economical antibody production levels (i.e., >20 μg/ml) in low serum or serum-free conditions, ion exchange and hydrophobic chroma-

TABLE 17–6 In-process Immunoglobulin Assays using Protein A
Chromatography and ELISA Methods[1]

Ion-Exchange Step	Protein A HPLC (g)	ELISA (g)	Intact IgG (%)
Pooled concentrates	177	345	51
Column 1 load (diluted)	170	335	51
Column 1 flow through	4	83	5
Column 1 elution product	172	258	67
Column 1 1 M NaCl	—	2	—
Column 2 flow through product	146	202	72
Column 2 1 M NaCl	10	12	83
Precipitation step sup.	10	70	14
(Ammonium Sulfate) ppt. (product)	134	133	100
Final purified: 133 g by OD			
Recovery of intact IgG_{280} (OD_{280} versus HPLC and ELISA)	75%	40%	

sup., Supernatant; ppt., precipitate (containing product); OD = Optical Density.
[1]Because the ELISA method detects free light chains as well as the intact antibody, this method overstates the quantity of IgG present for many of the process steps.
   The data show comparative in-process estimates of intact mouse antibody recovery using Protein A HPLC technique, versus ELISA that measures both the desirable product and free light chain. The hybridoma cell line produces a constant level of free light chain relative to intact antibody. The free light chain content at each step is reflected in the difference between the Protein A HPLC and ELISA values. Such inflated estimates of crude antibody by ELISA results in misleadingly low purification recoveries and can offset production schedules.

tography usually achieve the required levels of purification for producing acceptable clinical quality.

17.4 THE IMPACT OF CELL CULTURE TECHNIQUES ON PRODUCT INTEGRITY

Changes in cell culture conditions can often result in alterations to the physicochemical properties of a given product. Moellering et al. (1990) used identical procedures to isolate a monoclonal antibody from ascites, serum containing, and serum-free medium and found distinct differences in isoelectric mobilities, with subtle variation in tryptic mapping profiles. This observation was attributable to processes of deamidation that occurred in ascites and serum compared to the less degradative environment of serum-free conditions. The work emphasizes the importance of maintaining controlled cell culture conditions to produce consistent product without compromising the purification process.

The choice of bioreactor system can also have an impact on product quality. Perfusion systems for products secreted into the medium provide the opportunity for significant improvement in biological activity over that possible with batch or semibatch methods. Many processes can occur during the time that secreted proteins are exposed to the cell culture environment, which are detrimental to their integrity and biological function. Proteolytic enzymes, for example, may be present from serum supplements or from the cells, particularly if cells are allowed to die during extended production under batch operation. Secreted proteins, such as tPA, may also become irreversibly complexed with inhibitor molecules, which are contributed by the serum supplement or are naturally secreted by the cells. Quite often, cells that secrete active molecules also secrete inhibitors as part of an in vivo feedback control mechanism. In particular, tPA has been shown to form inhibitor complexes with alpha$_1$-antiplasmin, alpha$_1$-proteinase inhibitor, and alpha$_2$-macroglobulin, as described by Haggroth et al. (1984), Rijken et al. (1983), and Korninger et al. (1981). Other inactivating processes include aggregate formation and oxidation as described by Prior et al. (1988) and other possible chemical modifications of the desired protein structure. In all cases, the rate at which these inactivating processes occur is related to the temperature of the culture environment. By lowering the temperature, the inactivation process can be greatly reduced if not eliminated. Perfusion systems that separate the cell-conditioned medium that contains product from the cells allow the possibility of rapid temperature reduction and thereby preserve product integrity. As an example, the perfusion reactors operated at Invitron Corp. are computer automated and are arranged in 37°C incubator rooms immediately adjacent to 4°C cold rooms to facilitate rapid temperature reduction of perfused conditioned medium. Residence time for product at 37°C is one or two days for a 100 l perfusion culture system and only several hours for immobilized cell bioreactors. In contrast, batch production methods expose products accumulating in the reactor to degradative conditions throughout the length of the run, which can extend to weeks in many cases. The impact of such extended batch culture conditions on the biological activity of an IgM and IgG will be presented. Semibatch systems allow reduced contact time at higher temperatures but still involve several days at 37°C. Data will be presented showing that reduced product residency time inside the bioreactor does, in fact, improve the structural integrity of monoclonal IgM and rtPA (presented here as examples).

17.4.1 Human Recombinant Tissue Plasminogen Activator (rtPA)

Figure 17-7 shows growth curves of a recombinant CHO cell line used for production of tPA. The curve labeled III shows a batch production, the one labeled II a semibatch, and the I curve, perfusion production. The shorter product residency time in the perfusion vessel (hours), compared to semibatch (days to weeks) and batch (weeks), improved the specific activity of crude tPA in the conditioned medium (i.e., active versus inactive tPA mol-

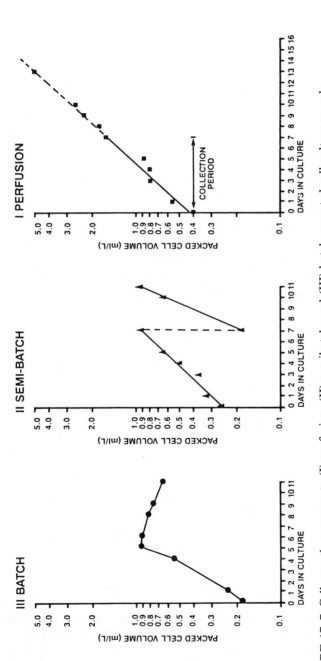

FIGURE 17–7 Cell growth response to (I) perfusion, (II) semibatch, and (III) batch operated cell culture vessels. Recombinant CHO cells producing human tPA were cultured at 37°C in medium containing 7.5% FBS under the three conditions described in the text. Cell growth was monitored by removing 25 ml aliquots of culture at the indicated times and determining packed cell volume (PCV) as described by Tolbert et al. (1980).

The batch (III) and semibatch (II) cultures were maintained in Invitron proprietary 14 L capacity spinner vessels and the perfusion culture (I) was maintained in an Invitron proprietary 100 l capacity perfusion culture system.

Conditioned medium was harvested from the three systems as indicated at a PCV of approximately 1 in all cases. The dotted line on the perfusion culture curve indicated that the PCV continued to rise beyond the media collection period.

ecules) and increased purification yield of active monomeric tPA as described by Prior et al. (1988). A summary of the relative purification recoveries of fully active material as a consequence of the three cell culture conditions (I, II, and III) is presented in Figure 17–8. This figure shows, in each case (I, II, III), the relative percent of active to inactive crude tPA in the conditioned media (left-hand column) and the corresponding percent of active tPA recovered from the purification process (right-hand column).

The longer product residency time in the semibatch and batch operated vessels resulted in complex formation with serum-derived inhibitor proteins and aggregation of tPA as a consequence of oxidation determined by nonreducing gel electrophoresis (Prior et al. 1988). This resulted in lower downstream recoveries compared to the perfusion case. These aggregates possess very low activity and could pose an immunogenic threat if they are co-purified as trace contaminants with product.

Figure 17–9 shows corresponding profiles of tPA eluted from the gel filtration column at the end of the purification process. Figure 17–10 shows an electrophoretic analysis of the individual column elution fractions using a 10% reduced SDS-PAGE. A major portion of the protein is irreversibly bound to inhibitor under batch conditions, a lesser amount in semibatch, and essentially no inhibitor complex is shown under perfusion. In addition,

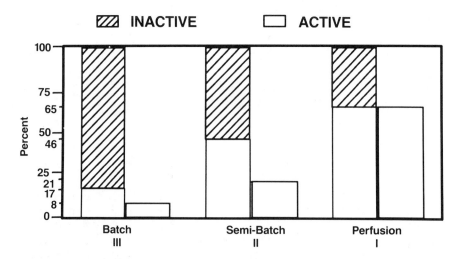

FIGURE 17–8 Recovery of active tPA molecules from cell culture conditions I, II, and III. The term "inactive" as used with respect to aggregated tPA refers to the substantially lowered specific activity of the aggregated form relative to that of the unaggregated form. For each case (I, II, and III), the right-hand column represents the percent recovery of final purified active antigen. The left-hand column (clear bar) represents the antigen in the conditioned media that accounts for 100% of the enzymatic activity. The striped bar represents the remaining portion of inactive antigen in the conditioned media.

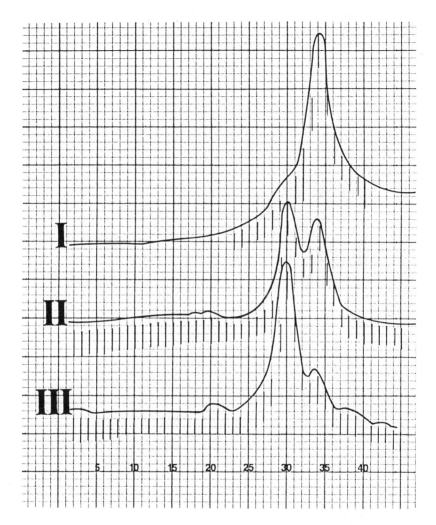

FIGURE 17–9 Gel filtration chromatography profiles of affinity purified tPA derived from (I) perfusion, (II) semibatch, and (III) batch operated cell culture vessels. Three separate tPA purification runs were performed on crude material derived from each of the three reactors (I, II, III). In each case, affinity purified tPA was subjected to gel filtration as a finishing step. The resin fractionates protein molecules by molecular weight and the rate of tPA elution was monitored by UV absorbance at 280 nm. tPA-associated complexes of high molecular weight (peak 1) were first to elute while monomeric molecules of lower molecular weight (peak 2) were retarded and eluted at later fraction volumes. All purification operations were performed at 4°C.

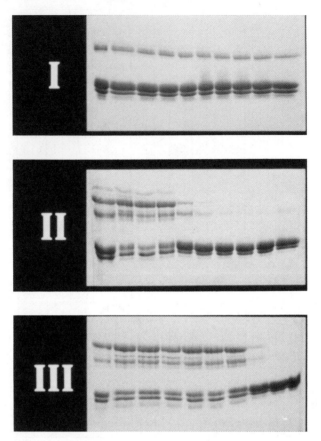

FIGURE 17–10 SDS-Page analysis of gel filtration chromatography profiles for cases I, II, and III. Forty micrograms of protein per lane was subjected to 10% SDS-PAGE (reducing) and stained with Coomassie brilliant blue. The first lane on the left-hand side of each panel represents the affinity purified tPA applied to the gel filtration column (load). The remaining lanes represent, from left to right, fractions 27 through 35, inclusive (see Figure 17–9).

the perfused product contains a significant portion of single chain tPA, indicating less proteolytic activity under serum-containing conditions. However, at high cell density ($>2 \times 10^7$ cells/ml) which normally occurs in a production perfusion chamber reactor, the media conditions are switched from serum to serum free with a subsequent conversion from the two chain form to the desired single chain product. As seen in Figure 17–11, perfusion technology delivers material with a single chain content >98% in serum-free conditions and without the use of protease inhibitors. Following final purification, the material contains <1.0 pg DNA/mg and <0.03 EU/mg of product.

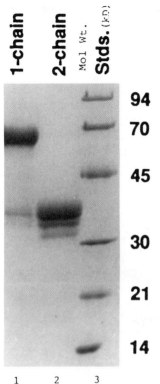

FIGURE 17–11 Ten percent SDS-PAGE (reducing) showing Invitron single chain and two chain human rtPA products. Lane 1 is the single chain product produced by perfusion technology in serum-free conditions without the use of protease inhibitors. Lane 2 is the two chain product produced in 5% FCS. The bands in lane 3 are the molecular weight standards.

In summary, perfusion systems, as compared to batch, improve the quality of the product in the conditioned media in terms of biological activity and structural integrity. This significantly increases downstream recovery and daily production output of final purified material. These findings have subsequently been corroborated by incubating cell-free conditioned medium spiked with active two chain tPA at 37°C and 4°C (control −40°C) and studying the rate of decay in specific activity at time points that simulate the product residency time of tPA in the various reactors. For purposes of comparison, the experiments were performed in serum and serum-free media and activity was measured using a fibrin dependent S-2251 assay as described by Verheijen et al. (1982). The results clearly demonstrate that at 4°C the tPA is remarkably stable in both media throughout a 17 day study period. This result confirms that tPA retains full activity when transferred to a 4°C environment. In contrast to the results at 4°C, the activity of the samples maintained at 37°C diminished considerably with time. In serum-free medium, 70% of the initial tPA activity was lost by day 17 (approximating batch conditions) while 40–50% of the initial activity was lost by five to six days (approximating semibatch conditions). In serum-containing conditions, most of the fibrin dependent tPA activity was lost beyond four days of incubation at 37°C.

17.4.2 Monoclonal IgM and IgG

A monoclonal IgM was produced and purified using identical procedures from a batch and perfusion culture. The purified samples from these cultures are shown on a reduced, silver stained SDS-PAGE, Figure 17–12. The first lane contained molecular weight standards and the second was antibody produced in batch culture in medium supplemented with 5% FBS. The third

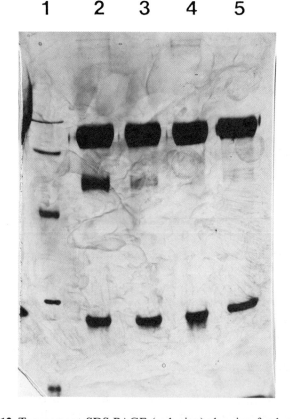

FIGURE 17–12 Ten percent SDS-PAGE (reducing) showing final purified monoclonal IgM derived from different cell culture conditions. Lane 1 contains low molecular weight standards. Lanes 2 through 5 show final purified IgM derived from batch cell culture reactor containing 5% FBS (lane 2); perfusion 5% FBS (lane 3); perfusion 2% FBS (lane 4); and perfusion serum free (lane 5). The band with an approximate molecular weight of 55,000 is a proteolytic fragment derived from the heavy chain as determined by immunoblot analysis.

lane contained antibody produced in perfusion culture in the same medium. The fourth and fifth lanes contained antibody produced in perfusion in the presence of 2% FBS and a serum-free medium, respectively. The extra band between the heavy and light chains in lane 2 is due to enzymatic degradation of the IgM molecule. Immunoblots suggest that all bands seen in lanes 2 through 5 are mouse-antibody related. The molecular weight markets (lane 1) serve as a negative control by not cross-reacting with the antimouse IgM polyclonal reagents. Perfusion in 5% serum greatly minimized this degradation compared to batch culture and eliminated it under reduced serum conditions.

Macmillan et al. (1987) studied the detrimental effects of extended time in batch culture on an IgG and IgM during the latter part of a traditional fermentation to completion. At this stage, as can be seen in Figure 17–13 (top panel), the culture enters a phase characterized by a decline in cell viability with consequent cell lysis and release of degradative enzymes. The data show that even though high IgG concentrations by mass, as measured by RID (bottom panel), may be achieved under these conditions, the specific biological activity of the crude product declines by day 4 in culture as measured by antigen binding in an ELISA (middle panel). The diverging assay results would be expected if fragmented antibody could not bind antigen in the ELISA reaction but could react with antimouse polyclonal antibodies in the less sensitive RID reaction, which is not dependent on the integrity of single reactive sites. The same workers investigated the structural integrity of an IgM with time in batch culture by HPLC. As seen in Figure 17–14, activity of the antibody measured by ELISA corresponded in molecular weight to the peak of an IgM (900,000 molecular weight) for the first few days in the batch culture run. At about day 3 in culture, however, activity was also detected in fractions corresponding to proteins with molecular weight of 150,000 molecular weight and less. Near the end of the culture period, antibody activity, though still associated with apparently intact IgM, was detectable over a broad area of the elution profile, indicating that small fragments still retained the ability to bind antigen. Antibody degradation over the course of the culture run was confirmed by analysis of individual peak fractions on subsequent gel filtration column chromatography. Although Macmillan et al. (1987) concluded that degradation only occurred to a limited extent, the regulatory issues relating to removal of breakdown products from material intended for use in humans may be significant.

An area of rapidly growing interest relates to the function of intracellular enzymes that modify carbohydrate structure of glycoproteins, such as those produced by mammalian cells, during processes of posttranslational modification in the endoplasmic reticulum and Golgi apparatus. Release of such enzymes from dead or dying cells during an extended period in batch culture could compromise the carbohydrate moiety and, consequently, the biological activity of the product. For a review on the effect of environmental

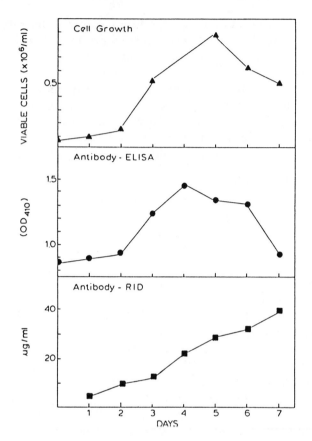

FIGURE 17–13 Kinetics of cell growth and monoclonal antibody (MAb) production in a batch reactor. The cell line produced an IgG1 MAb against staphylococcal enterotoxin. Antibody activity was measured in an ELISA using staphylococcal enterotoxin B as the antigen. Antibody concentration (RID) was measured with rabbit antimouse polyclonal antibody in the immobile phase.

conditions (i.e., during cell culture) on carbohydrate structure and biological activity, see Goochee and Monica (1990).

17.5 SUMMARY OF KEY POINTS

A primary objective of any biopharmaceutical manufacturing operation is to ensure that the desired protein secreted from a cell with full biological activity does, in fact, reach the final container in a "native" unadulterated state. In this effort, both cell culture and purification processes should employ the mildest conditions possible and be optimized to remove or reduce

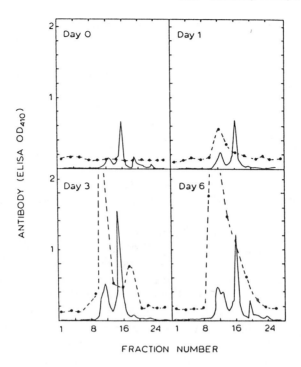

FIGURE 17-14 Elution patterns of an IgM monoclonal antibody produced in a 1 l batch reactor and analyzed by HPLC on Superose 6. Antibody in cell conditioned medium samples (50 ml), collected over an eight day period, was precipitated with $(HN_4)_2 SO_4$ (40%) and taken up in 0.15 M phosphate buffer, pH 7.8. Samples (50 μl) were applied to a column (bed volume, 25 cm^3) and eluted with phosphate buffer (0.5 ml/min); 1 ml fractions were collected. Protein was followed with a UV monitor (———). Antibody activity (-------) was measured as in Figure 17-13.

to an absolute minimum such potential contaminants as growth media and cellular proteins, aggregates, nucleic acids, and endotoxins. In addition, processes should have the ability to remove and inactivate adventitious agents using conditions not detrimental to the product. A critical aspect of achieving these demanding safety requirements is to ensure that the highest possible standards of sanitary operations are maintained. This requires that conditioned medium and freshly prepared buffers are filtered through 0.2 μm filters prior to column application. Adequate cleaning procedures should be in place to prevent processing equipment, particularly column resins, from contributing undesirable contaminants during the purification process itself. Therefore, column resins should be selected that are capable of withstanding either autoclave conditions or harsh chemical sanitization procedures on a repeated basis between individual purification runs. Data should be available as part of a manufacturer's MDF, describing the structural stability of the resin following repeated sterilization/sanitization techniques (i.e., leaching rates), particularly if procedures involve extreme temperatures and/or pH conditions. Testing procedures for measuring resin breakdown products with established levels of detection should be available. In terms of chemical stability, ion exchange resins are more suited to a pharmaceutical operation over the use of Protein A or immunoaffinity techniques. The latter are biological columns that cannot be autoclaved or sanitized as efficiently as the more inert ion exchange materials capable of withstanding treatment with 0.5–1.0 M NaOH. Also, leachables derived from biological columns, especially lectin columns, pose a greater immunogenic threat than those derived from conventional ion exchangers.

When comparing Protein A chromatography with ion exchange chromatography there are advantages and disadvantages associated with each. While Protein A can be a generic process for purifying antibodies, the concerns associated with cost, extreme pH conditions, copurification of contaminating antibodies, and sanitization renders Protein A less suitable than ion exchange chromatography for large-scale procedures. Protein A is often used in conjunction with ion exchange. However, the expense of the resin requires multiple recycling of product on a Protein A column of limited size. This creates multiple lots that must be individually analyzed for purity before being recombined. This can place an enormous workload on a quality control department and results in prolonged processing times. The user also has to invest in automation equipment and dedicated GMP work space. Protein A chromatography does serve as a very useful analytical tool for quantifying product and offers faster analysis than does ELISA. For the production of highly purified monoclonal antibodies for clinical application, ion exchange and hydrophobic chromatography are the more appropriate choice, especially when purifying from serum-free, low protein-containing medium.

With regard to choice of animal cell culture reactors and the impact on product quality, perfusion technology maintains cells more intact at high cell density than batch systems and effectively allows replacement of serum with serum-free conditions. Reducing the relative proportion of medium-derived contaminants to product substantially increases downstream recovery and improves final product quality. Perfusion systems do facilitate production of highly purified material low in DNA, endotoxin, and growth medium-derived proteins. Also, perfusion technology compared to batch and semibatch operated systems minimizes product residency time within the degradative 37°C environment of the bioreactor. This improves the structural integrity of biopharmaceuticals as described for production of monoclonal IgM, IgG, and one and two chain human rtPA.

These examples dramatically demonstrate reduced product degradation by perfusion methods when coupled with rapid cooling of the product stream. It is necessary to minimize injection of inactive, complexed, or aggregated materials, which could be significantly immunogenic to the patient. Therefore, the integrity, purity, and biological activities of these complex protein molecules are extremely important when they are to be used in human pharmaceutical applications.

REFERENCES

Abrams, P.G., Jeffrey, J.O., Steven L., et al. (1984) *J. Immunol.* 132, 1611–1617.

Bassin, R.H., Duran Troise, G., Gerwin, B.L., and Rein, A. (1978) *J. Virol.* 26, 306–315.

Bjorck, L., Erntell, M., and Myhre, E.B. (1987) European Patent Application No. 87850048.7 Publ. No. 0255497.

Burnouf, T., Martinache, L., and Goudermara, M. (1987) *Nouv. Rev. Fr. Hematot* 29, 93–96.

Center for Biologics Evaluation and Research (1987) *Points to Consider in the Manufacture of Monoclonal Antibody Products for Human Use*, Food and Drug Administration, Bethesda, MD.

Daher, K.A., Selsted, M.E., and Lehren, R.T. (1986) *J. Virol.* 60, 1068–1074.

Duffy, S.A., Moellering, B.J., Prior G.M., Doyle, K.D., and Prior, C.P. (1989) *Biopharm* 2, 34–47.

Ey, P.L., Prowse, J.S., and Jenkin, C.R. (1978) *Immunol. Chem.* 15, 429–436.

Feder, J., and Tolbert, W.R. (1983) *Sci. Am.* 248, 36–43.

Feder, J., and Tolbert, W.R., eds. (1985) *Large Scale Mammalian Cell Culture* Academic Press, London.

Fry, E.M. (1985) Process Validation: FDA Viewpoint—Drug and Chemical Industry, July, 46–51.

Goochee, C.F., and Monica, T. (1990) *Biotechnology* 5, 421–427.

Haggroth, L., Mattsson, C., and Friberg, T. (1984) *Thromb. Res.* 33, 583–594.

Hammen, R. (1988) *Biochromatography* 3(2), 54–59.

Hilfenhaus, J., and Weidmann, E. (1986) *Arzneim Forsch* 36, 621–625.

Hussain, S.F., Reveyemamu, M.M., Kaminjolo, J.S., Akhtar, A.S., and Mugera, G.M. (1980) *Zentralbt Veterinaermed Reihe B* 27, 233–242.

Ingerslev, J., Bukh, A., Wallevik, K., Moller, N.Ph., and Stenbjerg, S. (1987) *Thromb. Res.* 47, 175–182.

Knight, P. (1988) *Bio/Technology* 6, 1054–1058.

Korninger, C., Stassen, J.M., and Collen, D. (1981) *Thromb. Haemostat* 46, 658–661.

Laemmli, U.K. (1970) *Nature* 227, 680–684.

Lancz, G., and Sample, J. (1985) *Arch. Virol.* 84, 1–2.

Langone, J.J. (1982) *J. Immuno. Methods* 55, 227–296.

Lee, S.M., and Seaver, S.S., eds. (1987) *The Commercial Production of Monoclonal Antibodies: A Guide for Scale-Up*, pp. 199–216, Marcel Dekker, New York.

Macmillan, J.D., Velez, D., Miller, L., and Reuveny, S. (1987) *Monoclonal Antibody Production in Stirred Reactors in Large Scale Cell Culture Technology* (Lydersen, G.K., ed.), pp. 21–58, Hanser Verlag, Munich.

Manil, L., Motte, P., Pernas, P., et al. (1986) *J. Immunol. Methods* 90, 25–37.

Martin, L.S., McDougal, J.S., and Loskowski, S.L. (1985) *J. Infect. Dis.* 152, 400–403.

Mitra, G., and Wong, M. (1986) *Biotechnol. Bioeng.* 28, 297–300.

Moellering, B.J., and Prior, C.P. (1990) *Biopharm* 1, 34–38.

Moellering, B.J., Tedesco, J.L., Townsend, R.R., et al. (1990) *Biopharm* 2, 30–38.

Ng, P.K., and Dobkin, M.B. (1985) *Thromb. Res.* 39, 439–448.

Parenteral Drug Association (1985) *Technical Report No. 7* Parenteral Drug Association Inc., Philadelphia.

Petricciani, J.C. (1985) in *Large-Scale Mammalian Cell Culture* (Feder. J., and Tolbert, W.R., eds.), pp. 79–86, Academic Press, London.

Pharmacia, Inc. (1986) *Downstream News and Views for Process Biotechnologists, No. 2,* Pharmacia, Inc., Piscataway, N.J.

Prior, C.P., Prior, G.M., and Hope, J.A. (1988) *Am. Biotechnol. Lab.* April, pp. 25–31.

Repligen (1988) *Immobilized rProtein A™ Product Brochure,* Cambridge, MA.

Rigby, P.W.J., Dieckmann, M., Rhodes, C., and Berg, P. (1977) *J. Mol. Biol.* 113, 237–241.

Rijken, D.C., Vague-Juhon, I., and Collen, D.J. (1983) *Lab. Clin. Med.* 101, 285–294.

Schrezenmeier, H., and Fleischer, B. (1987) *J. Immunol. Methods* 105, 133–137.

Stegmann, T., Booy, F.P., and Wilschut, J. (1987) *J. Biol. Chem.* 262, 17744–17749.

Tolbert, W.R., Srigley, W.R., and Prior, C.P. (1988) *Animal Cell Biotechnology* Vol. 3 (Griffiths, G.B., and Spier, R.E., eds.), pp. 374–393, Academic Press, London.

Tolbert, W.R., Hitt, M.M., and Feder, J. (1980) *Anal. Biochem.* 106, 109–113.

Truskey, G., Gabler, R., DiLeo, A., and Manter, T. (1987) *J. Parenteral Sci. Technol.* 41(6), 180–193.

Verheijen, J.H., Nieuwenhuizen, W., and Wijngaards, G. (1982) *Thromb. Res.* 27, 377–385.

Whitehouse, R.L., and Clegg, J. (1963) *Dairy Res.* 30, 315–322.

Wickerhauser, M., and Williams, C. (1987) *Vox Sang* 53(3), 188–189.

Spier, R.E., and Griffiths, G.B., eds. (1988) *Animal Cell Biotechnology* Vol. 3, Academic Press, London.